Real-Time Embedded Systems with Open-Source Operating Systems

This book aims to provide readers with hands-on knowledge about real-time operating systems and their possible application in the embedded systems domain to streamline, simplify, and make software development more efficient, without requiring any significant previous experience with them. A thorough presentation of operating system-based programming techniques is especially important because they enjoy an ever-increasing popularity in the embedded systems domain but are often misunderstood, because they still lack comprehensive support in the scientific and technical literature.

The book analyzes in detail three realistic case studies of increasing complexity, of which the first one requires only a commonly available PC or laptop, while the other two involve low-cost, open-source hardware platforms readily available to the majority of readers. They serve as starting points and running examples while introducing theoretical concepts, as well as real-time operating systems' operations and interfaces. A set of exercises and their solutions completes the book, to enable readers to self-assess their knowledge as they proceed. Moreover, the source code developed for the case studies is freely available for download and further experimentation.

- Provides hands-on description of the most important real-time operating system concepts
- Includes case studies of practical interest to experiment with while reading the book
- Provides an in-depth, but accessible presentation of real-time scheduling theory
- A balanced mix of operating system theory, exercises, and case studies in a single book

The use cases involve inexpensive hardware boards readily available on the market

Together, the topics covered by this book help embedded system designers understand the benefits and shortcomings of real-time operating systems and then decide whether it may be worth adopting one of them for their next project instead of relying on more traditional, but less powerful, techniques. At the same time, students will acquire all the knowledge and skills they need to take part in real-world embedded software development without sacrificing a proper theoretical foundation. In this context, the case studies play the crucial role of underlining the strong relationship between operating system theory and application, along with the relevance of theoretical concepts in day-to-day project design and implementation.

Real-Time Embedded Systems with Open-Source Operating Systems

Second Edition

Ivan Cibrario Bertolotti and Gabriele Manduchi

CRC Press
Taylor & Francis Group
Boca Raton London New York

CRC Press is an imprint of the
Taylor & Francis Group, an **informa** business

Designed cover image: Shutterstock

Second edition published 2026
by CRC Press
2385 NW Executive Center Drive, Suite 320, Boca Raton FL 33431

and by CRC Press
4 Park Square, Milton Park, Abingdon, Oxon, OX14 4RN

CRC Press is an imprint of Taylor & Francis Group, LLC

© 2026 Ivan Cibrario Bertolotti and Gabriele Manduchi

First edition published by CRC Press 2012

ISBN: 978-1-032-97371-5 (hbk)
ISBN: 978-1-032-97651-8 (pbk)
ISBN: 978-1-003-59341-6 (ebk)

DOI: 10.1201/9781003593416

Typeset in Nimbus Roman font
by KnowledgeWorks Global Ltd.

Publisher's note: This book has been prepared from camera-ready copy provided by the authors.

To Maria Cristina, Samuele, and Guglielmo
— ICB

To Ornella, Silvia, and Laura
— GM

Contents

SECTION I   Concurrent Programming Concepts

SECTION II   Real-Time Scheduling Analysis

SECTION III   *Case Studies*

Foreword

Real-time embedded systems have become an integral part of our technological and social space. But is the engineering profession equipped with the right knowledge to develop those systems in ways dictated by the economic and safety aspects? Likely yes. But the knowledge is fragmented and scattered among different engineering disciplines and computing sciences. Seldom anyone of us has the clear picture of the whole. If so, then parts of it are at an abstract level. That poses a question whether the academic system provides education in a way holistic enough to prepare graduates to embark on the development of real-time embedded systems, frequently complex and imposing safety requirements. How many electrical and computer engineering departments offer subjects focusing on the application-dependent specialized communication networks used to connect embedded nodes in distributed real-time systems. If so, then the discussion is confined to the Controller Area Network (CAN), or sometimes FlexRay, in the context of automotive applications—usually a small unit of an embedded systems subject. (The impression might be that specialized communication networks are mostly used in automotive applications.) The requirement for the underlying network technology to provide real-time guarantees for message transmissions is central to proper functioning of real-time systems. Most of computer engineering streams teach operating systems. But real-time aspects are scantly covered. Computer science students, on the other hand, have very little, if any, exposure to the "physicality" of the real systems the real-time operating systems are intended to interact with. Does this put computer science graduates in a disadvantaged position? In the late 1990s and early 2000s, I was involved in the Sun Microsystems lead initiative to develop real-time extensions for the Java language. The working group comprised professionals mostly from industry with backgrounds largely in computing sciences. I was taken aback by the slow pace of the process. On reflection, the lack of exposure to the actual real-time systems in different application areas and their physicality was likely to be behind difficulties to identify generic functional requirements to be implemented by the intended extensions.

In the second part of 1980s, I was teaching digital control to the final year students of the electrical engineering course. The lab experiments to illustrate different control algorithms were designed around the, at that time, already antiquated Data General microNOVA MP/200 minicomputer, running one of the few real-time operating systems commercially available at that time—QNX, if I remember correctly. Showing things work was fun. But students' insight into the working of the whole system stopped at the system-level commands of the operating systems. The mystery had to be revealed by discussing hypothetical implementations of the system level calls and interaction with the operating system kernel—of course, at the expense of the digital control subject. At that time, seldom any electrical engineering curriculum had a separate subject dedicated to operating systems. Of frustration and to avoid the "black box" approach to illustrating control systems in action, I have written in C a simple multitasking real-time executive for MS-DOS-based platforms, to be run on an IBM PC (Intel 8088). Students were provided with the implementation documentation in addition to the theoretical background; quite a lot of pages to study. But the reward was substantial: they were now in full "control." With the support of an enterprising post-graduate student, the executive was intended to be grown into more robust RTOS with a view for commercialization. But it was never to be. Academic life has other priorities. Around 1992, I decided to harness the MINIX operating system, which I then taught to the final-year graduate students, to run my real-time control lab experiments to illustrate control algorithms in their supporting real-time operating system environment. But soon after that came the Linux kernel.

If you are one of those professionals with the compartmented knowledge, particularly with the electrical and computer engineering or software engineering background, with not much theoretical knowledge of and practical exposure to real-time operating systems, this book is certainly an invaluable help to "close the loop" in your knowledge, and to develop an insight into how things work in the realm of real-time systems. Readers with a background in computer science will benefit from the hands-on approach, and a comprehensive overview of the aspects of control theory and signal processing relevant to the real-time systems. The book also discusses a range of advanced topics which will allow computer science professionals to stay up-to-date with the recent developments and emerging trends.

The book was written by two Italian researchers from the Italian National Research Council (CNR) actively working in the area of real-time (embedded) operating systems, with a considerable background in control and communication systems, and a history of the development of actual real-time systems. Both authors are also involved in teaching several courses related to these topics at Politecnico di Torino and University of Padova.

The book has been written with a remarkable clarity, which is particularly appreciated whilst reading the section on real-time scheduling analysis. The presentation of real-time scheduling is probably the best in terms of clarity I have ever read in the professional literature. Easy to understand, which is important for busy professionals keen to acquire (or refresh) new knowledge without being bogged down in a convoluted narrative and an excessive detail overload. The authors managed to largely avoid theoretical-only presentation of the subject, which frequently affects books on operating systems. Selected concepts are illustrated by practical programming examples developed for the Linux and FreeRTOS operating systems. As the authors stated: Linux has a potential to evolve in a fully fledged real-time operating system; FreeRTOS, on the other hand, gives a taste of an operating system for small footprint applications typical of most of embedded systems. Irrespective of the rationale for this choice, the availability of the programming examples allows the reader to develop insight into the generic implementation issues transferrable to other real-time (embedded) operating systems.

This book is an indispensable addition to the professional library of anyone who wishes to gain a thorough understanding of real-time systems from the operating systems perspective, and to stay up to date with the recent trends and actual developments of the open-source real-time operating systems.

Richard Zurawski
ISA Group, San Francisco, California

Preface

This book is the outcome of more than 10 years of research and teaching activity in the field of real-time operating systems and real-time control systems. During this time, we have been positively influenced by many other people we came in contact with, both from academy and industry. They are too numerous to mention individually, but we are nonetheless indebted to them for their contribution to our professional growth.

A special thank you goes to our university students, who first made use of the lecture notes this book is based upon. Their questions, suggestions, and remarks were helpful to make the book clearer and easier to read.

We would also like to express our appreciation to our coworkers for their support and patience while we were busy with the preparation of the manuscript. A special mention goes to one of Ivan's past teachers, Albert Werbrouck, who first brought his attention to the wonderful world of embedded systems.

Last, but not least, we are grateful to Richard Zurawski, who gave us the opportunity to write the first edition of this book. We are also indebted to the CRC Press publishing and editorial staff, Nora Konopka and Sifat Kaur Keer in particular. Without their help, the book would probably not exist.

Authors

Ivan Cibrario Bertolotti earned the Laurea degree (*summa cum laude*) in computer science from the University of Torino, Turin, Italy, in 1996. Since then, he has been a researcher with the National Research Council of Italy (CNR). Currently, he is senior researcher with the Istituto di Elettronica e di Ingegneria dell'Informazione e delle Telecomunicazioni (IEIIT), Turin, Italy.

His research interests include real-time operating system design and implementation, industrial communication systems and protocols, and formal methods for security and dependability analysis of distributed systems. Along the years, he published more than 100 peer-reviewed articles on these subjects, while also co-authoring several book chapters and three books. His contribution encompasses both theoretical work and practical applications, carried out in cooperation with leading Italian and international companies.

He has taught various courses on real-time operating systems at Politecnico di Torino, Turin, Italy, starting in 2003, as well as a PhD degree course at the University of Padova in 2009. He regularly serves as a technical referee for the main international conferences and journals on industrial informatics, factory automation, and communication. He has been an IEEE member since 2006.

Gabriele Manduchi obtained the Laurea Degree (*summa cum laude*) in Electronic Engineering from the University of Padua, Italy, in 1987. Since then, he has been a researcher with the National Research Council (CNR). He is currently with the Istituto Gas Ionizzati (IGI), Padua, Italy.

Since 1989 he has been involved in the design and development of acquisition and real-time control systems for large physics experiments. In particular, he is one of the developers of MDSplus, an open-source data acquisition system widely adopted in nuclear fusion experiments. He is also one of the developers of the real-time control system of the RFX, a nuclear fusion experiment in the framework of the coordinated European Fusion Development Agreement. He has been also involved as deputy project leader in the International Tokamak Modeling European task aiming at building a comprehensive simulation framework for thermonuclear fusion. Since 2021 he has been the project leader of the Control and Data Acquisition task for the nuclear fusion experiment DTT in Rome, Italy.

His current research interests include real-time system design and implementation, large data management and supervision for mission-critical systems.

Since 1995 Gabriele Manduchi has been teaching introductory and advanced courses on computer architectures at the University of Padua. In recent years, he has also taught PhD courses on Software Design Patterns and on Concurrent and Real-time Programming in the same University.

Section I

Concurrent Programming Concepts

1 Introduction

This book addresses three different topics: *Embedded Systems*, *Real-Time Systems*, and *Open Source Operating Systems*. Even if every single topic can well represent the argument of a whole book, they are normally intermixed in practical applications. This is in particular true for the first two topics: very often industrial or automotive applications, implemented as embedded systems, must provide timely responses in order to perform the required operation. Further, in general, real-time requirements typically refer to applications that are expected to react to the events of some kind of controlled process.

Often in the literature, real-time embedded systems are presented and analyzed in terms of abstract concepts such as tasks, priorities, and concurrence. However, in order to be of practical usage, such concepts must be then eventually implemented in *real* programs, interacting with *real* operating systems, to be executed for the control of *real* applications.

Traditionally, textbooks concentrate on specific topics using different approaches. Scheduling theory is often presented using a formal approach based on a set of assumptions for describing a computer system in a mathematical framework. This is fine, provided that the reader has enough experience and skills to understand how well real systems fit into the presented models, and this may not be the case when the textbook is used in a course or, more in general, when the reader is entering this area as a primer. Operating system textbooks traditionally make a much more limited usage of mathematical formalism and take a more practical approach, but often lack practical programming examples in the main text (some provide specific examples in appendices), as the presented concepts apply to a variety of real-world systems.

A different approach is taken here: after a general presentation of the basic concepts in the first chapters, the remaining ones make explicit reference to two specific operating systems: *Linux* and *FreeRTOS*. Linux represents a full-fledged operating system with a steadily growing user base and, what is more important from the perspective of this book, is moving toward real-time responsiveness, and is becoming a feasible choice for the development of real-time applications. FreeRTOS represents somewhat the opposite extreme in complexity. FreeRTOS is a minimal system with a very limited footprint in system resources and can therefore be used in very small applications such as microcontrollers. At the same time, FreeRTOS supports a multithreading programming model with primitives for thread synchronization that are not far from what larger systems offer.

In this second edition, two new case studies have been introduced, in addition to the one originally presented. These three use cases (two based on Linux and the third one on FreeRTOS) present nontrivial applications and provide a deeper insight into the presented concepts. For this edition, a companion GitHub repository (`https://github.com/minimap-xl/RTOS_Book`) has been developed, providing all the presented source code and, more in general, all the material that is required

DOI: 10.1201/9781003593416-1

to replicate the presented case studies, including detailed guides for the components installation.

The first case study, presented in this chapter, addresses the programming of a camera device in Linux, and it will be used to refresh several Operating Systems (OS) concepts whose knowledge is important in order to develop efficient distributes systems. No additional hardware is required, just the camera embedded in any laptop, and therefore it can be easily replicated by students at home.

The second case study, discussed in Chapter 16, shows how FreeRTOS can be used to implement a simple gateway between an Ethernet-based UDP/IP network and a Controller Area Network (CAN), a network technology mainly used in the automotive and industrial automation domains. This case study gives us the opportunity to show how a fully functional firmware can be designed, implemented, and deployed on a low-cost embedded controller board, albeit on a small scale. At the same time, contrasting the second and the third case study provides valuable information on how software development for embedded systems differs from the way it is typically done on more general-purpose operating systems like Linux.

The third case study, presented in Chapter 17, presents a more complex system, that is fast Analog to Digital Conversion (ADC) and Digital to Analog conversion (DAC) implemented in a System on Chip (SoC) architecture. Here, the required FPGA components are introduced and the development of a Linux driver explained. The presented system is based on RedPitaya, a cheap and powerful hardware architecture based on the Zynq SoC. All the steps that are required to use the fast ADC and DAC in software applications are presented in the chapters. While the presentation of the FPGA components does not enter in full detail, being FPGA programming outside the scope of this book, the Linux driver is discussed in depth and may represent a useful baseline for the development of other Linux drivers. All the software components that are required to replicate the presented application on a RedPitaya STEMlab 125-14 (the most popular RedPitaya board that has been used here) are available in the GitHub repository.

If, on the one side, the choice of specific case studies may leave some details of other widespread operating systems uncovered, on the other one, it presents to the reader a complete conceptual path from general concepts of concurrence and synchronization down to their specific implementation, including dealing with the unavoidable idiosyncrasies of specific application programming interfaces. Here, most code examples are not collected in appendices, but presented in the book chapters to stress the fact that concepts cannot be fully grasped unless undertaking the "dirty job" of writing, debugging, and running programs.

The same philosophy has been adopted in the chapters dealing with scheduling theory. It is not possible, of course, to get rid of some mathematical formalism, nor to avoid mathematical proofs (which can, however, be skipped without losing the main conceptual flow). However, thanks to the fact that such chapters follow the presentation of concurrence-related issues in operating systems, it has been possible to provide a more practical perspective to the presented results and to better describe how the used formalism maps onto real-world applications.

This book differs from other textbooks in two further aspects:

- The presentation of a case study at the beginning of the book, rather than at its end. This choice may sound bizarre as case studies are normally used to summarize presented concepts and results. However, the purpose of the case study here is different: rather than providing a final example, it is used to summarize prerequisite concepts on computer architectures that are assumed to be known by the reader afterward. Readers may, in fact, have different backgrounds: less experienced ones may find the informal description of computer architecture details useful to understand more in-depth concepts that are presented later in the book such as task context switch and virtual memory issues. The more experienced will likely skip details on computer input/output or memory management but may, nevertheless, have some interest in the presented application, handling online image processing over a stream of frames acquired by a digital camera.
- The presentation of the basic concepts of control theory and Digital Signal Processing in a nutshell. Traditionally, control theory and Digital Signal Processing are not presented in textbooks dealing with concurrency and schedulability, as this kind of knowledge is not strictly related to operating systems issues. However, the practical development of embedded systems is often not restricted to the choice of the optimal operating system architecture and task organization, but also requires analyzing the system from different perspectives, finding proper solutions, and finally implementing them. Different engineering disciplines cover the various facets of embedded systems: control engineers develop the optimal control strategies in the case the embedded system is devoted to process control; electronic engineers will develop the front-end electronics, such as sensor and actuator circuitry, and finally software engineers will define the computing architecture and implement the control and supervision algorithms. Active involvement of different competencies in the development of a control or monitoring system is important in order to reduce the risk of missing major functional requirements or, on the opposite end, of an overkill, that is ending in a system which is more expensive than what is necessary. Not unusual is the situation in which the system proves both incomplete in some requirements and redundant in other aspects.

  Even if several competencies may be required in the development of embedded systems, involving specialists for every system aspect is not always affordable. This may be true with small companies or research groups, and in this case different competencies may be requested by the same developer. Even when this is not the case (e.g., in large companies), a basic knowledge of control engineering and electronics is desirable for those software engineers involved in the development of embedded systems. Communication in the team can, in fact, be greatly improved if there is some overlap in competencies, and this may reduce the risk of flaws in the system due to the lack of communication within the development team. In large projects, different components are developed by different teams, possibly in different companies, and clear interfaces must be defined in the system's architecture to allow the proper component integration, but it is always possible that some misunderstanding could occur even with the most accurate interface definition. If there is no competence overlap among development teams, this risk may become a reality, as it happened in the development of the trajectory control system of the NASA Mars Climate Orbiter, where a software component developed by an external company was working in pounds force, while the spacecraft expected values in newtons. As a result, the \$125 million Mars probe miserably crashed when it reached the Mars atmosphere [71].

As a final remark, in the title an explicit reference is made to open source systems, and two open source systems are taken as example through the book. This choice should not mislead the reader in assuming that open source systems are the common solution in industrial or automotive applications, but it is the authors' opinion that open source solutions are going to share a larger and larger portion of applications in the near future.

The book is divided into three sections: *Concurrent Programming Concepts*, *Real-Time Scheduling Analysis*, and *Case Studies*. The first section presents the basic concepts about processes and synchronization, and it is introduced by a case study represented by a nontrivial application for vision-based control. Along the example, the basic concepts of computer architectures and in particular of input/output management are introduced, as well as the terminology used in the rest of the book.

After the case study presentation, the basic concepts of concurrent programming are introduced. This is done in two steps: first, the main concepts are presented in a generic context without referring to any specific platform and therefore without detailed code examples. Afterward, the same concepts are described, with the aid of several code examples, in the context of the two reference systems: Linux and FreeRTOS. This section also includes a chapter on network communication; even if not explicitly addressing network communication, a topic which deserves by itself a whole book, some basic concepts about network concepts and network programming are very often required when developing embedded applications.

The chapters of this section are the following:

- Chapter 2: A Case Study: Vision Control. Here, an application is presented that acquires a stream of images from a Web camera and detects online the center of a circular shape in the acquired images. This represents a complete example of an embedded application. Both theoretical and practical concepts are introduced here, such as the input/output architecture in operating systems and the video capture application programming interface for Linux.
- Chapter 3: Real-Time Concurrent Programming Principles. From this chapter onwards, an organic presentation of concurrent programming concepts is provided. Here, the concept of parallelism and its consequences, such as

race conditions and deadlocks, are presented. Some general implementation issues of multiprocessing, such as process context and states, are discussed.

- Chapter 4: Deadlock. This chapter focuses on deadlock, arguably one of the most important issues that may affect a concurrent application. After defining the problem in formal terms, several solutions of practical interest are presented, each characterized by a different trade-off between ease of application, execution overhead, and conceptual complexity.

- Chapter 5: Interprocess Communication Based on Shared Variables. The chapter introduces the notions of Interprocess Communication (IPC), and it concentrates on the shared memory approach, introducing the concepts of lock variable, mutual exclusion, semaphore and monitors, which represent the basic mechanisms for process coordination and synchronization in concurrent programming.

- Chapter 6: Interprocess Communication Based on Message Passing. An alternate way for achieving interprocess communication, based on the exchange of messages, is discussed in this chapter. As in the previous two chapters, the general concepts are presented and discussed without any explicit reference to any specific operating system.

- Chapter 7: Interprocess Communication Primitives in POSIX/Linux. This chapter introduces several examples showing how the general concurrent programming concepts presented before are then mapped into Linux and POSIX. The presented information lies somewhere between a user guide and a reference for Linux/POSIX IPC primitives.

- Chapter 8: Interprocess Communication Primitives in FreeRTOS. The chapter presents the implementation of the above concurrent programming concepts in FreeRTOS, the other reference operating system for this book. The same examples of the previous chapter are used, showing how the general concepts presented in Chapters 3–6 can be implemented both on a full-fledged system and a minimal one.

- Chapter 9: Network Communication. Although not covering concepts strictly related to concurrent programming, this chapter provides important practical concepts for programming network communication using the socket abstraction. Network communication also represents a possible implementation of the message-passing synchronization method presented in Chapter 6. Several examples are provided in the chapter: although they refer to Linux applications, they can be easily ported to other systems that support the socket programming layer, such as FreeRTOS.

The second section, Real-Time Scheduling Analysis, presents several theoretical results that are useful in practice for building systems which are guaranteed to respond within a maximum, given delay. It is worth noting now that "real-time" does not always mean "fast." Rather, a real-time system is a system whose timely response can be trusted, even if this may imply a reduced overall throughput. This section introduces the terminology and the main results in scheduling theory. They are initially presented using a simplified model which, if on the one side it allows the formal derivation of many useful properties, on the other it is still too far from real-world applications to use the above results as they are. The last two chapters of this section will progressively extend the model to include facts occurring in real applications, so that the final results can be used in practical applications.

The chapters of this section are the following:

- Chapter 10: Real-Time Scheduling Based on the Cyclic Executive. This chapter introduces the basic concepts and the terminology used throughout the second section of the book. In this section, the concepts are presented in a more general way, assuming that the reader, after reading the first section of the book, is now able to use the generic concepts presented here in practical systems. A first and simple approach to real-time scheduling, cyclic executive, is presented here and its implications discussed.
- Chapter 11: Real-Time, Task-Based Scheduling. After introducing the general concepts and terminology, this chapter addresses real-time issues in the concurrent multitask model, widely described in the first section. The chapter presents two important results with immediate practical consequences: Rate Monotonic (RM) and Earliest Deadline First (EDF), which represent the optimal scheduling for fixed and variable task priority systems, respectively.
- Chapter 12: Schedulability Analysis Based on Utilization. While the previous chapter presented the optimal policies for real-time scheduling, this chapter addresses the problem of stating whether a given set of tasks can be schedulable under real-time constraints. The outcome of this chapter is readily usable in practice for the development of real-time systems.
- Chapter 13: Schedulability Analysis Based on Response Time Analysis. This chapter provides a refinement of the results presented in the previous one. Although readily usable in practice, the results of Chapter 12 provide a conservative approach, which can be relaxed with the procedures presented in this chapter, at the cost of a more complex schedulability analysis. The chapter also takes into account sporadic tasks, whose behavior cannot be directly described in the general model used so far, but which nevertheless describe important facts, such as the occurrence of exceptions that happen in practice.
- Chapter 14: Process Interactions and Blocking. This chapter and the next provide the concepts that are required to map the theoretical results on scheduling analysis presented up to now onto real-world applications, where the tasks cannot anymore be described as independent processes, but interact with each other. In particular, this chapter addresses the interference among tasks due to the sharing of system resources, and introduces the priority inheritance and priority ceiling procedures, which are of fundamental importance in the implementation of real-world applications.
- Chapter 15: Self-Suspension and Schedulability Analysis. This chapter addresses another fact which differentiates real systems from the model used to derive the theoretical results in schedulability analysis, that is, the suspension of tasks due, for instance, to I/O operations. The implications of this fact, and the quantification of its effects, are discussed here.

The last section will cover other aspects of embedded systems and the two new case studies. Unlike the first two sections, where concepts are introduced step by step to provide a comprehensive understanding of concurrent programming and real-time systems, the chapters of the last section cover separate, self-consistent arguments. The chapters of this section are the following:

- Chapter 16: General-Purpose IoT/Embedded Controller. In this case study we use a typical embedded controller board to build a gateway between an Ethernet-based UDP/IP network, briefly described in Chapter 9, and a Controller Area Network (CAN), a network technology mainly used in the automotive and industrial automation domains. Overall, this case study gives us the opportunity to show how a fully functional firmware can be designed, implemented, and deployed on a low-cost embedded controller board.
- Chapter 17: Real-Time, High-Performance Data Acquisition. In this chapter the development from scratch of the required components for ADC acquisition and DAC generation is presented. Whilst the parts not related to the topics presented in this book, namely the FPGA components, are just introduced but not analyzed in detail, the basic concepts of Linux drivers are presented here in detail. The general approach taken in this book is also followed in this chapter, that is, introducing concepts step-by-step along with the presentation of a test case application.
- Chapter 18: Control Theory and Digital Signal Processing Primer. This chapter provides a quick tour of the most important mathematical concepts for control theory and digital signal processing, using two case studies: the control of a pump and the development of a digital low-pass filter. The only mathematical background required of the reader corresponds to what is taught in a basic math course for engineering, and no specific previous knowledge in control theory and digital signal processing is assumed.

In the second edition of this book, an "Exercises" section has been added to every chapter. For the chapters dealing with theory, the solutions of the proposed exercises are given in the appendix. However, for the remaining chapters dealing with code, listing the lengthy source code of the solutions in the appendix would represent a waste of pages. For this reason, the solutions of the proposed exercises for such chapters are provided in the companion GitHub repository at `https://github.com/minimap-xl/RTOS_Book` from where they can be easily copied.

The short bibliography at the end of the book has been compiled with less experienced *readers* in mind. For this reason, we did not provide an exhaustive list of references, aimed at acknowledging each and every author who contributed to the rather vast field of real-time systems.

Rather, the bibliography is meant to point to a limited number of additional sources of information, which readers can and should actually use as a starting point to seek further information, without getting lost. There, readers will also find more, and more detailed, references to continue their quest.

2 A Case Study: Vision Control

This chapter describes a case study consisting of an embedded application performing online image processing. Both theoretical and practical concepts are introduced here: after an overview of basic concepts in computer input/output, some important facts on operating systems (OS) and software complexity will be presented here. Moreover, some techniques for software optimization and parallelization will be presented and discussed in the framework of the presented case study. The theory and techniques that are going to be introduced do not represent the main topic of this book. They are necessary, nevertheless, to fully understand the remaining chapters, which will concentrate on more specific aspects such as multithreading and process scheduling.

The presented case study consists of a Linux application that acquires a sequence of images (frames) from a video camera device. The data acquisition program will then perform some elaboration on the acquired images in order to detect the coordinates of the center of a circular shape in the acquired images.

This chapter is divided into four main sections. In the first section general concepts in computer input/output (I/O) are presented. The second section will discuss how I/O is managed by operating systems, in particular Linux, while in the third one the implementation of the frame acquisition is presented. The fourth section will concentrate on the analysis of the acquired frames to retrieve the desired information; after presenting two widespread algorithms for image analysis, the main concepts about software complexity will be presented, and it will be shown how the execution time for those algorithms can be reduced, sometimes drastically, using a few optimization and parallelization techniques.

Embedded systems carrying out online analysis of acquired images are becoming widespread in industrial control and surveillance. In order to acquire the sequence of the frames, the video capture application programming interface for Linux (V4L2) will be used. This interface supports most commercial USB webcams, which are now ubiquitous in laptops and other PCs. Therefore this sample application can be easily reproduced by the reader, using for example his/her laptop with an integrated webcam.

2.1 INPUT OUTPUT ON COMPUTERS

Every computer does input/output (I/O); a computer composed only of a processor and the memory would do barely anything useful, even if containing all the basic components for running programs. I/O represents the way computers interact with

DOI: 10.1201/9781003593416-2

9

the outside environment. There is a great variety of I/O devices: A personal computer will input data from the keyboard and the mouse, and output data to the screen and the speakers while using the disk, the network connection, and the USB ports for both input and output. An embedded system typically uses different I/O devices for reading data from sensors and writing data to actuators, leaving user interaction be handled by remote clients connected through the local area network (LAN).

2.1.1   ACCESSING THE I/O REGISTERS

In order to communicate with I/O devices, computer designers have followed two different approaches: *dedicated I/O bus* and *memory-mapped I/O*. Every device defines a set of registers for I/O management. *Input registers* will contain data to be read by the processor; *output registers* will contain data to be outputted by the device and will be written by the processor; *status registers* will contain information about the current status of the device; and finally *control registers* will be written by the processor to initiate or terminate device activities.

When a dedicated bus is defined for the communication between the processor and the device registers, it is also necessary that specific instructions for reading or writing device register are defined in the set of machine instructions. In order to interact with the device, a program will read and write appropriate values onto the I/O bus locations (i.e., at the addresses corresponding to the device registers) via specific I/O Read and Write instructions.

In memory-mapped I/O, devices are seen by the processor as a set of registers, but no specific bus for I/O is defined. Rather, the same bus used to exchange data between the processor and the memory is used to access I/O devices. Clearly, the address range used for addressing device registers must be disjoint from the set of addresses for the memory locations. Figures 2.1 and 2.2 show the bus organization for computers using a dedicated I/O bus and memory-mapped I/O, respectively. Memory-mapped architectures are more common nowadays, but connecting all the external I/O devices directly to the memory bus represents a somewhat simplified solution with several potential drawbacks in reliability and performance. In fact, since speed in memory access represents one of the major bottlenecks in computer performance, the memory bus is intended to operate at a very high speed, and therefore it has very strict constraints on the electrical characteristics of the bus lines, such as capacity, and in their dimension. Letting external devices be directly connected to the memory bus would increase the likelihood that possible malfunctions of the connected devices would seriously affect the function of the whole system and, even if that were not the case, there would be the concrete risk of lowering the data throughput over the memory bus.

In practice, one or more separate buses are present in the computer for I/O, even with memory-mapped architectures. This is achieved by letting a *bridge* component connect the memory bus with the I/O bus. The bridge presents itself to the processor as a device, defining a set of registers for programming the way the I/O bus is mapped onto the memory bus. Basically, a bridge can be programmed to define one or more

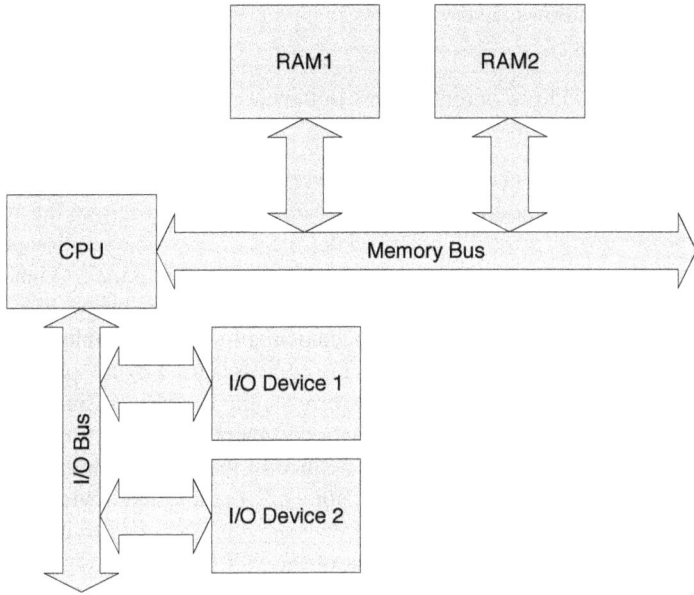

Figure 2.1 Bus architecture with a separate I/O bus.

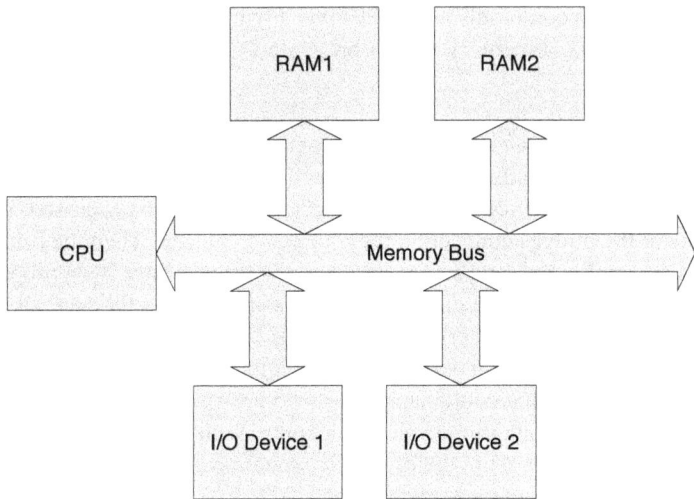

Figure 2.2 Bus architecture for Memory Mapped I/O.

address mapping windows. Every address mapping window is characterized by the following parameters:

1. Start and end address of the window in the memory bus
2. Mapping address offset

Once the bridge has been programmed, for every further memory access performed by the processor whose address falls in the selected address range, the bridge responds in the bus access protocol and translates the read or write operation performed in the memory bus into an equivalent read or write operation in the I/O bus. The address used in the I/O bus is obtained by adding the preprogrammed address offset for that mapping window. This simple mechanism allows to decouple the addresses used by I/O devices over the I/O bus from the addresses used by the processor.

A common I/O bus in computer architectures is the Peripheral Component Interconnect (PCI) bus, widely used in personal computers for connecting I/O devices. Normally, more than one PCI segment is defined in the same computer board. The PCI protocol, in fact, poses a limit in the number of connected devices and, therefore, in order to handle a larger number of devices, it is necessary to use PCI to PCI bridges, which connect different segments of the PCI bus. The bridge will be programmed in order to define map address windows in the primary PCI bus (which sees the bridge as a device connected to the bus) that are mapped onto the corresponding address range in the secondary PCI bus (for which the bridge is the master, i.e., leads bus operations). Following the same approach, new I/O buses, such as the Small Computer System Interface (SCSI) bus for high-speed disk I/O, can be integrated into the computer board by means of bridges connecting the I/O bus to the memory bus or, more commonly, to the PCI bus. Figure 2.3 shows an example of bus configuration defining a memory to PCI bridge, a PCI to PCI bridge, and a PCI to SCSI bridge.

One of the first actions performed when a computer boots is the configuration of the bridges in the system. Firstly, the bridges directly connected to the memory bus are configured, so that the devices over the connected buses can be accessed, including the registers of the bridges connecting these to new I/O buses. Then the bridges over these buses are configured, and so on. When all the bridges have been properly configured, the registers of all the devices in the system are directly accessible by the processor at given addresses over the memory bus. Properly setting all the bridges in the system may be tricky, and a wrong setting may make the system totally unusable. Suppose, for example, what could happen if an address map window for a bridge on the memory bus were programmed with an overlap with the address range used by the RAM memory. At this point the processor would be unable to access portions of memory and therefore would not anymore be able to execute programs.

Bridge setting, as well as other very low-level configurations are normally performed before the operating system starts, and are carried out by the Basic Input/Output System (BIOS), a code which is normally stored on ROM and executed as soon as the computer is powered. So, when the operating system starts, all the device registers

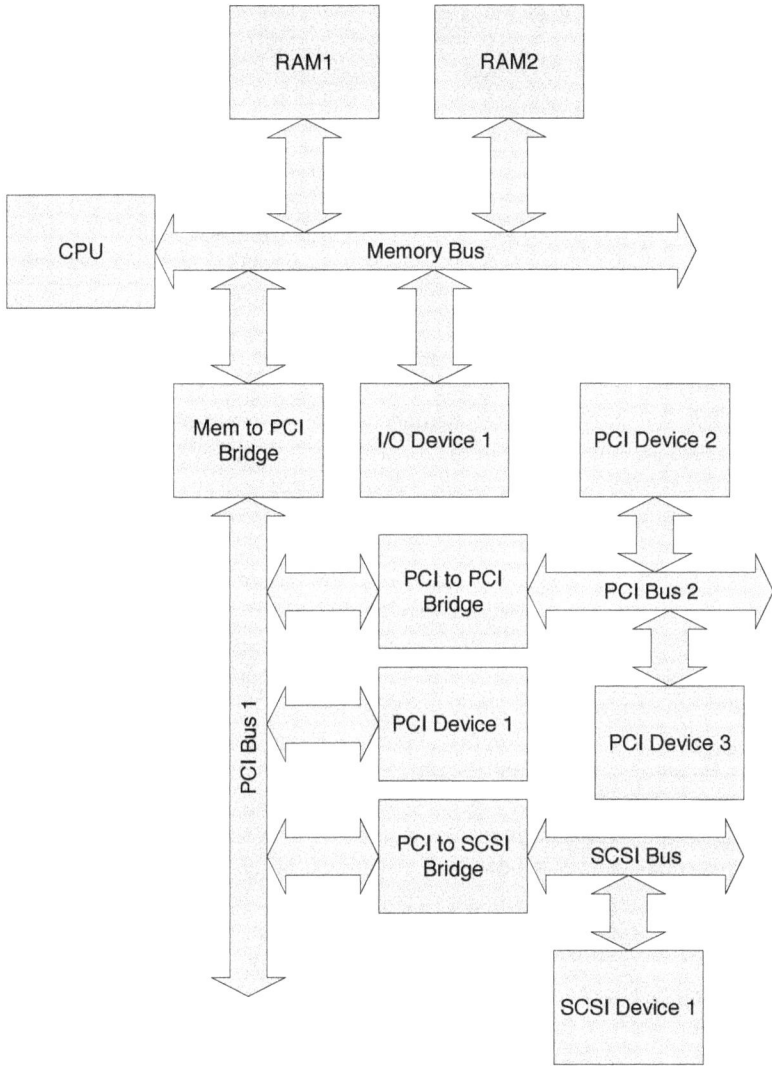

Figure 2.3 Bus architecture with two PCI buses and one SCSI bus.

are available at proper memory addresses. This is, however, not the end of the story: in fact, even if device registers are seen by the processor as if they were memory locations, there is a fundamental difference between devices and RAM blocks. While RAM memory chips are expected to respond in a time frame on the order of nanoseconds, the response time of devices largely varies and in general can be much longer. It is therefore necessary to synchronize the processor and the I/O devices.

2.1.2   SYNCHRONIZATION IN I/O

Consider, for example, a serial port with a baud rate of 9600 bit/s, and suppose that an incoming data stream is being received; even if ignoring the protocol overhead, the maximum incoming byte rate is 1200 byte/s. This means that the computer has to wait 0.83 milliseconds between two subsequent incoming bytes. Therefore, a sort of synchronization mechanism is needed to let the computer know when a new byte is available to be read in a data register for readout. The simplest method is *polling*, that is, repeatedly reading a status register that indicates whether new data is available in the data register. In this way, the computer can synchronize itself with the actual data rate of the device. This comes, however, at a cost: no useful operation can be carried out by the processor when synchronizing to devices in polling. If we assume that 100 ns are required on average for memory access, and assuming that access to device registers takes the same time as a memory access (a somewhat simplified scenario since we ignore here the effects of the memory cache), acquiring a data stream from the serial port would require more than 8000 read operations of the status register for every incoming byte of the stream – that is, wasting 99.99% of the processor power in useless accesses to the status register. This situation becomes even worse for slower devices; imagine the percentage of processor power for doing anything useful if polling were used to acquire data from the keyboard!

Observe that the operations carried out by I/O devices, once programmed by a proper configuration of the device registers, can normally proceed in parallel with the execution of programs. It is only required that the device should notify the processor when an I/O operation has been completed, and new data can be read or written by the processor. This is achieved using Interrupts, a mechanism supported by most I/O buses. When a device has been started, typically by writing an appropriate value in a command register, it proceeds on its own. When new data is available, or the device is ready to accept new data, the device raises an interrupt request to the processor (in most buses, some lines are dedicated to interrupt notification) which, as soon as it finishes executing the current machine instruction, will serve the interrupt request by executing a specific routine, called Interrupt Service Routine (ISR), for the management of the condition for which the interrupt has been generated.

Several facts must be taken into account when interrupts are used to synchronize the processor and the I/O operations. First of all, more than one device could issue an interrupt at the same time. For this reason, in most systems, a priority is associated with interrupts. Devices can, in fact, be ranked based on their importance, where important devices require a faster response. As an example, consider a system controlling a nuclear plant: An interrupt generated by a device monitoring the temperature of a nuclear reactor core is for sure more important than the interrupt generated by a printer device for printing daily reports. When a processor receives an interrupt request with a given associated priority level N, it will soon respond to the request only if it is not executing any service routine for a previous interrupt of priority M, $M \geq N$. In this case, the interrupt request will be served as soon as the previous Interrupt Service Routine has terminated and there are no pending interrupts with priority greater or equal to the current one.

When a processor starts serving an interrupt, it is necessary that it does not lose information about the program currently in execution. A program is fully described by the associated memory contents (the program itself and the associated data items), and by the content of the processor registers, including the Program Counter (PC), which records the address of the current machine instruction, and the Status Register (SR), which contains information on the current processor status. Assuming that memory locations used to store the program and the associated data are not overwritten during the execution of the interrupt service routine, it is only necessary to preserve the content of the processor registers. Normally, the first actions of the routine are to save in the stack the content of the registers that are going to be used, and such registers will be restored just before its termination. Not all the registers can be saved in this way; in particular, the PC and the SR are changed just before starting the execution of the interrupt service routine. The PC will be set to the address of the first instruction of the routine, and the SR will be updated to reflect the fact that the process is starting to service an interrupt of a given priority. So it is necessary that these two register are saved by the processor itself and restored when the interrupt service routine has finished (a specific instruction to return from ISR is defined in most computer architectures). In most architectures the SR and PC registers are saved on the stack, but others, such as the ARM architecture, define specific registers to hold the saved values.

A specific interrupt service routine has to be associated with every possible source of interrupt, so that the processor can take the appropriate actions when an I/O device generates an interrupt request. Typically, computer architectures define a vector of addresses in memory, called a *Vector Table*, containing the start addresses of the interrupt service routines for all the I/O devices able to generate interrupt requests. The offset of a given ISR within the vector table is called the *Interrupt Vector Number*. So, if the interrupt vector number were communicated by the device issuing the interrupt request, the right service routine could then be called by the processor. This is exactly what happens; when the processor starts serving a given interrupt, it performs a cycle on the bus called the *Interrupt Acknowledge Cycle* (IACK) where the processor communicates the priority of the interrupt being served, and the device which issued the interrupt request at the specified priority returns the interrupt vector number. In case two different devices issued an interrupt request at the same time with the same priority, the device closest to the processor in the bus will be served. This is achieved in many buses by defining a bus line in *Daisy Chain* configuration, that is, which is propagated from every device to the next one along the bus, only in cases where it did not answer to an IACK cycle. Therefore, a device will answer to an IACK cycle only if both conditions are met:

1. It has generated a request for interrupt at the specified priority
2. It has received a signal over the daisy chain line

Note that in this case it will not propagate the daisy chain signal to the next device.

The offset returned by the device in an IACK cycle depends on the current organization of the vector table and therefore must be a programmable parameter in the

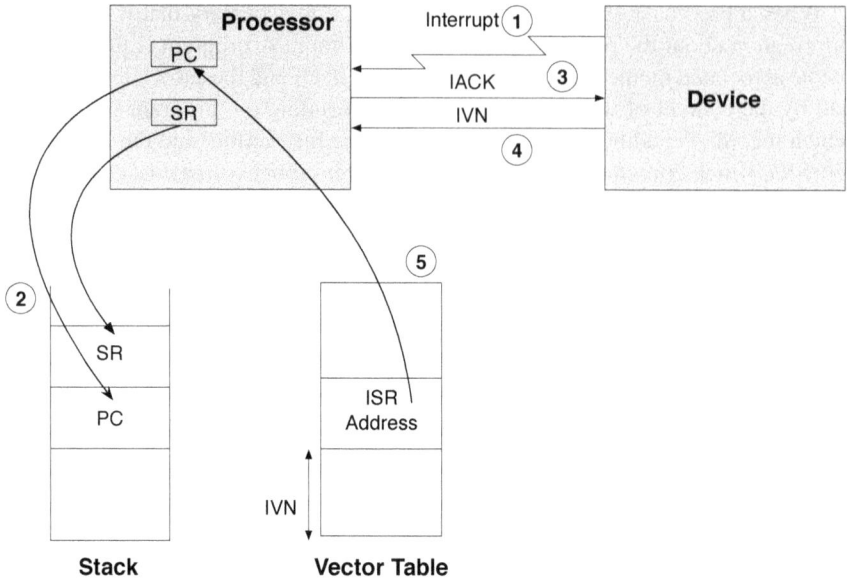

Figure 2.4 The interrupt sequence.

device. Typically, all the devices which are able to issue an interrupt request have two registers for the definition of the interrupt priority and the interrupt vector number, respectively. The sequence of actions is shown in Figure 2.4, highlighting the main steps of the sequence:

1. The device issues an interrupt request;
2. The processor saves the context, i.e., puts the current values of the PC and of the SR on the stack;
3. The processor issues an interrupt acknowledge cycle (IACK) on the bus;
4. The device responds by putting the interrupt vector number (IVN) over the data lines of the bus;
5. The processor uses the IVN as an offset in the vector table and loads the interrupt service routine address in the PC.

Programming a device using interrupts is not a trivial task, and it consists of the following steps:

1. The interrupt service routine has to be written. The routine can assume that the device is ready at the time it is called, and therefore no synchronization (e.g., polling) needs to be implemented;
2. During system boot, that is when the computer and the connected I/O devices are configured, the code of the interrupt service routine has to be loaded in memory, and its start address written in the vector table at, say, offset N;

3. The value N has to be communicated to the device, usually written in the interrupt vector number register;
4. When an I/O operation is requested by the program, the device is started, usually by writing appropriate values in one or more command registers. At this point the processor can continue with the program execution, while the device operates. As soon as the device is ready, it will generate an interrupt request, which will be eventually served by the processor by running the associated interrupt service routine.

In this case it is necessary to handle the fact that data reception is asynchronous. A commonly used techniques is to let the program continue after issuing an I/O request until the data received by the device is required. At this point the program has to suspend its execution waiting for data, unless not already available, that is, waiting until the corresponding interrupt service routine has been executed. For this purpose the interprocess communication mechanisms described in Chapter 5 will be used.

2.1.3   DIRECT MEMORY ACCESS (DMA)

The use of interrupts for synchronizing the processor and the connected I/O devices is ubiquitous, and we will see in the next chapters how interrupts represent the basic mechanism over which operating systems are built. Using interrupts clearly spares processor cycles when compared with polling; however, there are situations in which even interrupt-driven I/O would require too much computing resources. To better understand this fact, let's consider a mouse which communicates its current position by interrupting the processor 30 times per second. Let's assume that 400 processor cycles are required for the dispatching of the interrupt and the execution of the interrupt service routine. Therefore, the number of processor cycles which are dedicated to the mouse management per second is $400 * 30 = 12000$. For a 1 GHz clock, the fraction of processor time dedicated to the management of the mouse is $12000/10^9$, that is, 0.0012% of the processor load. Managing the mouse requires, therefore, a negligible fraction of processor power.

Consider now a hard disk that is able to read data with a transfer rate of 4 MByte/s, and assume that the device interrupts the processor every time 16 bytes of data are available. Let's also assume that 400 clock cycles are still required to dispatch the interrupt and execute the associated service routine. The device will therefore interrupt the processor 250000 times per second, and 10^8 processor cycles will be dedicated to handle data transfer every second. For a 1 GHz processor this means that 10% of the processor time is dedicated to data transfer, a percentage clearly no more acceptable.

Very often data exchanged with I/O devices are transferred from or to memory. For example, when a disk block is read it is first transferred to memory so that it is later available to the processor. If the processor itself were in charge of transferring the block, say, after receiving an interrupt request from the disk device to signal the block availability, the processor would repeatedly read data items from the device's data register into an internal processor register and write it back into memory. The net effect is that a block of data has been transferred from the disk into memory, but it has been obtained at the expense of a number of processor cycles that could have been used to do other jobs if the device were allowed to write the disk block into memory by itself. This is exactly the basic concept of *Direct Memory Access* (DMA), which is letting the devices read and write memory by themselves so that the processor will handle I/O data directly in memory. In order to put this simple concept in practice it is, however, necessary to consider a set of facts. First of all, it is necessary that the processor can "program" the device so that it will perform the correct actions, that is, reading/writing a number N of data items in memory, starting from a given memory address A. For this purpose, every device able to perform DMA provides at least the following registers:

- A Memory Address Register (MAR) initially containing the start address in memory of the block to be transferred;
- A Word Count register (WC) containing the number of data items to be transferred.

So, in order to program a block read or write operation, it is necessary that the processor, after allocating a block in memory and, in case of a write operation, filling it with the data to be output to the device, writes the start address and the number of data items in the MAR and WC registers, respectively. Afterward the device will be started by writing an appropriate value in (one of) the command register(s). When the device has been started, it will operate in parallel with the processor, which can proceed in the execution of the program. However, as soon as the device is ready to transfer a data item, it will require the memory bus used by the processor to exchange data with memory, and therefore some sort of bus arbitration is needed since it is not possible that two devices read or write the memory at the same time on the same bus (note however that nowadays memories often provide multiport access, that is, allow simultaneous access to different memory addresses). At any time one, and only one, device (including the processor) connected to the bus is the *master*, i.e., can initiate a read or write operation. All the other connected devices at that time are slaves and can only answer to a read/write bus cycle when they are addressed. The memory will be always a slave in the bus, as well as the DMA-enabled devices when they are not performing DMA. At the time such a device needs to exchange data with the memory, it will ask the current master (normally the processor, but it may be another device performing DMA) the ownership of the bus. For this purpose the protocol of every bus able to support ownership transfer is to define a cycle for the bus owner-ship transfer. In this cycle, the potential master raises a request line and the current master, in response, relinquishes the mastership, signaling this over another bus line, and possibly waiting for the termination of a read/write operation in progress. When a device has taken the bus ownership, it can then perform the transfer of the data item and will remain the current master until the processor or another device asks to become the new master. It is worth noting that the bus ownership transfers are han-dled by the bus controller components and are carried out entirely in hardware. They are, therefore, totally transparent to the programs being executed by the processor, except for a possible (normally very small) delay in their execution.

Every time a data item has been transferred, the MAR is incremented and the WC is decremented. When the content of the WC becomes zero, all the data have been transferred, and it is necessary to inform the processor of this fact by issuing an interrupt request. The associated Interrupt Service Routine will handle the block transfer termination by notifying the system of the availability of new data. This is normally achieved using the interprocess communication mechanisms described in Chapter 5.

2.2   INPUT/OUTPUT OPERATIONS AND THE OPERATING SYSTEM

After having seen the techniques for handling I/O in computers, the reader will be convinced that it is highly desirable that the complexity of I/O should be handled by the operating system and not by user programs. Not surprisingly, this is the case for most operating systems, which offer a unified interface for I/O operations despite the large number of different devices, each one defining a specific set of registers and requiring a specific I/O protocol. Of course, it is not possible that operating systems could include the code for handling I/O in every available device. Even if it were the case, and the developers of the operating system succeed in the titanic effort of providing the device specific code for every known device, the day after the system release there will be tens of new devices not supported by such an operating system. For this reason, operating systems implement the generic I/O functionality, but leave the details to a device-specific code, called the *Device Driver*. In order to be integrated into the system, every device requires its software driver, which depends not only on the kind of hardware device but also on the operating system. In fact, every operating system defines its specific set of interfaces and rules a driver must adhere to in order to be integrated. Once installed, the driver becomes a component of the operating system. This means that a failure in the device driver code execution becomes a failure of the operating system, which may lead to the crash of the whole system. (At least in monolithic operating systems such as Linux and Windows; this may be not true for other systems, such as microkernel-based ones.) User programs will never interact directly with the driver as the device is accessible only via the Application Programming Interface (API) provided by the operating system. In the following we shall refer to the Linux operating systems and shall see how a uniform interface can be adapted to the variety of available devices. A complete example if Linux driver will be provided in Chapter 17. The other operating systems adopt a similar architecture for I/O, which typically differ only by the name and the arguments of the I/O systems routines, but not on their functionality.

2.2.1   USER AND KERNEL MODES

We have seen how interacting with I/O devices means reading and writing into device registers, mapped at given memory addresses. It is easy to guess what could happen if user programs were allowed to read and write at the memory locations corresponding to device registers. The same consideration holds also for the memory structures used by the operating system itself. If user programs were allowed to freely access the whole addressing range of the computer, an error in a program causing a memory access to a wrong address (something every C programmer experiences often) may lead to the corruption of the operating system data structures, or to an interference with the device operation, leading to a system crash.

For this reason most processors define at least two levels of execution: *user mode* and *kernel* (or *supervisor*) *mode*. When operating in user mode, a program is not allowed to execute some machine instructions (called *Privileged Instructions*) or to access sets of memory addresses. Conversely, when operating in kernel mode, a program has full access to the processor instructions and to the full addressing range. Clearly, most of the operating system code will be executed in kernel mode, while user programs are kept away from dangerous operations and are intended to be executed in user mode. Imagine what would happen if the HALT machine instruction for stopping the processor were available in user mode, possibly on a server with tens of connected users.

A first problem arises when considering how a program can switch from user to kernel mode. If this were carried out by a specific machine instruction, would such an instruction be accessible in user mode? If not, it would be useless, but if it were, the barrier between kernel mode and user mode would be easily circumvented, and malicious programs could easily take the whole system down.

So, how to solve the dilemma? The solution lies in a new mechanism for the invocation of software routines. In the normal routine invocation, the calling program copies the arguments of the called routine over the stack and then puts the address of the first instruction of the routine into the Program Counter register, after having copied on the stack the return address, that is, the address of the next instruction in the calling program. Once the called routine terminates, it will pick the saved return address from the stack and put it into the Program Counter, so that the execution of the calling program is resumed. We have already seen, however, how the interrupt mechanism can be used to "invoke" an interrupt service routine. In this case the sequence is different, and is triggered not by the calling program but by an external hardware device. It is exactly when the processor starts executing an Interrupt Service routine that the current execution mode is switched to kernel mode. When the interrupt service routine returns and the interrupted program resumes its execution, unless not switching to a new interrupt service routine, the execution mode is switched to user mode. It is worth noting that the mode switch is not controlled by the software, but it is the processor which only switches to kernel mode when servicing an interrupt.

This mechanism makes sense because interrupt service routines interact with devices and are part of the device driver, that is, of a software component that is integrated in the operating system. However, it may happen that user programs have to do I/O operations, and therefore they need to execute some code in kernel mode. We have claimed that all the code handling I/O is part of the operating system and therefore the user program will call some system routine for doing I/O. However, how do we switch to kernel mode in this case where the trigger does not come from an hardware device? The solution is given by *Software Interrupts*. Software interrupts are not triggered by an external hardware signal, but by the execution of a specific machine instruction. The interrupt mechanism is quite the same: The processor saves the current context, picks the address of the associated interrupt service routine from the vector table and switches to kernel mode, but in this case the Interrupt Vector number is not obtained by a bus IACK cycle; rather, it is given as an argument to the machine instruction for the generation of the software interrupt.

The net effect of software interrupts is very similar to that of a function call, but the underlying mechanism is completely different. This is the typical way the operating system is invoked by user programs when requesting system services, and it represents an effective barrier protecting the integrity of the system. In fact, in order to let any code to be executed via software interrupts, it is necessary to write in the vector table the initial address of such code but, not surprisingly, the vector table is not accessible in user mode, as it belongs to the set of data structures whose integrity is essential for the correct operation of the computer. The vector table is typically initialized during the system boot (executed in kernel mode) when the operating system initializes all its data structures.

To summarize the above concepts, let's consider the execution story of one of the most used C library function: `printf()`, which takes as parameter the (possibly formatted) string to be printed on the screen. Its execution consists of the following steps:

1. The program calls routine `printf()`, provided by the C run time library. Arguments are passed on the stack and the start address of the `printf` routine is put in the program counter;

2. The `printf` code will carry out the required formatting of the passed string and the other optional arguments, and then calls the operating system specific system service for writing the formatted string on the screen;

3. The system routine executes initially in user mode, makes some preparatory work and then needs to switch in kernel mode. To do this, it will issue a software interrupt, where the passed interrupt vector number specifies the offset in the Vector Table of the corresponding ISR routine to be executed in kernel mode;

4. The ISR is eventually activated by the processor in response to the software interrupt. This routine is provided by the operating system and it is now executing in kernel mode;

5. After some work to prepare the required data structures, the ISR routine will interact with the output device. To do this, it will call specific routines of the device driver;

6. The activated driver code will write appropriate values in the device registers to start transferring the string to the video device. In the meantime the calling process is put in wait state (see Chapter 3 for more information on processes and process states);

7. A sequence of interrupts will be likely generated by the device to handle the transfer of the bytes of the string to be printed on the screen;

8. When the whole string has been printed on the screen, the calling process will be resumed by the operating system and `printf` will return.

Software interrupts provide the required barrier between user and kernel mode, which is of paramount importance in general purpose operating systems. This comes, however, at a cost: the activation of a kernel routine involves a sequence of actions, such as saving the context, which is not necessary in a direct call. Many embedded systems are then not intended to be of general usage. Rather, they are intended to run a single program for control and supervision or, in more complex systems involving multitasking, a well-defined set of programs developed ad hoc. For this reason several real-time operating systems do not support different execution levels (even if the underlying hardware could), and all the software is executed in kernel mode, with full access to the whole set of system resources. In this case, a direct call is used to activate system routines. Of course, the failure of a program will likely bring the whole system down, but in this case it is assumed that the programs being executed have already been tested and can therefore be trusted.

2.2.2 INPUT/OUTPUT ABSTRACTION IN LINUX

Letting the operating system manage input/output on behalf of the user is highly desirable, hiding as far as possible the communication details and providing a simple and possibly uniform interface for I/O operations. We shall learn how a simple Application Programming Interface for I/O can be effectively used despite the great variety of devices and of techniques for handling I/O. Here we shall refer to Linux, but the same concepts hold for the vast majority of the other operating systems.

In Linux every I/O device is basically presented to users as a *file*. This may seem at a first glance a bit surprising since the similarity between files and devices is not so evident, but the following considerations hold:

- In order to be used, a file must be open. The open() system routine will create a set of data structures that are required to handle further operations on that file. A file identifier is returned to be used in the following operations for that file in order to identify the associated data structures. In general, every I/O device requires some sort of initialization before being used. Initialization will consist of a set of operations performed on the device and in the preparation of a set of support data structures to be used when operating on that device. So an open() system routine makes sense also for I/O devices. The returned identifier (actually an integer number in Linux) is called a *Device Descriptor* and uniquely identifies the device instance in the following operations. When a file is no more used, it is closed and the associated data structures deallocated. Similarly, when a I/O device is no more used, it will be closed, performing cleanup operations and freeing the associated resources.
- A file can be read or written. In the read operation, data stored in the file are copied in the computer memory, and the converse holds for write operations. Regardless of the actual nature of a I/O device, there are two main categories of interaction with the computer: read and write. In read operation, data from the device is copied into the computer memory to be used by

the program. In write operations, data in memory will be transferred to the device. Both `read()` and `write()` system routines will require the target file or device to be uniquely identified. This will be achieved by passing the identifier returned by the `open()` routine.

However, due to the variety of hardware devices that can be connected to a computer, it is not always possible to provide a logical mapping of the device's functions exclusively into read-and-write operations. Consider, as an example, a network card: actions such as receiving data and sending data over the network can be mapped into read-and-write operations, respectively, but others, like the configuration of the network address, require a different interface. In Linux this is achieved by providing an additional routine for I/O management: `ioctl()`. In addition to the device descriptor, `ioctl()` defines two more arguments: the first one is an integer number and is normally used to specify the kind of operation to be performed; the second one is a pointer to a data structure that is specific to the device and the operation. The actual meaning of the last argument will depend on the kind of device and on the specified kind of operation. It is worth noting that Linux does not make any use of the last two `ioctl()` arguments, passing them as they are to the device-specific code, i.e., the device driver.

The outcome of the device abstraction described above is deceptively simple: the functionality of all the possible devices connected to the computers is basically carried out by the following four routines:

- `open()` to initialize the device;
- `close()` to close and release the device;
- `read()` to get data from the device;
- `write()` to send data to the device;
- `ioctl()` for all the remaining operations of the device.

The evil, however, hides in the details, and in fact, all the complexity in the device/-computer interaction has been simply moved to `ioctl()`. Depending on the device's nature, the set of operations and of the associated data structures may range from a very few and simple configurations to a fairly complex set of operations and data structures, described by hundreds of user manual pages. This is exactly the case of the standard driver for the camera devices that will be used in the subsequent sections of this chapter for the presented case study.

The abstraction carried out by the operating system in the application programming interface for device I/O is also maintained in the interaction between the operating system and the device-specific driver. We have already seen that, in order to integrate a device in the systems, it is necessary to provide a device-specific code assembled into the device driver and then integrated into the operating system. Basically, a device driver provides the implementation of the open, close, read, write, and ioctl operations. So, when a program opens a device by invoking the `open()` system routine, the operating system will first carry out some generic operations common to all devices, such as the preparation of its own data structures for handling the device, and will then call the device driver's `open()` routine to carry out the required device specific actions. The actions carried out by the operating system may involve the management of the calling process. For example, in a read operation, the operating system, after calling the device-specific read routine, may suspend the current process (see Chapter 3 for a description of the process states) in the case the required data are not currently available. When the data to be read becomes available, the system will be notified of it, say, with an interrupt from the device, and the operating system will wake the process that issued the `read()` operation, which can now terminate the `read()` system call.

2.3   ACQUIRING IMAGES FROM A CAMERA DEVICE

So far, we have learned how input/output operation are managed by Linux. Here we shall see in detail how the generic routines for I/O can be used for a real application, that is, acquiring images from a video camera device. A wide range of camera devices is available, ranging from $10 USB Webcams to $100K cameras for ultra-fast image recording. The number and the type of configuration parameters varies from device to device, but it will always include at least:

- Device capability configuration parameters, such as the ability of supporting data streaming and the supported pixel formats;
- Image format definition, such as the width and height of the frame, the number of bytes per line, and the pixel format.

Due to the large number of different camera devices available on the market, having a specific driver for every device, with its own configuration parameters and `ioctl()` protocol (i.e., the defined operations and the associated data structures), would complicate the life of the programmers quite a lot. Suppose, for example, what would happen if in an embedded system for on-line quality control based on image analysis the type of used camera is changed, say, because a new better device is available. This would imply re-writing all the code which interacts with the device. For this reason, a unified interface to camera devices has been developed in the Linux community. This interface, called V4L2 (Video for Linux Two), defines a set of ioctl operations and associated data structures that are general enough to be adapted for all the available camera devices of common usage. If the driver of a given camera device adheres to the V4L2 standards, the usability of such device is greatly improved and it can be quickly integrated into existing systems. V4L2 improves also interchangeability of camera devices in applications. To this purpose, an important feature of V4L2 is the availability of query operations for discovering the supported functionality of the device. A well-written program, first querying the device capabilities and then selecting the appropriate configuration, can be reused for a different camera device with no change in the code.

As V4L2 in principle covers the functionality of all the devices available on the market, the standard is rather complicated because it has to foresee even the most exotic functionality. Here we shall not provide a complete description of V4L2 interface, which can be found in [81], but will illustrate its usage by means of two examples. In the first example, a camera device is inquired in order to find out the supported formats and to check whether the YUYV format is supported. If this format is supported, camera image acquisition is started using the `read()` system routine. YUYV is a format to encode pixel information expressed by the following information:

- Luminance (Y)
- Blue Difference Chrominance (C_b)
- Red Difference Chrominance (C_r)

Y, C_b, and C_r represent a way to encode *RGB* information in which red (R), green (G), and blue (B) light are added together to reproduce a broad array of colors for image pixels, and there is a precise mathematical relationship between R, G, B, and Y, C_b, and C_r parameters, respectively. The luminance Y represents the brightness in an image and can be considered alone if only a grey scale representation of the image is needed. In our case study we are not interested in the colors of the acquired images, rather we are interested in retrieving information from the shape of the objects in the image, so we shall consider only the component Y.

The YUYV format represents a compressed version of the Y, C_b, and C_r. In fact, while the luminance is encoded for every pixel in the image, the chrominance values are encoded for every two pixels. This choice stems from the fact that the human eye is more sensitive to variation of the light intensity, rather than of the colors components. So in the YUYV format, pixels are encoded from the topmost image line and from the left to the right, and four bytes are used to encode two pixels with the following pattern: $Y_i, C_{bi}, Y_{i+1}, C_{ri}$. To get the grey scale representation of the acquired image, our program will therefore take the first byte of every pair.

2.3.1 SYNCHRONOUS READ FROM A CAMERA DEVICE

This first example shows how to read from a camera device using synchronous frame readout, that is, using the `read()` function for reading data from the camera device. Its code is listed below;

```
#include <fcntl.h>
#include <stdio.h>
#include <stdlib.h>
#include <string.h>
#include <errno.h>
#include <linux/videodev2.h>
#include <asm/unistd.h>
#include <sys/ioctl.h>
#include <unistd.h>
#include <poll.h>

#define MAX_FORMAT 100
#define FALSE 0
#define TRUE 1
#define CHECK_IOCTL_STATUS(message)  \\
if(status == -1)                     \\
{                                    \\
    perror(message);                 \\
    exit(EXIT_FAILURE);              \\
```

```
}

main (int argc, char *argv[])
{
    int fd, idx, status;
    int pixelformat;
    int imageSize;
    int width, height;
    int yuyvFound;

    struct v4l2_capability cap;      //Query Capability structure
    struct v4l2_fmtdesc fmt;         //Query Format Description structure
    struct v4l2_format format;       //Query Format structure
    char *buf;                       //Image buffer
    fd_set fds;                      //Select descriptors
    struct timeval tv;               //Timeout specification structure

/* Step 1: Open the device */
    fd = open("/dev/video1", O_RDWR);

/* Step 2: Check read/write capability */
    status = ioctl(fd, VIDIOC_QUERYCAP, &cap);
    CHECK_IOCTL_STATUS("Error Querying capability")
    if(!(cap.capabilities & V4L2_CAP_READWRITE))
    {
        printf("Read I/O NOT supported\n");
        exit(EXIT_FAILURE);
    }

/* Step 3: Check supported formats */
    yuyvFound = FALSE;
    for(idx = 0; idx < MAX_FORMAT; idx++)
    {
        fmt.index = idx;
        fmt.type = V4L2_BUF_TYPE_VIDEO_CAPTURE;
        status = ioctl(fd, VIDIOC_ENUM_FMT, &fmt);
        if(status != 0) break;
        if(fmt.pixelformat == V4L2_PIX_FMT_YUYV)
        {
            yuyvFound = TRUE;
            break;
        }
    }
    if(!yuyvFound)
    {
        printf("YUYV format not supported\n");
        exit(EXIT_FAILURE);
    }

/* Step 4: Read current format definition */
    memset(&format, 0, sizeof(format));
    format.type = V4L2_BUF_TYPE_VIDEO_CAPTURE;
    status = ioctl(fd, VIDIOC_G_FMT, &format);
    CHECK_IOCTL_STATUS("Error Querying Format")

/* Step 5: Set format fields to desired values: YUYV coding,
   480 lines, 640 pixels per line */
    format.fmt.pix.width = 640;
    format.fmt.pix.height = 480;
    format.fmt.pix.pixelformat = V4L2_PIX_FMT_YUYV;

/* Step 6: Write desired format and check actual image size */
    status = ioctl(fd, VIDIOC_S_FMT, &format);
    CHECK_IOCTL_STATUS("Error Setting Format")
    width = format.fmt.pix.width;              //Image Width
    height = format.fmt.pix.height;            //Image Height
```

```
        //Total image size in bytes
        imageSize = (unsigned int)format.fmt.pix.sizeimage;

  /* Step 7: Start reading from the camera */
        buf = malloc(imageSize);
        FD_ZERO(&fds);
        FD_SET(fd, &fds);
        tv.tv_sec = 20;
        tv.tv_usec = 0;
        for(;;)
        {
            status = select(1, &fds, NULL, NULL, &tv);
            if(status == -1)
            {
                perror("Error in Select");
                exit(EXIT_FAILURE);
            }
            status = read(fd, buf, imageSize);
            if(status == -1)
            {
                perror("Error reading buffer");
                exit(EXIT_FAILURE);
            }
  /* Step 8: Do image processing */
            processImage(buf, width, height, imageSize);
        }
}
```

The first action (step 1)in the program is opening the device. System routine open() looks exactly as an open call for a file. As for files, the first argument is a path name, but in this case such a name specifies the device instance. In Linux the names of the devices are all contained in the /dev directory. The files contained in this directory do not correspond to real files (a Webcam is obviously different from a file), rather, they represent a rule for associating a unique name with each device in the system. In this way it is also possible to discover the available devices using the ls command to list the files contained in a directory. By convention, camera devices have the name /dev/video<n>, so the command ls /dev/video* will show how many camera devices are available in the system. The second argument given to system routine open() specifies the protection associated with that device. In this case the constant O_RDWR specifies that the device can be read and written. The returned value is an integer value that uniquely specifies within the system the *Device Descriptor*, that is the set of data structures held by Linux to manage this device. This number is then passed to the following ioctl() calls to specify the target device. Step 2 consists in checking whether the camera device supports read/write operation. The attentive reader may find this a bit strange—how could the image frames be acquired otherwise?—but we shall see in the second example that an alternative way, called *streaming*, is normally (and indeed most often) provided. This query operation is carried out by the following line:

```
    status = ioctl(fd, VIDIOC_QUERYCAP, &cap);
```

In the above line the ioctl operation code is given by constant VIDIOC_QUERYCAP (defined, as all the other constants used in the management of the video device, in linux/videodev2.h), and the associated data structure for the pointer argument is of type `v4l2_capability`. This structure, documented in the V4L2 API specification, defines, among others, a capability field containing a bit mask specifying the supported capabilities for that device.

Line

```
if(cap.capabilities & V4L2_CAP_READWRITE)
```

will let the program know whether read/write ability is supported by the device.

In step 3 the device is queried about the supported pixel formats. To do this, `ioctl()` is repeatedly called, specifying `VIDIOC_ENUM_FMT` operation and passing the pointer to a `v4l2_fmtdesc` structure whose fields of interest are:

- `index`: to be set before calling `ioctl()` in order to specify the index of the queried format. When no more formats will be available, that is, when the index is greater or equal the number of supported indexes, `ioctl()` will return an error.
- `type`: specifies the type of the buffer for which the supported format is being queried. Here, we are interested in the returned image frame, and this is set to `V4L2_BUF_TYPE_VIDEO_CAPTURE`
- `pixelformat`: returned by `ioctl()`, specifies supported format at the given index

If the pixel format YUYV is found (this is the normal format supported by all Webcams), the program proceeds in defining an appropriate image format. There are many parameters for specifying such information, all defined in structure `v4l2_format` passed to `ioctl` to get (operation `VIDIOC_G_FMT`) or to set the format (operation `VIDIOC_S_FMT`). The program will first read (step 4) the currently defined image format (normally most default values are already appropriate) and then change (step 5) the formats of interest, namely, image width, image height, and the pixel format. Here, we are going to define a 640 x 480 image using the YUYV pixel format by writing the appropriate values in fields `fmt.pix.width`, `fmt.pix.height`, and `fmt.pix.pixelformat` of the format structure. Observe that, after setting the new image format, the program checks the returned values for image width and height. In fact, it may happen that the device does not support exactly the requested image width and height, and in this case the format structure returned by `ioctl` contains the appropriate values, that is, the supported width and height that are closest to the desired ones. Fields `pix.sizeimage` will contain the total length in bytes of the image frame, which in our case will be given by 2 times width times height (recall that in YUYV format four bytes are used to encode two pixels).

At this point the camera device is configured, and the program can start acquiring image frames. In this example a frame is acquired via a `read()` call whose arguments are:

- The device descriptor;
- The data buffer;
- The dimension of the buffer.

Function `read()` returns the number of bytes actually read, which is not necessarily equal to the number of bytes passed as argument. In fact, it may happen that at the time the function is called, not all the required bytes are available, and the program has to manage this properly. So, it is necessary to make sure that when `read()` is called, a frame is available for readout. The usual technique in Linux to synchronize read operation on device is the usage of the `select()` function, which allows a program to monitor multiple device descriptors, waiting until one or more devices become "ready" for some class of I/O operation (e.g., input data available). A device is considered ready if it is possible to perform the corresponding I/O operation (e.g., read) without blocking. Observe that the usage of select is very useful when a program has to deal with several devices. In fact, since `read()` is blocking, that is, it suspends the execution of the calling program until some data are available, a program reading on multiple devices may suspend in a `read()` operation regardless the fact that some other device may have data ready to be read. The arguments passed to `select()` are

- The number of involved devices;
- The read device mask;
- The write device mask
- The mask of devices to be monitored for exceptions;
- The wait timeout specification.

The devices masks have are of type `fd_set`, and there is no need to know its definition since macros FD˙ZERO and FD˙SET allow resetting the mask and adding a device descriptor to it, respectively. When the select has not to monitor a device class, the corresponding mask is NULL, as in the above example for the write and exception mask. The timeout is specified using the structure `timeval`, which defines two fields, `tv_sec` and `tv_usec`, to specify the number of seconds and microseconds, respectively.

The above example will work fine, provided the camera device supports direct `read()` operation, as far as it is possible to guarantee that `read()` routine is called as often as the frame rate. This is, however, not always the case because the process running the program may be preempted by the operating system in order to assign the processor to other processes. Even if we can guarantee that, on average, the read rate is high enough, it is in general necessary to handle the occasional cases in which the reading process is late and the frame may be lost. Several chapters of this book will discuss this fact, and we shall see several techniques to ensure real-time behavior, that is, making sure that a given action will be executed within a given amount of time. If this were the case, and we could ensure that the `read()` operation for the current frame will be always executed before a new frame is acquired, there would be no risk of losing frames. Otherwise it is necessary to handle occasional delays in frame readout. The common technique for this is *double buffering*, that is using two buffers for the acquired frames. As soon as the driver is able to read a frame, normally in response to an interrupt indicating that the DMA transfer for that frame has terminated, the frame is written in two alternate memory buffers. The process acquiring such frames can then copy from one buffer while the driver is filling in the other one. In this case, if T is the frame acquisition period, a process is allowed to read a frame with a delay up to T. Beyond this time, the process may be reading a buffer that at the same time is being written by the driver, producing inconsistent data or losing entire frames. The double buffering technique can be extended to multiple buffering by using N buffers linked to form a circular chain. When the driver has filled the nth buffer, it will use buffer $(n+1)modN$ for the next acquisition. Similarly, when a process has read a buffer it will proceed to the next one, selected in the same way as above. If the process is fast enough, the new buffer will not be yet filled, and the process will be blocked in the select operation. When `select()` returns, at least one buffer contains valid frame data. If, for any reason, the process is late, more than one buffer will contain acquired frames not yet read by the program. With N buffers, for a frame acquisition period of T, the maximum allowable delay for the reading process is $(N-1)T$. In the next example, we shall use this technique, and we shall see that it is no more necessary to call function `read()` to get data, as one or more frames will be already available in the buffers that have been set before by the program. Before proceeding with the discussion of the new example, it is, however, necessary to introduce the *virtual memory* concept.

2.3.2   VIRTUAL MEMORY

Virtual memory, supported by most general-purpose operating systems, is a mechanism by which the memory addresses used by the programs running in user mode do not correspond to the addresses the CPU uses to access the RAM memory in the same instructions. The address translation is performed by a component of the processor called the *Memory Management Unit* (MMU). The details of the translation may vary, depending on the computer architecture, but the basic mechanism always relies on a data structure called the *Page Table*. The memory address managed by the user program, called *Virtual Address* (or *Logical Address*) is translated by the MMU, first dividing its N bits into two parts, the first one composed of the K least significant bits and the other one composed of the remaining $N-K$ bits, as shown in Figure 2.5. The most significant $N-K$ bits are used as an index in the *Page Table*, which is composed of an array of numbers, each L bits long. The entry in the page table corresponding to the given index is then paired to the least significant K bits of the virtual address, thus obtaining a number composed of $L+K$ bits that represents the *physical address*, which will be used to read the physical memory. In this way it is also possible to use a different number of bits in the representation of virtual and physical addresses. If we consider the common case of 32 bit architectures, where 32 bits are used to represent virtual addresses, the top $32-K$ bits of virtual addresses are used as the index in the page table. This corresponds to providing a logical organization of the virtual address rage in a set of *memory pages*, each 2^K bytes long. So the most significant $32-K$ bits will provide the memory page number, and the least significant K bits will specify the *offset* within the memory page. Under this perspective, the page table provides a page number translation mechanism, from the logical page number into the physical page number. In fact, also the physical memory can be

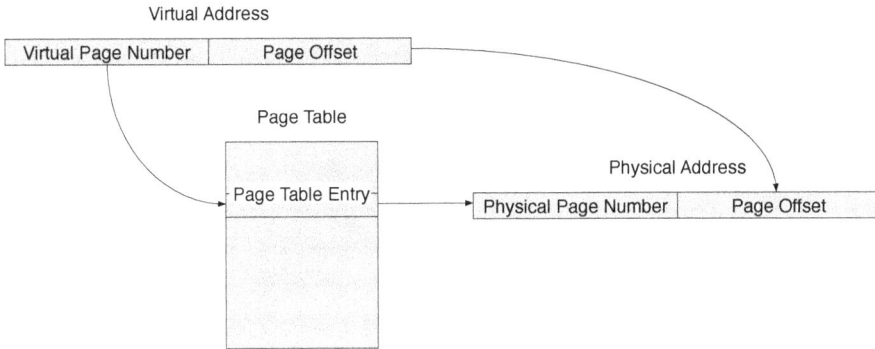

Figure 2.5 The virtual memory address translation.

considered divided into pages of the same size, and the offset of the physical address within the translated page will be the same of the original logical page.

Even if virtual memory may seem at a first glance a method merely invented to complicate the engineer's life, the following example should convince the skeptics of its convenience. Consider two processes running the same program: This is perfectly normal in everyday's life, and no one is, in fact, surprised by the fact that two Web browsers or editor programs can be run by different processes in Linux (or tasks in Windows). Recalling that a program is composed of a sequence of machine instructions handling data in processor registers and in memory, if no virtual memory were supported, the two instances of the same program run by two different processes would interfere with each other since they would access the same memory locations (they are running the *same* program). This situation is elegantly solved, using the virtual memory mechanism, by providing two different mappings to the two processes so that the same virtual address page is mapped onto two different physical pages for the two processes, as shown in Figure 2.6. Recalling that the address translation is driven by the content of the page table, this means that the operating systems, whenever it assigns the processor to one process, will also set accordingly the corresponding page table entries. The page table contents become therefore part of the set of information, called *Process Context*, which needs to be restored by the operating system in a context switch, that is whenever a process regains the usage of the processor. Chapter 3 will describe process management in more detail; here it suffices to know that virtual address translation is part of the process context.

Virtual memory support complicates quite a bit the implementation of an operating system, but it greatly simplifies the programmer's life, which does not need concerns about possible interferences with other programs. At this point, however, the reader may be falsely convinced that in an operating system not supporting virtual memory it is not possible to run the same program in two different processes, or that, in any case, there is always the risk of memory interferences among programs executed by different processes. Luckily, this is not the case, but memory consistence

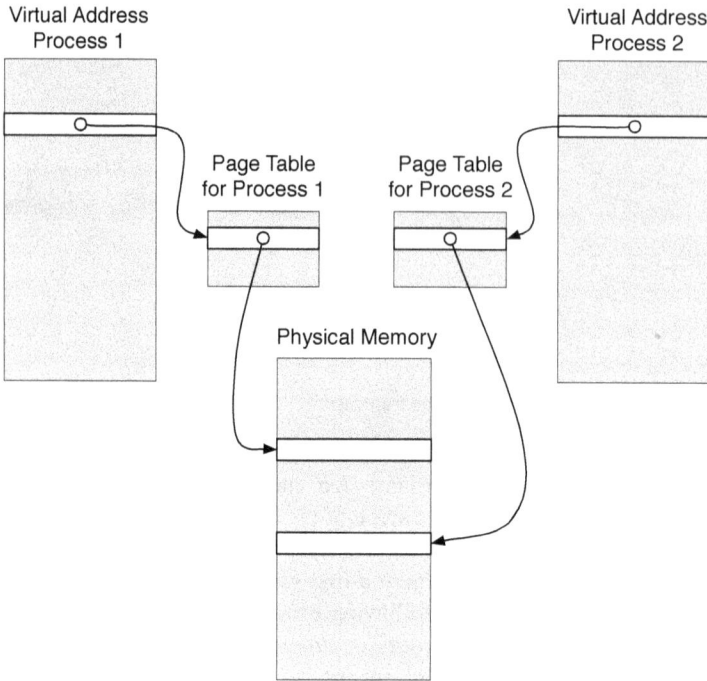

Figure 2.6   The usage of virtual address translation to avoid memory conflicts.

can be obtained only by imposing a set of rules for programs, such as the usage of the stack for keeping local variables. Programs which are compiled by a C compiler normally use the stack to contain local variables (i.e., variables which are declared inside a program block without the `static` qualifier) and the arguments passed in routine calls. Only static variables (i.e., local variables declared with the `static` qualifier or variables declared outside program blocks) are allocated outside the stack. A separate stack is then associated with each process, thus allowing memory insulation, even on systems supporting virtual memory. When writing code for systems without virtual memory, it is therefore important to pay attention in the usage of static variables, since these are shared among different processes, as shown in Figure 2.7. This is not necessarily a negative fact, since a proper usage of static data structures may represent an effective way for achieving interprocess communication. Interprocess communication, that is, exchanging data among different processes, can be achieved also with virtual memory, but in this case it is necessary that the operating system is involved so that it can set-up the content of the page table in order to allow the sharing of one or more physical memory pages among different processes, as shown in Figure 2.8.

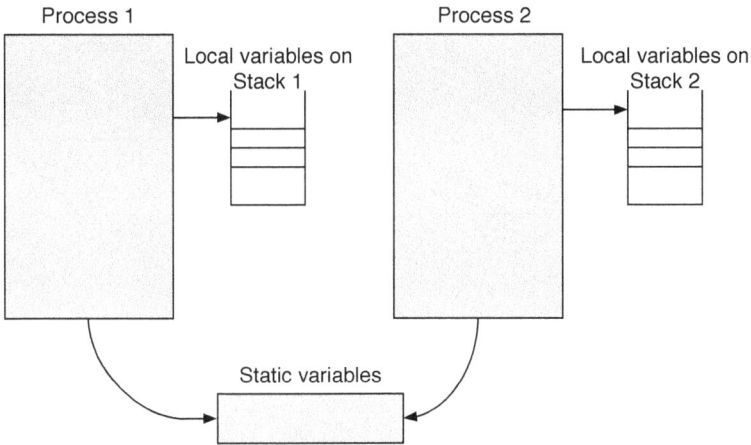

Figure 2.7 Sharing data via static variable on systems which do not support virtual addresses.

2.3.3 HANDLING DATA STREAMING FROM THE CAMERA DEVICE

Coming back to the acquisition of camera images using double buffering, we face the problem of properly mapping the buffers filled by the driver, running in Kernel mode, and the process running the frame acquisition program, running in User mode. When operating in Kernel mode, Linux uses, in fact, direct physical addresses (the operating system must have a direct access to every computer resource), so the buffer addresses seen by the driver will be different from the addresses of the same memory areas seen by the program. To cope with such a situation, Linux provides the mmap() system call. In order to understand how mmap() works, it is necessary to recall the file model adopted by Linux to support device I/O programming. In this conceptual model, files are represented by a contiguous space corresponding to the bytes stored in the file on the disk. A current address is defined for every file, representing the index of the current byte into the file. So address 0 refers to the first byte of the file, and address N will refer to the Nth byte of the file. Read-and-write operations on files implicitly refer to the current address in the file. When N bytes are read or written, they are read or written starting from the current address, which is then incremented by N. The current address can be changed using the lseek() system routine, taking as argument the new address within the file. When working with files, mmap() routine allows to map a region in the file onto a region in memory. The arguments passed to mmap() will include the relative starting address of the file region and the size in bytes of the region, and mmap() will return the (virtual) start address in memory of the mapped region. Afterward, reading and writing in that memory area will correspond to reading and writing into the corresponding region in the file. The concept of current file address cannot be exported as it is when using the same abstraction to describe I/O devices. For example, in a network device the current address is meaningless: read operations will return the bytes that have just

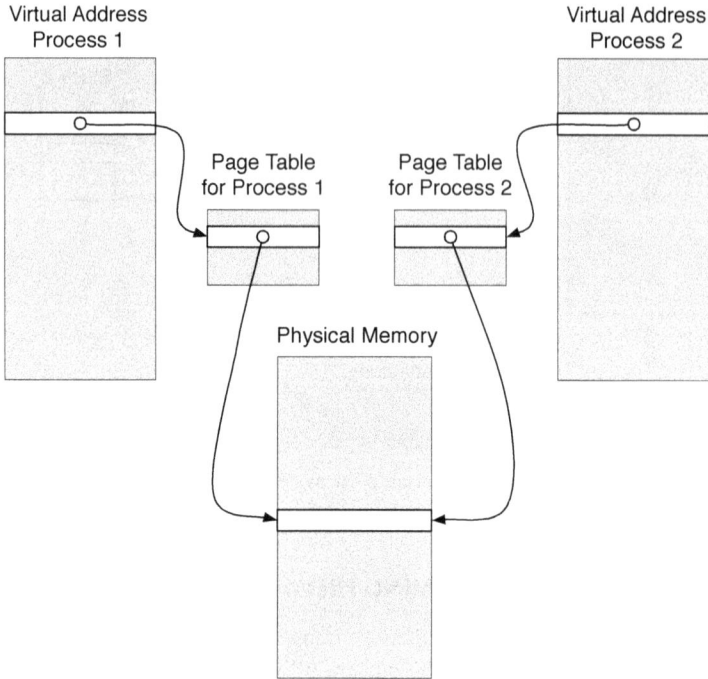

Figure 2.8 Using the Page Table translation to map possibly different virtual addresses onto the same physical memory page.

been received, and write operations will send the passed bytes over the network. The same holds for a video device, and read operation will get the acquired image frame, not read from any "address." However, when handling memory buffers in double buffering, it is necessary to find some way to map region of memory used by the driver into memory buffers for the program. mmap() can be used for this purpose, and the preparation of the shared buffers is carried out in two steps:

1. The driver allocates the buffers in its (physical) memory space, and returns (in a data structure passed to ioctl()) the unique *address* (in the driver context) of such buffers. The returned addresses may be the same physical address of the buffers, but in any case they are seen outside the driver as *addresses referred to the conceptual file model*.
2. The user programs calls mmap() to map such buffers in its virtual memory onto the driver buffers, passing as arguments the *file addresses* returned in the previous ioctl() call. After the mmap() call the memory buffers are shared between the driver, using physical addresses, and the program, using virtual addresses.

The code of the program using multiple buffering for handling image frame streaming from the camera device is listed below.

```
#include <fcntl.h>
#include <stdio.h>
#include <stdlib.h>
#include <string.h>
#include <errno.h>
#include <linux/videodev2.h>
#include <asm/unistd.h>
#include <poll.h>
#include <sys/mman.h>


#define MAX_FORMAT 100
#define FALSE 0
#define TRUE 1
#define CHECK_IOCTL_STATUS(message) \\
if(status == -1)                    \\
{                                   \\
    perror(message);                \\
    exit(EXIT_FAILURE);             \\
}

main (int argc, char *argv[])
{
    int fd, idx, status;
    int pixelformat;
    int imageSize;
    int width, height;
    int yuyvFound;

    struct v4l2_capability cap;        //Query Capability structure
    struct v4l2_fmtdesc fmt;           //Query Format Description structure
    struct v4l2_format format;         //Query Format structure
    struct v4l2_requestbuffers reqBuf; //Buffer request structure
    struct v4l2_buffer buf;            //Buffer setup structure
    enum v4l2_buf_type bufType;        //Used to enqueue buffers

    typedef struct {                   //Buffer descriptors
        void *start;
        size_t length;
    } bufferDsc;
    int idx;
    fd_set fds;                        //Select descriptors
    struct timeval tv;                 //Timeout specification structure

        bufferDsc *buffers;


/* Step 1: Open the device */
    fd = open("/dev/video1", O_RDWR);

/* Step 2: Check streaming capability */
    status = ioctl(fd, VIDIOC_QUERYCAP, &cap);
    CHECK_IOCTL_STATUS("Error querying capability")
    if(!(cap.capabilities & V4L2_CAP_STREAMING))
    {
        printf("Streaming NOT supported\n");
        exit(EXIT_FAILURE);
    }

/* Step 3: Check supported formats */
    yuyvFound = FALSE;
    for(idx = 0; idx < MAX_FORMAT; idx++)
    {
        fmt.index = idx;
        fmt.type = V4L2_BUF_TYPE_VIDEO_CAPTURE;
        status = ioctl(fd, VIDIOC_ENUM_FMT, &fmt);
```

```
        if(status != 0) break;
        if(fmt.pixelformat == V4L2_PIX_FMT_YUYV)
        {
            yuyvFound = TRUE;
            break;
        }
    }
    if(!yuyvFound)
    {
        printf("YUYV format not supported\n");
        exit(EXIT_FAILURE);
    }

/* Step 4: Read current format definition */
    memset(&format, 0, sizeof(format));
    format.type = V4L2_BUF_TYPE_VIDEO_CAPTURE;
    status = ioctl(fd, VIDIOC_G_FMT, &format);
    CHECK_IOCTL_STATUS("Error Querying Format")

/* Step 5: Set format fields to desired values: YUYV coding,
    480 lines, 640 pixels per line */
    format.fmt.pix.width = 640;
    format.fmt.pix.height = 480;
    format.fmt.pix.pixelformat = V4L2_PIX_FMT_YUYV;

/* Step 6: Write desired format and check actual image size */
    status = ioctl(fd, VIDIOC_S_FMT, &format);
    CHECK_IOCTL_STATUS("Error Setting Format");
    width = format.fmt.pix.width;                   //Image Width
    height = format.fmt.pix.height;                 //Image Height
    //Total image size in bytes
    imageSize = (unsigned int)format.fmt.pix.sizeimage;

/* Step 7: request for allocation of 4 frame buffers by the driver */
    reqBuf.count = 4;
    reqBuf.type = V4L2_BUF_TYPE_VIDEO_CAPTURE;
    reqBuf.memory = V4L2_MEMORY_MMAP;
    status = ioctl(fd, VIDIOC_REQBUFS, &reqBuf);
    CHECK_IOCTL_STATUS("Error requesting buffers")
/* Check the number of returned buffers. It must be at least 2 */
    if(reqBuf.count < 2)
    {
        printf("Insufficient buffers\n");
        exit(EXIT_FAILURE);
    }

/* Step 8: Allocate a descriptor for each buffer and request its
    address to the driver. The start address in user space and the
    size of the buffers are recorded in the buffers descriptors. */
    buffers = calloc(reqBuf.count, sizeof(bufferDsc));
    for(idx = 0; idx < reqBuf.count; idx++)
    {
        buf.type = V4L2_BUF_TYPE_VIDEO_CAPTURE;
        buf.memory = V4L2_MEMORY_MMAP;
        buf.index = idx;
/* Get the start address in the driver space of buffer idx */
        status = ioctl(fd, VIDIOC_QUERYBUF, &buf);
        CHECK_IOCTL_STATUS("Error querying buffers")
/* Prepare the buffer descriptor with the address in user space
    returned by mmap() */
        buffers[idx].length = buf.length;
        buffers[idx].start = mmap(NULL, buf.length,
            PROT_READ | PROT_WRITE,MAP_SHARED,
            fd, buf.m.offset);
        if(buffers[idx].start == MAP_FAILED)
        {
```

```
                perror("Error mapping memory");
                exit(EXIT_FAILURE);
            }
        }

/* Step 9: request the driver to enqueue all the buffers
   in a circular list */
    for(idx = 0; idx < reqBuf.count; idx++)
    {
        buf.type = V4L2_BUF_TYPE_VIDEO_CAPTURE;
        buf.memory = V4L2_MEMORY_MMAP;
        buf.index = idx;
        status = ioctl(fd, VIDIOC_QBUF, &buf);
        CHECK_IOCTL_STATUS("Error enqueuing buffers")
    }

/* Step 10: start streaming */
    bufType = V4L2_BUF_TYPE_VIDEO_CAPTURE;
    status = ioctl(fd, VIDIOC_STREAMON, &bufType);
    CHECK_IOCTL_STATUS("Error starting streaming")

/* Step 11: wait for a buffer ready */
    FD_ZERO(&fds);
    FD_SET(fd, &fds);
    tv.tv_sec = 20;
    tv.tv_usec = 0;
    for(;;)
    {
        status = select(1, &fds, NULL, NULL, &tv);
        if(status == -1)
        {
            perror("Error in Select");
            exit(EXIT_FAILURE);
        }
/* Step 12: Dequeue buffer */
        buf.type = V4L2_BUF_TYPE_VIDEO_CAPTURE;
        buf.memory = V4L2_MEMORY_MMAP;
        status = ioctl(fd, VIDIOC_DQBUF, &buf);
        CHECK_IOCTL_STATUS("Error dequeuing buffer")

/* Step 13: Do image processing */
        processImage( buffers[buf.index].start, width, height, imageSize);

/* Step 14: Enqueue used buffer */
        status = ioctl(fd, VIDIOC_QBUF, &buf);
        CHECK_IOCTL_STATUS("Error enqueuing buffer")
    }
}
```

Steps 1–6 are the same of the previous program, except for step 2, where the streaming capability of the device is now checked. In Step 7, the driver is asked to allocate four image buffers. The actual number of allocated buffers is returned in the count field of the v4l2_requestbuffers structure passed to ioctl(). At least two buffers must have been allocated by the driver to allow double buffering. In Step 8 the descriptors of the buffers are allocated via the calloc() system routine (every descriptor contains the dimension and a pointer to the associated buffer). The actual buffers, which have been allocated by the driver, are queried in order to get their address in the driver's space. Such an address, returned in field m.offset of the v4l2_buffer structure passed to ioctl(), cannot be used directly in the program since it refers to a different address space. The actual address in the user address space is returned by the following mmap() call. When the program arrives at Step

9, the buffers have been allocated by the driver and also mapped to the program address space. They are now enqueued by the driver, which maintains a linked queue of available buffers. Initially, all the buffers are available: every time the driver has acquired a frame, the first available buffer in the queue is filled. Streaming, that is, frame acquisition, is started at Step 10, and then at Step 11 the program waits for the availability of a filled buffer, using the `select()` system call. Whenever `select()` returns, at least one buffer contains an acquired frame. It is dequeued in Step 12, and then enqueued in Step 13, after it has been used in image processing. The reason for dequeuing and then enqueuing the buffer again is to make sure that the buffer will not be used by the driver during image processing.

Finally, image processing will be carried out by routine `processImage()`, which will first build a byte buffer containing only the luminance, that is, taking the first byte of every 16 bit word of the passed buffer, coded using the YUYV format.

2.4   EDGE DETECTION

In the following text we shall proceed with the case study by detecting, for each acquired frame, the center of a circular shape in the acquired image. In general, image elaboration is not an easy task, and its results may not only depend on the actual shapes captured in the image, but also on several other factors, such as illumination and angle of view, which may alter the information retrieved from the image frame. Nowadays this class of problems is solved using Artificial Intelligence (AI) techniques, in this case a Convolutional Neural Network (CNN). However, we shall adopt here a more traditional approach because we are not interested in the solution per se, but we are using it to introduce and discuss several programming techniques. Center coordinates detection will be performed here in two steps. Firstly, the edges in the acquired image are detected. This first step allows reducing the size of the problem, since for the following analysis it suffices to take into account the pixels representing the edges in the image. Edge detection is carried out by computing the approximation of the gradients in the X (L_x) and Y (L_y) directions for every pixel of the image, selecting, then, only those pixels for which the gradient magnitude, computed as $|\nabla L| = \sqrt{L_x^2 + L_y^2}$, is above a given threshold. In fact, informally stated, an edge corresponds to a region where the brightness of the image changes sharply, the gradient magnitude being an indication of the "sharpness" of the change. Observe that in edge detection we are only interested in the luminance, so in the YUYV pixel format, only the first byte of every two will be considered. The gradient is computed using a *convolution matrix filter*. Image filters based on convolution matrix filters are very common in image elaboration and, based on the matrix used for the computation, often called *kernel*, can perform several types of image processing. Such a matrix is normally a 3 x 3 or 5 x 5 square matrix, and the computation is carried out by considering, for each pixel image $P(x, y)$, the pixels surrounding the considered one and multiplying them for the corresponding coefficient of the kernel matrix K. Here we shall use a 3 x 3 kernel matrix, and therefore the computation of the filtered pixel value $P^f(x,y)$ is

$$P^f(x,y) = \sum_{i=0}^{2} \sum_{j=0}^{2} K(i,j) P(x+i-1, y+j-1) \tag{2.1}$$

Here, we use the *Sobel Filter* for edge detection, which defines the following two kernel matrixes:

$$\begin{bmatrix} -1 & 0 & 1 \\ -2 & 0 & 2 \\ -1 & 0 & 1 \end{bmatrix} \tag{2.2}$$

for the gradient along the X direction, and

$$\begin{bmatrix} 1 & 2 & 1 \\ 0 & 0 & 0 \\ -1 & -2 & -1 \end{bmatrix} \tag{2.3}$$

for the gradient along the Y direction.

The C source code for the gradient detection is listed below:

```c
#define THRESHOLD 100
/* Sobel matrixes */
static int GX [3][3];
static int GY [3][3];
/* Initialization of the Sobel matrixes, to be called before
Sobel filter computation */
static void initG ()
{
/* 3x3 GX Sobel mask. */
    GX [0][0] = -1; GX [0][1] = 0; GX [0][2] = 1;
    GX [1][0] = -2; GX [1][1] = 0; GX [1][2] = 2;
    GX [2][0] = -1; GX [2][1] = 0; GX [2][2] = 1;

/* 3x3 GY Sobel mask. */
    GY [0][0] =  1; GY [0][1] =  2; GY [0][2] =  1;
    GY [1][0] =  0; GY [1][1] =  0; GY [1][2] =  0;
    GY [2][0] = -1; GY [2][1] = -2; GY [2][2] = -1;
}

/* Sobel Filter computation for Edge detection. */
static void makeBorder (char *image, char *border, int cols, int rows)
/* Input image is passed in the byte array image (cols x rows pixels)
   Filtered image is returned in byte array border */
{
    int x,y, i, j, sumX, sumY, sum;

    for (y = 0; y <= (rows-1); y++)
    {
        for (x = 0; x <= (cols-1); x++)
        {
            sumX = 0;
            sumY = 0;
/* handle image boundaries */
            if (y == 0 || y == rows-1)
                sum = 0;
            else if (x == 0 || x == cols-1)
                sum = 0;
```

```
/* Convolution starts here */
          else
          {
/* X Gradient */
              for (i = -1; i <= 1; i++)
              {
                  for (j =- 1; j <= 1; j++)
                  {
                      sumX = sumX + (int)( (*(image + x + i +
                          (y + j)*cols)) * GX[i+1][j+1]);
                  }
              }
/* Y Gradient */
              for (i = -1; i <= 1; i++)
              {
                  for (j = -1; j <= 1; j++)
                  {
                      sumY = sumY + (int)( (*(image + x + i +
                          (y + j)*cols)) * GY[i+1][j+1]);
                  }
              }
/* Gradient Magnitude approximation to avoid square root operations */
              sum = abs(sumX) + abs(sumY);
          }

          if(sum > 255) sum = 255;
          if(sum < THRESHOLD) sum = 0;
          *(border + x + y*cols) = 255 - (unsigned char)(sum);
      }
   }
}
```

Routine makeBorder() computes a new image representing the edges of the scene in the image. Only such pixels will then be considered in the following computation for detecting the center of a circular shape in the image.

2.4.1 OPTIMIZING THE CODE

Before proceeding, it is worth to consider the performance of such algorithm. In fact, if we intend to use the edge detection algorithm in an embedded system with real-time constraints, we must ensure that its execution time will be bound to a given value, short enough to guarantee that the system will meet its requirements. First of all we observe that for every pixel of the image, 2*3*3 multiplications and sums are performed to compute the X and Y gradients, not considering the operation on the matrix indices. This means that, supposing a square image of size N is considered, the number of operation is proportional to N*N, and we say that the algorithm has complexity $\mathcal{O}(N^2)$. This notation is called the *big-O notation* and provides an indication of the complexity for computer algorithms. More formally, given two functions $f(x)$ and $g(x)$, if a value M and a value x_0 exist for which the following condition holds:

$$| f(x) | \leq M \, | g(x) | \tag{2.4}$$

for every $x > x_0$, then we say that $f(x)$ is $\mathcal{O}(g(x))$.

Informally stated, the above notation states that, for very large values of x the two functions tend to become proportional. For example, if $f(x) = 3x$ and $g(x) = 100x + 1000$, then we can find a pair M, x_0 for which 2.4 holds, and therefore $f(x)$

is $\mathcal{O}(g(x))$. However, if we consider $f(x) = 3x^2$ instead, it is not possible to find such a pair M, x_0. In fact, $f(x)$ grows faster than every multiple of $g(x)$. Normally, when expressing the complexity of an algorithm, the variable x used above represents the "dimension" of the problem. For example, in a sorting algorithm, the dimension of the problem is represented by the dimension of the vector to be sorted. Often some simplifying assumption must be done in order to provide a measure of the dimension to be used in the big-O notation. In our edge detection problem, we make the simplifying assumption that the image is represented by a square pixel matrix of size N, and therefore we can state that the Sobel filter computation is $\mathcal{O}(N^2)$ since the number of operations is proportional to N^2.

The big-O notation provides a very important measurement of the efficiency for computer algorithms, which normally become unmanageable when the dimension of the problem increases. Take as an example the algorithms for sorting a given array of values. Elementary sorting algorithms such as Bubble Sort or Insertion Sort require a number of operation that is proportional to N^2, where N is the dimension of the array to be sorted and therefore are $\mathcal{O}(N^2)$. Other sorting algorithms, such as Shell Sort and Quick Sort are instead $\mathcal{O}(Nlog(N))$. This implies that for very large arrays, only the latter algorithms can be used in practice because the number of operations becomes orders of magnitude lower in this case.

Even if the big-O notation is very important in the classification of algorithms and in determining their applicability when the dimension of the problem grows, it does not suffice for providing a complete estimate of the computation time. To convince ourselves of this fact, it suffices to consider two algorithms for a problem of dimension N, the first one requiring $f(N)$ operations, and the second one requiring exactly $100f(N)$. Of course, we would never choose the second one; however they are equivalent in the big-O notation, being both $\mathcal{O}(f(N))$.

Therefore, in order to assess the complexity of a given algorithm and to optimize it, other techniques must be considered, in addition to the choice of the appropriate algorithm. This the case of our application: given the algorithm, we want to make its computation as fast as possible.

First of all, we need to perform a measurement of the time the algorithm takes. A crude but effective method is to use the system routines for getting the current time, and measure the difference between the time measured first and after the computation of the algorithm. The following code snippet makes a raw estimation of the time procedure `makeBorder()` takes in a Linux system.

```
#define ITERATIONS 1000
struct time_t beforeTime, afterTime;
int executionTime;
....
gettimeofday(&beforeTime, NULL);
for(i = 0; i < ITERATIONS; i++)
    makeBorder(image, border, cols, rows);
gettimeofday(&afterTime, NULL);
/* Execution time is expressed in microseconds */
executionTime = (afterTime.tv_sec - beforeTime.tv_sec) * 1000000
    + afterTime.tv_usec - beforeTime.tv_usec;
executionTime /= ITERATIONS;
...
```

The POSIX routine `gettimeofday()` reads the current time from the CPU clock and stores it in a `time_t` structure whose fields define the number of seconds (`tv_sec`) and microseconds (`tv_usec`) from the *Epoch*, that is, a reference time which, for POSIX, is assumed to be 00:00:00 UTC, January 1, 1970.

The execution time measured in this way can be affected by several factors, among which can be the current load of the computer. In fact, the process running the program may be interrupted during execution by other processes in the system. Even after setting the priority of the current process as the highest one, the CPU will be interrupted many times for performing I/O and for the operating system operation. Nevertheless, if the computer is not loaded, and the process running the program has a high priority, the measurement is accurate enough.

We are now ready to start the optimization of our edge detection algorithm. The first action is the simplest one: let the compiler do it. Modern compilers perform very sophisticated optimization of the machine code that is produced when parsing the source code. It is easy to get an idea of the degree of optimization by comparing the execution time when compiling the program without optimization (compiler flag -00) and with the highest degree of optimization (compiler flag -03), which turns out to be 5–10 times shorter for the edge detection routine. The optimization performed by the compiler addresses the following aspects:

- *Code Reduction*: Reducing the number of machine instructions makes the program execution faster. Very often in programs the same information is computed many times in different parts. So the compiler can reuse the value computed before, instead of executing again a sequence of machine instructions leading to the same result. The compiler tries also to carry out computation in advance, rather than producing the machine instructions for doing it. For example, if an expression formed by constant values is present in the source code, the compiler can produce the result at compile time, rather than doing it during the program execution. The compiler also moves away from loops the computation that does not depend on loop variable, and which therefore would produce the same result at every loop iteration.

  Observe that code reduction does not mean reduction in the size of the produced program; rather, it reduces the number of instruction actually executed during the program. For example, whenever the number N of loop iterations can be deduced at compile time (i.e., does not depend on run-time information) and N is not too high, compilers often replace the conditional jump instruction by concatenating N segments of machine instruction, each corresponding to the loop body. The resulting executable program is longer, but the number of instructions actually performed is lower since the conditional jumps instruction and the corresponding condition evaluation are avoided. For the same reason, compilers can also perform inline expansion when a routine is called in the program. Inserting the code of the routine again makes the size of the executable program bigger, but avoids the overhead due to the routine invocation and the passage of the arguments.

- *Instruction Selection*: Even if several operations defined in the source code, such as multiplications, can be directly executed by machine instruction, this choice does not represent the most efficient one. Consider, for example, a multiplication by two: this can be performed either with a multiplication (MUL) or with an addition (ADD) machine instruction. Clearly, the second choice is preferable since in most computer architectures addition is performed faster than multiplication. Therefore, the compiler selects the most appropriate sequence of machine instructions for carrying out the required computation. Observe that again this may lead to the generation of a program with a larger number of machine instructions, where some operations for which a direct machine instruction exists are instead implemented with a longer sequence of faster machine instruction. In this context, a very important optimization carried out by the compiler is the recognition of induction variables in loops and the replacement of operations on such variables with simpler ones. As an example, consider the following loop:

```
for (i = 0; i < 10; i++)
{
    a = 15 * i;
    . . . . .
}
```

Variable i is an induction variable, that is, a variable which gets increased or decreased by a fixed amount on every iteration of a loop, or which is a linear function of another induction variable. In this case, it is possible to replace the multiplication with an addition, getting the equivalent loop:

```
a = 0;
for (i = 0; i < 10; i++)
{
    a = a + 15;
    . . . . .
}
```

The compiler recognizes, then, induction variables and replaces more complex operations with additions. This optimization is particularly useful for the loop variables used as indexes in arrays; in fact, many computer architectures define memory access operations (arrays are stored in memory and are therefore accessed via memory access machine instructions such as LOAD or STORE), which increment the passed memory index by a given amount in the same memory access operation.

- *Register Allocation*: Every computer architecture defines a number of registers that can store temporary information during computation. Registers are implemented within the processor, and therefore reading or writing to a register is much faster than reading and writing from memory. For this reason the compiler will try to use processor registers as far as possible, for example, using registers to hold the variables defined in the program. The number of registers is, however, finite (up to some tents), and therefore it is not possible to store all the variables into registers. Memory locations must be used, too. Moreover, when arrays are used in the program, they are stored in memory, and access to array elements in the program normally requires an access to memory in run time. The compiler uses a variety of algorithms to optimize the use of registers, and to maximize the likelihood that a variable access will be performed by a register access. For example, if a variable stored in memory is accessed for the second time, and it has not been changed since its first access (something which can be detected under certain conditions by the compiler), then the compiler will temporarily hold a copy of the variable on a register so that the second time it is read from the register instead from memory.

- *Machine-Dependent Optimization*: the above optimizations hold for every computer. In fact, reducing the number and the complexity of instructions executed in run time will always reduce execution time, as well as optimizing the usage of registers. There are, however, other optimizations that depend on specific computer architecture. A first set of optimizations addresses the pipeline. All modern processors are pipelined, that is the execution of machine instructions is implemented as a sequence of stages. Each stage is carried out by a different processors components. For example, a processor may define the following stages for a machine instruction:

  1. Fetch: read the instruction from memory;
  2. Decode: decode the machine instruction;
  3. Read arguments: load the arguments of the machine instruction (from registers or from memory);
  4. Execute: do what the instruction specifies;
  5. Store results: store the results of the execution (to registers or to memory).

A separate hardware processor component, called the pipeline stage, will carry out every stage. So, when the first stage has terminated fetching the instruction N, it can start fetching instruction $N+1$ while instruction N is being decoded. After a startup time, under ideal conditions, the K stages of the pipeline will all be busy, and the processor is executing K instruction in parallel, reducing the average execution time of a factor of K. There are, however, several conditions that may block the parallel execution of the pipeline stages, forcing a stage to wait for some clock cycle before resuming operation. One such condition is given by the occurrence of two consecutive instructions, say, instructions N and $N+1$ in the program, where the latter uses as input the results of the former. In this case, when instruction $N+1$ enters its third stage (Read arguments), instructions N enters the execute phase. However, instruction $N+1$ cannot proceed in reading the arguments, since they have not yet been reported by the previous instruction. Only when instruction N finishes its execution (and its results have been stored) execution $N + 1$ can resume its execution, thus producing a delay in the execution of two clock cycles, assuming that every pipeline stage is executed in one clock period. This condition is called *Data Hazard* and depends on the existence of sequences of two or more dependent consecutive instructions.

If the two instruction were separated by at least two independent instructions in the program sequence, no data hazard would occur and no time would be spent with the pipeline execution partially blocked. The compiler, therefore, tries to separate dependent instruction in the program. Of course, instructions cannot be freely moved in the code, and code analysis is required to figure out which instruction sequence rearrangement are legal, that is, which combination maintain the program correct. This kind of analysis is also performed by the compiler to take advantage of the availability of multiple execution units in superscalar processors. In fact, instructions can be executed in parallel only when they are independent from each other.

At this point we may be tempted to state that the all the possible optimizations in the edge detection program have been carried out by the compiler, and there is no need to further analyze the program for reducing its execution time. This is, however, not completely true: while compilers are very clever in optimizing code, very often achieving a better optimization than what can be achieved with manual optimization, there is one aspect of the program in which compilers cannot exploit extreme optimization—that is, memory access via pointers. We have already seen that a compiler can often maintain in a register a copy of a variable stored in memory so that the register copy can be used instead. However, it is not possible to store in a register a memory location accessed via a pointer and reuse it afterward in spite of the memory location, because it is not possible to make sure that the memory address has not been modified in the meantime. In fact, while in many cases the compiler can analyze in advance how variables are used in the program, in general it cannot do the same for memory location accessed via pointers because the pointer values, that is, the memory addresses, are normally computed *run time*, and cannot therefore be foreseen during program compilation.

As we shall see shortly, there is still room for optimization in the edge detection routine, but it is necessary to introduce first some concepts of memory caching.

In order to speed memory accesses computers use *memory caches*. A memory cache is basically a fast memory that is much faster that the RAM memory used by the processor, and which holds data recently accessed by the computer. The memory cache does not correspond to any fixed address in the addressing space of the processor, and therefore contains only *copies* for memory locations stored in the RAM. The caching mechanism is based on a common fact in programs: *locality in memory access*. Informally stated, memory access locality expresses the fact that if a processor makes a memory access, say, at address K, the next access in memory is likely to occur at an address that is close to K. To convince ourselves of this fact, consider the two main categories of memory data access in a program execution: fetching program instructions and accessing program data. Fetching memory instructions (recall that a processor has to read the instruction from memory in order to execute it) is clearly sequential in most cases. The only exception is for the Jump instructions, which, however, represent a small fraction of the program instructions. Data is mostly accessed in memory when the program accesses array elements, and arrays are normally (albeit not always) accessed in loops using some sort of sequential indexing.

Cache memory is organized in blocks (called also *cache lines*), which can be up to a few hundred bytes large. When the processor tries to access a memory location for reading or writing a data item at a given address, the cache controller will first check if a cache block containing that location is currently present in the cache. If it is found in the cache memory, fast read/write access is performed in the cached copy of the data item. Otherwise, a free block in the cache is found (possibly copying in memory an existing cache block if the cache is full), and a block of data located around that memory address is first copied from memory to the cache. The two cases are called *Cache Hit* and *Cache Miss*, respectively. Clearly, a cache miss incurs in a penalty in execution time (the copy of a block from memory to cache), but, due to memory access locality, it is likely that further memory accesses will hit the cache, with a significant reduction in data access time.

The gain in performance due to the cache memory depends on the program itself: the more local is memory access, the faster will be program execution. Consider the following code snippet, which computes the sum of the elements of a M x N matrix.

```
double a[M][N];
double sum = 0;
for(i = 0; i < M, i++)
    for(j = 0; j < N; j++)
        sum += a[i][j];
```

In C, matrixes are stored in row first order, that is, rows are stored sequentially. In this case a[i][j] will be adjacent in memory to a[i][j+1], and the program will access matrix memory sequentially. The following code is also correct, differing from the previous one only for the exchange of the two for statements.

```
double a[M][N];
double sum = 0;
for(j = 0; j < N; j++)
    for(i = 0; i < M, i++)
        sum += a[i][j];
```

However in this case memory access is not sequential since matrix elements a[i][j] and a[i+1][j] are stored in memory locations that are N elements far away. In this case, the number of cache misses will be much higher than in the former case, especially for large matrixes, affecting the execution time of that code.

Coming back to routine makeBorder(), we observe that it is accessing memory in the right order. In fact, what the routine basically does is to consider a 3 x 3 matrix sweeping along the 480 rows of the image. The order of access is therefore row first, corresponding to the order in which bytes are stored in the image buffer. So, if bytes are being considered in a "cache friendly" order, what can we do to improve performance? Recall that the compiler is very clever in optimizing access to information stored in program variables, but is mostly blind as regard the management of information stored in memory (i.e., in arrays and matrixes). This fact suggests to us a possible strategy: move the current 3 x 3 portion of the image being considered in the Sobel filter into 9 variables. Filling this set of 9 variables the first time a line is considered will require reading 9 values from memory, but at the following iterations, that is, moving the 3 x 3 matrix one position left, only three new values will be read from memory, the others already being stored in program variables. Moreover, the nine multiplications and summations required to compute the value of the current output filter can be directly expressed in the code, without defining the 3 x 3 matrixes GX and GY used in the program listed above. The new implementation of makeBorder() is listed below, using the new variables c_{11}, c_{12}, ..., c_{33} to store the current portion of the image being considered for every image pixel.

```
void makeBorder(char *image, char *border, int cols, int rows)
{
    int x, y, sumX, sumY, sum;
/* Variables to hold the 3x3 portion of the image used in the computation
of the Sobel filter output */
    int c11,c12,c13,c21,c22,c23,c31,c32,c33;

    for(y = 0; y <= (rows-1); y++)
    {
    /* First image row: the first row of cij is zero */
        if(y == 0)
        {
            c11 = c12 = c13 = 0;
        }
        else
    /* First image column: the first column of cij matrix is zero */
        {
            c11=0;
            c12 = *(image + (y - 1) * cols);
            c13 = *(image + 1 + (y - 1)*cols);
        }
        c21 = 0;
        c22 = *(image + y*cols);
        c23 = *(image + 1 + y*cols);
        if(y == rows - 1)
    /* Last image row: the third row of cij matrix is zero */
        {
            c31 = c32 = c33 = 0;
        }
        else
        {
            c31=0;
            c32 = *(image + (y + 1)*cols);
            c33 = *(image + 1 + (y + 1)*cols);
        }
/* The 3x3 matrix corresponding to the first pixel of the current image
   row has been loaded in program variables.
   The following iterations will only load
   from memory the rightmost column of such matrix */
        for(x = 0; x <= (cols-1); x++)
        {
            sumX = sumY = 0;
/* Skip image boundaries */
            if(y == 0 || y == rows-1)
                sum = 0;
            else if(x == 0 || x == cols-1)
                sum = 0;
/* Convolution starts here.
   GX and GY parameters are now "cabled" in the code */
            else
```

```
          {
              sumX = sumX - c11;
              sumY = sumY + c11;
              sumY = sumY + 2*c12;
              sumX = sumX + c13;
              sumY = sumY + c13;
              sumX = sumX - 2 * c21;
              sumX = sumX + 2*c23;
              sumX = sumX - c31;
              sumY = sumY - c31;
              sumY = sumY - 2*c32;
              sumX = sumX + c33;
              sumY = sumY - c33;
              if(sumX < 0) sumX = -sumX; //Abs value
              if(sumY < 0) sumY = -sumY;
              sum = sumX + sumY;
          }
/* Move one pixel on the right in the current row.
   Update the first/last row only if not in the first/last image row */
          if(y > 0)
          {
              c11 = c12;
              c12 = c13;
              c13 = *(image + x + 2 + (y - 1) * cols);
          }
          c21 = c22;
          c22 = c23;
          c33 = *(image + x +2 + y * cols);
          if(y < cols - 1)
          {
              c31 = c32;
              c32 = c33;
              c33 = *(image + x + 2 + (y + 1) * cols);
          }
          if(sum > 255) sum = 255;
          if(sum < THRESHOLD) sum=0;
/* Report the new pixel in the output image */
          *(border + x + y*cols) = 255 - (unsigned char)(sum);
      }
   }
}
```

The resulting code is for sure less readable then the previous version, but, when compiled, it produces a code that is likely faster because the compiler has now more chance for optimizing the management of information, being memory access limited to the essential cases.

In general code optimization is not a trivial task and requires ingenuity and a good knowledge of the optimization strategies carried out by the compiler. Very often, in fact, the programmer experiences the frustration of getting no advantage after working hard in optimizing his/her code, simply because the foreseen optimization had already been carried out by the compiler. Since optimized source code is often much less readable that a nonoptimized one, implementing a given algorithm taking care also of possible code optimization, may be an error-prone task. For this reason, implementation should be done in two steps:

1. Provide a first implementation with no regard to efficiency, but concentrating on a clearly readable and understandable implementation. At this level, the program should be fully debugged to make sure that no errors are present in the code, preparing also a set of test cases that fully covers the different aspects of the algorithm.

2. Starting from the previous implementation, and using the test cases prepared in the first step, perform optimization, possibly in steps, in order to
   address separately possible sources of inefficiency. At every try (not all the
   tentatives will actually produce a faster version) check the correctness of
   the new code and the amount of gained performance.

2.5 FINDING THE CENTER COORDINATES OF A CIRCULAR SHAPE

After the edge detection stage, a much reduced number of pixels has to be taken into
consideration to compute the final result of image analysis in our case study: locating
the coordinates of the center of a circular shape in the image. To this purpose, the
Hough transform will be used, a technique for feature extraction in images. In the
original image, every element of the image matrix brings information on the luminance of the corresponding pixel (we are not considering colors here). The Hough
transform procedure converts pixel luminance information into a set of parameters,
so that a voting procedure can be defined in the parameter space to derive the desired
feature, even in the case of imperfect instances of objects in the input image.

   The Hough transform was originally used to detect lines in images. In this case,
the parameter space components are r and θ, where every line in the original image
is represented by a (r, θ) pair, as shown in Figure 2.9. Using parameters r and θ, the
equation of a line in the x, y plane is expressed as:

$$y = -(\frac{\cos\theta}{\sin\theta})x + (\frac{r}{\sin\theta}) \tag{2.5}$$

   Imagine an image containing one line. After edge detection, the pixels associated
with the detected edges may belong to the line, or to some other element of the scene
represented by the image. Every such pixel at coordinates (x_0, y_0) is assumed by the
algorithm as belonging to a potential line, and the (infinite) set of lines passing for
(x_0, y_0) is considered. For all such lines, the associated parameters r and θ obey to
the following relation:

$$r = x_0\cos\theta + y_0\sin\theta \tag{2.6}$$

that is, a sinusoidal law in the plane (r, θ). Suppose now that the considered pixel effectively belongs to the line, and consider another pixel at position (x_1, y_1), belonging

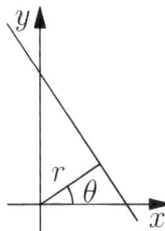

Figure 2.9 r and θ representation of a line.

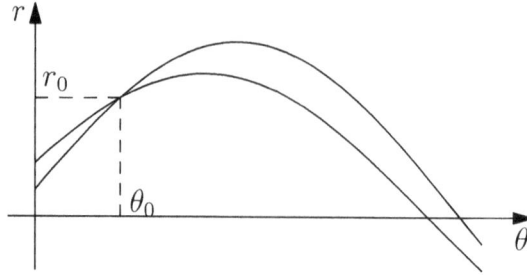

Figure 2.10 (r, θ) relationship for points (x_0, y_0) and (x_1, y_1).

to the same line. Again, for the set of lines passing through (x_1, y_1), their r and θ will obey the law:

$$r = x_1 \cos \theta + y_1 \sin \theta \tag{2.7}$$

Plotting (2.5) and (2.7) in the (r, θ) (Figure 2.10) we observe that the two graphs intersect in (r_0, θ_0), where r_0 and θ_0 are the parameters of the line passing through (x_0, y_0) and (x_1, y_1). Considering every pixel on that line, all the corresponding curves in place (r, θ) will intersect in (r_0, θ_0). This suggests a voting procedure for detecting the lines in an image. We must consider, in fact, that in an image spurious pixels are present, in addition to those representing the line. Moreover, the (x, y) position of the line pixels may lie not exactly in the expected coordinates for that line. So, a matrix corresponding to the (r, θ) plane, initially set to 0, is maintained in memory. For every edge pixel, the matrix elements corresponding to all the pairs (r, θ) defined by the associated sinusoidal relation are incremented by one. When all the edge pixels have been considered, supposing a single line is represented in the image, the matrix element at coordinates (r_0, θ_0) will hold the highest value, and therefore it suffices to choose the matrix element with the highest value, whose coordinates will identify the recognized line in the image.

A similar procedure can be used to detect the center of a circular shape in the image. Assume initially that the radius R of such circle is known. In this case, a matrix with the same dimension of the image is maintained, initially set to 0. For every edge pixel (x_0, y_0) in the image, the circle of radius R centered in (x_0, y_0) is considered, and the corresponding elements in the matrix incremented by 1. All such circles intersect in the center of the circle in the image, as shown in Figure 2.11. Again, a voting procedure will allow discovery of the center of the circle in edge image, even in presence of spurious pixels, and the approximate position of the pixels representing the circle edges. If the radius R is not known in advance, it is necessary to repeat the above procedure for different values of R and choose the radius value that yields the maximum count value for the candidate center. Intuitively, this holds, because only when the considered radius is the right one will all the circles built around the border pixels of the original circle intersect in a single point.

Observe that even if the effective radius of the circular object to be detected in the image is known in advance, the radius of its shape in the image may depend on

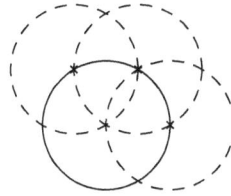

Figure 2.11   Circles drawn around points over the circumference intersect in the circle center.

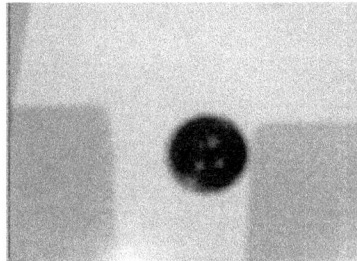

Figure 2.12   A sample image with a circular shape.

several factors, such as its distance from the camera, or even from the illumination of the scene, which may yield slightly different edges in the image, so in practice it is always necessary to consider a range of possible radius values.

The overall detection procedure is summarized in Figures 2.12–2.15. The original image and the detected edges are shown in Figures 2.12 and 2.13, respectively. Figure 2.14 is a representation of the support matrix used in the detection procedure. It can be seen that most of the circles in the image intersect in a single point (the others are circles drawn around the other edges of the image), reported then in the original image in Figure 2.15.

The code of routine `findCenter()` is listed below. Its input arguments are the radius of the circle, the buffer containing the edges of the original image (created by

Figure 2.13   The image of 2.12 after edge detection.

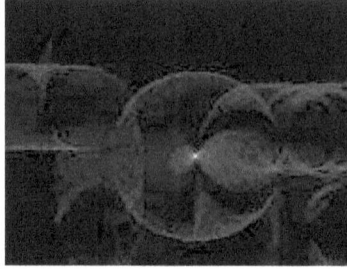

Figure 2.14 The content of the voting matrix generated from the edge pixels of 2.13.

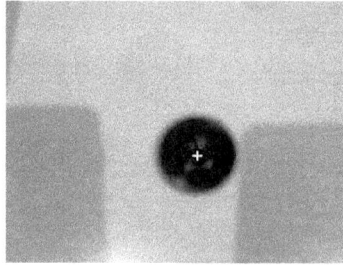

Figure 2.15 The detected center in the original image.

routine `makeBorder()`), and the number of rows and columns. The routine returns the position of the detected center and a quality indicator, expressed as the normalized maximum value in the matrix used for center detection. The buffer for such a matrix is passed in the last argument.

```
/* Black threshold:
   a pixel value less than the threshold is considered black. */
#define BLACK_LIMIT 10
void findCenter(int radius,unsigned char *buf, int rows, int cols,
    int *retX, int *retY, int *retMax, unsigned char *map)
{
    int x, y, l, m, currCol, currRow, maxCount = 0;
    int maxI  = 0, maxJ = 0;
/* Square roots needed for computation are computed only once
   and maintained in array sqr */
    static int sqr[2 * MAX_RADIUS];
    static int sqrInitialized = 0;
/* Hit counter, used to normalize the returned quality indicator */
    double totCounts = 0;
/* The matrix is initially set to 0 */
    memset(map, 0, rows * cols);
/* If square root values not yet initialized, compute them */
    if(!sqrInitialized)
    {
        sqrInitialized = 1;
        for(l = -radius; l <= radius; l++)
/*integer approximation of sqrt(radius^2 - l^2) */
            sqr[l+radius] = sqrt(radius*radius - l*l) + 0.5;
    }
    for(currRow = 0; currRow < rows; currRow++)
```

```
    {
        for(currCol = 0; currCol < cols; currCol++)
        {
/* Consider only pixels corresponding to borders of the image
   Such pixels are set by makeBorder as dark ones */
            if(buf[currRow*cols + currCol] <= BLACK_LIMIT)
            {
                x = currCol;
                y = currRow;
/* Increment the value of the pixels in map buffer which corresponds to
   a circle of the given radius centered in (currCol, currRow) */
                for(l = x - radius; l <= x+radius; l++)
                {
                    if(l < 0 || l >= cols)
                        continue; // Out of image X range
                    m = sqr[l-x+radius];
                    if(y-m < 0 || y+m >= rows)
                        continue; //Out of image Y range
                    map[(y-m)*cols + l]++;
                    map[(y+m)*cols + l]++;
                    totCounts += 2;        //Two more pixels incremented
/* Update current maximum */
                    if(maxCount < map[(y+m)*cols + l])
                    {
                        maxCount = map[(y+m)*cols + l];
                        maxI = y + m;
                        maxJ = l;
                    }
                    if(maxCount < map[(y-m)*cols + l])
                    {
                        maxCount = map[(y-m)*cols + l];
                        maxI = y - m;
                        maxJ = l;
                    }
                }
            }
        }
    }
/* Return the (X,y) position in the map which yields the largest value */
    *retX = maxJ;
    *retY = maxI;
/* The returned quality indicator is expressed as maximum pixel
   value in map matrix */
    *retMax = maxCount;
}
```

As stated before, due to small variations of the actual radius of the circular shape in the image, routine findCenter() will be iterated for a set of radius values, ranging between a given minimum and maximum value.

When considering the possible optimization of the detection procedure, we observe that every time routine findCenter() is called, it is necessary to compute the square root values that are required to select the map elements which lie on a circumference centered on the current point. Since the routine is called for a fixed range of radius values, we may think of removing the square root calculation at the beginning of the routine, and to pass on an array of precomputed values, which are prepared in an initialization phase for all the considered radius values. This improvement would, however, bring very little improvement in speed: in fact, only few tens of square root computations (i.e., the pixel dimension of the radius) are carried out every time findCenter() is called, a very small number of operations if compared with the total number of operations actually performed. A much larger improvement can be obtained by observing that it is possible to execute `findCenter()` for the different radius values in parallel instead of in a sequence. The following code uses POSIX threads, described in detail in Chapter 7, to launch a set of thread, each computing the center coordinates for a given value of the radius. Every thread can be considered an independent flow of execution for the passed routine. In a multicore processor, threads can run on different cores, thus providing a drastic reduction of the execution time because code is executed effectively in parallel. A new thread is created by POSIX routine `pthread_create()`, which takes as arguments the routine to be executed and the (single) parameter to be passed. As `findCenter()` accepts multiple input and output parameters, it cannot be passed directly as argument to `pthread_create()`. The normal practice is to allocate a data structure containing the routine-specific parameters and to pass its pointer to `pthread_create()` using a support routine (`doCenter()` in the code below).

After launching the threads, it is necessary to wait for their termination before selecting the best result. This is achieved using POSIX routine `pthread_join()`, which suspends the execution of the calling program until the specified thread terminates, called in a loop for every created thread. When the loop exits, all the centers have been computed, and the best candidate can be chosen using the returned arguments stored in the support argument structures.

```
#include <pthreads.h>
/* Definition of a structure to contain the arguments to be
   exchanged with findCenter() */
struct arguments{
    unsigned char *edges;    //Edge image
    int rows, cols;          //Rows and columns if the image
    int r;                   //Current radius
    int retX, retY;          //Returned center position
    int retMax;              //Returned quality factor
    unsigned char *map;      //Buffer memory for the voting matrix
};
struct arguments *args;
/* Initialization of the support structure. initCenter()
   will be called once and will allocate the required memory */
void initCenter(unsigned char *edges,
                int minR, int maxR, int rows, int cols)
{
    int i;
    args = (struct arguments *)
        malloc((maxR - minR + 1)*sizeof(struct arguments));
    for(i = 0; in <= maxR - minR; i++)
    {
        args[i].edges = edges;
        args[i].r = minR + i;
        args[i].rows = rows;
        args[i].cols = cols;
        args[i].map = (unsigned char *)malloc(rows * cols);
    }
}

/* Routine executed by the thread. It receives the pointer to the
   associated arguments structure */
static void *doCenter(void *ptr)
{
    struct arguments *arg = (struct arguments *)ptr;
/* Take arguments from the passed structure */
    findCenter(arg->r, arg->borders, arg->rows, arg->cols,
        &arg->retX, &arg->retY, &arg->max, arg->map);
```

```
       return NULL;
}
/* Parallel execution of multiple findCenter() routines for radius
   values ranging from minR to maxR */
static void parallelFindCenter(unsigned char *borders, int minR,
  int maxR, int rows, int cols, int *retX, int *retY,
  int *retRadius, unsigned char *map)
{
    int i;
    double currMax = 0;
/* Dummy thread return value (not used) */
    void *retVal;
/* Array of thread indentifiers */
    pthread_t trs[maxR - minR];

/* Create the threads. Each thread will receive the pointer of the
   associated argument structure */
    for(i = 0; i <= maxR - minR; i++)
        pthread_create(&trs[i], NULL, doCenter, &args[i]);
/* Wait the termination of all threads */
    for(i = 0; i < maxR - minR; i++)
        pthread_join(trs[i], &retVal);
/* All threads are now terminated: select the best radius and return
   the detected center for it */
    for(i = 0; i < maxR - minR; i++)
    {
        if(args[i].max > currMax)
        {
            currMax = args[i].max;
            *retX = args[i].retX;
            *retY = args[i].retY;
            *retRadius = args[i].r;
        }
    }
}
```

2.6 SUMMARY

In this chapter a case study has been used to introduce several important facts about embedded systems. In the first part, the I/O architecture of computers has been presented, introducing basilar techniques such as polling, interrupts and Direct Memory Access.

The interface to I/O operations provided by operating systems, in particular Linux, has then been presented. The operating system shields all the internal management of I/O operations, offering a very simple interface, but nonetheless knowledge in the I/O techniques is essential to fully understand how I/O routines can be used. The rather sophisticated interface provided by the library V4L2 for camera devices allowed us to learn more concepts such as virtual memory and multiple buffer techniques for streaming.

The second part of the chapter concentrates on image analysis, introducing some basic concepts and algorithms. In particular, the important problem of code optimization is discussed, presenting some optimization techniques carried out by compilers and showing how to "help" compilers in producing more optimized code. Finally, an example of code parallelization has been presented, to introduce the basic concepts of threads activation and synchronization.

We are ready to enter the more specific topics of the book. As explained in the introduction, embedded systems represent a field of application with many aspects, only few of which can be treated in depth in a reasonably sized text. Nevertheless, the general concepts we met so far will hopefully help us in gaining some understanding of the facets not "officially" covered by this book.

2.7 EXERCISES

You can find all the solutions of the following exercises in the companion GitHub repository at `https://github.com/minimap-xl/RTOS_Book`.

EXERCISE 1

In the presented use case the YUYV format has been used in order to retrieve the frames as pixel matrices that could be then used by the image processing programs. However, for many other applications such as frame display, the jpeg format, supported by almost all webcams, is preferred for a variety of reasons, among which the reduced memory requirements. Starting from the presented code based on V4L2, modify it to handle frame acquisition in jpeg format.

Hint: the required changes in the code are minimal. It suffices replacing the requested pixelformat to `V4L2_PIX_FMT_JPEG` when checking the supported formats and when setting the format via `VIDIOC_S_FMT ioctl` call. In addition, you have to consider the fact that the size in bytes of the acquired frame may change from frame to frame because of jpeg compression. This information can be retrieved from field `bytesused` of the `struct v4l2_buffer` argument passed to `ioctl` call for dequeueing buffers during frame acquisition

EXERCISE 2

Modify the frame acquisition program so that acquired frames are stored on disk as .jpg files, taking into account possible delays in disk saving times.

Hint: even if buffering is already performed at the driver level, the limited number of buffers may prove insufficient if a longer occasional delay occurs when writing files on disk. For this reason it is convenient to decouple disk write operation on a thread that is separated from the main thread performing frames acquisition and to use a dynamically allocated queue between the two threads. In this case no frames are lost, provided of course that the average disk throughput is not lower than the frame throughput.

EXERCISE 3

Modify the frame acquisition program so that acquired frames are sent as UDP datagrams over the network.

Hint: as for the previous exercise, it is necessary to take into account possible delays in transmission and therefore it is convenient to decouple on a separate thread the transmission of UDP datagrams. A quick test can be then performed at the receiver side with the following python program using matplotlib to display the received frames:

```python
import socket
import numpy as np
import sys
import io
from PIL import Image
import matplotlib.pyplot as plt

if len(sys.argv) != 2:
        print('Usage: python udp_display <port>')
        sys.exit(0)

port = int(sys.argv[1])

# Max UDP packet size. It is assumed here that the jpg image
# size does not exceeds the maximum UDP datagram size
BUFFER_SIZE = 65507  # Max UDP packet size.
sock = socket.socket(socket.AF_INET, socket.SOCK_DGRAM)
sock.bind(("0.0.0.0", port))
isFirst = True
while True:
        data, addr = sock.recvfrom(BUFFER_SIZE)
        print(f"Received image from {addr}, size: {len(data)} bytes")
        try:
                image = Image.open(io.BytesIO(data))
                # Display image using Matplotlib, clear it first
                if not isFirst:
                        plt.clf()
                isFirst = False
                plt.imshow(image)
                plt.axis("off")
                plt.show(block=False)
                plt.pause(0.01)
        except Exception as e:
                print(f"Failed to decode image: {e}")
```

A limitation in using UDP datagrams is the maximum size that datagrams can have, i.e. 65527 bytes (the UDP header itself is 8 bytes). In general, when using webcams mounted on laptops this is the case, otherwise a different approach, using TCP/IP instead, must be adopted. In this case you may send over the TCP socket first the size of the frame, followed by the frame itself.

EXERCISE 4

You may have noted that developing the code for the previous two exercises, some operation are the same, that is, the management of the threads and of the queue used to exchange information between them. A good software engineering practice is to avoid duplicating code, integrating common functionality in classes that can then be reused. For the queue management, one could define the following class organization in C++:

```cpp
<include pthread.h>
class QueueItem
{
        public:
        char *buf;
        int size;
```

```
        QueueItem *nxt;
//Constructor
        QueueItem(char *buf, int size):buf(buf), size(size), nxt(NULL){}
};

class Queue
{
        pthread_mutex_t lock;
        pthread_cond_t cond;
        QueueItem *first, *last;
        public:
//Constructor
                Queue();
//Destrutor
                ~Queue();
//Enqueue a new item
                void addItem(char *buf, int size);
//Get the oldest item in the queue
                char *getItem(int &size);
};
```

The code implementing class Queue dynamically allocates QueueItem objects and
enqueue then in a linked list, providing the required mechanisms to avoid race conditions and to synchronize access (this will be the subject of the following chapters).
In this way the management of the queue and of the required synchronization is carried out by the Queue object. It is possible however to abstract more, and to embed
also the thread management in a further QueueManager class:

```
#include "queue.h"
#include <pthread.h>
class QueueManager {
//The queue internally used to enqueue data iitems
        Queue queue;
//The decoupled thread
        pthread_t thread;
        bool started;
        public:
                QueueManager():started(false){};
                ~QueueManager(){};
                void produce(char *buf, int size);
                void start();
                void stop();
                Queue *getQueue() {return &queue;}
                virtual void consume(char *buf, int size) = 0;
};
```

Using QueueManager a program can enqueue data, generically represented as an
array of bytes via produce() method. The companion thread will be activated by
method start() taking enqueued messages and calling method consume(). As
we are developing a general purpose class handling thread decoupling and queue
management, the semantics of method consume() will depend on the specific application. For exercise 2, method consume()will store the passed buffer on a file,
while in for exercise 3 it will send data over the network. This is why method consume is declared as virtual: it will be implemented by specific classes inheriting from
QueueManager. You may rewrite the programs proposed in exercises 2 and 3 greatly
reducing the amount of required code, just providing the specific implementation of
method consume.

3 Real-Time Concurrent Programming Principles

This chapter lays the foundation of real-time concurrent programming theory by introducing what is probably its most central concept, that is, the definition of *process* as the abstraction of an executing program. This definition is also useful to clearly distinguish between *sequential* and *concurrent* programming, and to highlight the pitfalls of the latter.

3.1 THE ROLE OF PARALLELISM

Most contemporary computers are able to perform more than one activity at the same time, at least apparently. This is particularly evident with personal computers, in which users ordinarily interact with many different applications at the same time through a graphics user interface. In addition, even if this aspect is often overlooked by the users themselves, the same is true also at a much finer level of detail. For example, contemporary computers are usually able to manage user interaction while they are reading and writing data to the hard disk, and are actively involved in network communication. In most cases, this is accomplished by having peripheral devices *interrupt* the current processor activity when they need attention. Once it has finished taking care of the interrupting devices, the processor goes back to whatever it was doing before.

A key concept here is that all these activities are not performed in a fixed, predetermined *sequence*, but they all seemingly proceed in *parallel*, or *concurrently*, as the need arises. This is particularly useful to enhance the user experience since it would be very awkward, at least by modern standards, to be constrained to have only one application active on a personal computer at any given time and to have to quit one application before switching to the next. Similarly, having a computer stop doing anything else only because a hard disk operation is in progress would seem strange, to say the least.

Even more importantly, the ability of carrying out multiple activities "at once" helps in fulfilling any timing constraint that may be present in the system in an efficient way. This aspect is often of concern whenever the computer interacts with the outside world. For example, network interfaces usually have a limited amount of space to buffer incoming data. If the system as a whole is unable to remove them from the buffer and process them within a short amount of time—on the order of a few milliseconds for a high-speed network coupled with a low-end interface—the buffer will eventually overflow, and some data will be lost or will have to be retransmitted. In more extreme cases, an excessive delay will also trigger higher-level errors, such as network communication timeouts.

DOI: 10.1201/9781003593416-3

59

In this particular situation, a sequential implementation would be tricky, because all applications would have to voluntarily suspend whatever they were doing, at predetermined instants, to take care of network communication. In addition, deciding in advance when and how often to perform this activity would be difficult because the exact arrival time of network data and their rate are often hard to predict.

Depending on the hardware characteristics, the apparent execution parallelism may correspond to a true parallelism at the hardware level. This is the case of *multiprocessor* and *multicore* systems, in which either multiple processors or a single processor with multiple execution cores share a common memory, and each processor or core is able to carry out its own sequential flow of instructions.

The same end result can be obtained when a single processor or core is available, or when the number of parallel activities exceeds the number of available processors or cores, by means of software techniques implemented at the operating system level, known as *multiprogramming*, that repeatedly switch the processor back and forth from one activity to another. If properly implemented, this *context switch* is completely transparent to, and independent from, the activities themselves, and they are usually unaware of its details. The term *pseudo* parallelism is often used in this case, to contrast it with the real hardware-supported parallelism discussed before, because technically the computer is still executing exactly one activity at any given instant of time.

The notion of *sequential process* (or *process* for short) was born, mainly in the operating system community, to help programmers express parallel activities in a precise way and keep them under control. It provides both an abstraction and a conceptual model of a running program.

3.2   DEFINITION OF PROCESS

The concept of process was first introduced in the seminal work of Dijkstra [25]. In this model, any concurrent system, regardless of its nature or complexity, is represented by, and organized as, a set of processes that execute in parallel. Therefore, the process model encompasses both the application programs and the operating system itself. Each process is autonomous and holds all the information needed to represent the evolving execution state of a sequential program. This necessarily includes not only the program instructions but also the state of the processor (program counter, registers) and memory (variables).

Informally speaking, each process can be regarded as the execution of a sequential program by "its own" processor even if, as shown in Figure 3.1, in a multiprogrammed system the physical processors may actually switch from one process to another. The abstract view of the system given by the process model is shown on the left side of the figure, where we see three independent processes, each with its own control flow and state information. For the sake of simplicity, both of them have been depicted with an arrow, representing in an abstract way how the execution proceeds with time.

On the other hand, the right side of the figure shows one of the many possible sequences of operations performed by a single-processor system to execute them. The

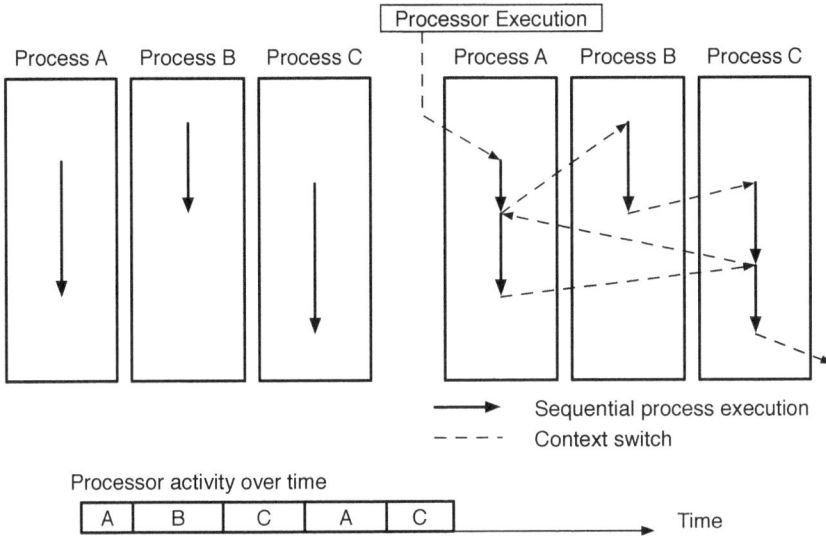

Figure 3.1 Multiprogramming: abstract model of three sequential processes (left) and their execution on a single-processor system (right).

solid lines represent the execution of a certain process, whereas the dashed lines represent context switches. The multiprogramming mechanism ensures, in the long run, that all processes will make progress even if, as shown in the time line of processor activity over time at the bottom of the figure, the processor indeed executes only one process at a time.

Comparing the left and right sides of Figure 3.1 explains why the adoption of the process model simplifies the design and implementation of a concurrent system: By using this model, the system design is carried out at the process level, a clean and easy to understand abstraction, without worrying about the low-level mechanisms behind its implementation. In principle, it is not even necessary to know whether the system's hardware is really able to execute more than one process at a time or not, or the degree of such a parallelism, as long as the execution platform actually provides multiprogramming.

The responsibility of choosing which processes will be executed at any given time by the available processors, and for how long, falls on the operating system and, in particular, on an operating system component known as *scheduler*. Of course, if a set of processes must cooperate to solve a certain problem, not all possible choices will produce meaningful results. For example, if a certain process P makes use of some values computed by another process Q, executing P before Q is probably not a good idea.

Therefore, the main goal of *concurrent programming* is to define a set of inter-process communication and synchronization primitives. When used appropriately, these primitives ensure that the *results* of the concurrent program will be correct by

Wrong results (P executed before Q)

(a) | R | P | Q | → Time

Correct results (P completes after 40 ms, R after 60 ms)

(b) | Q | P | R | → Time

Correct results (P completes after 60 ms, R after 50 ms)

(c) | Q | R | P | → Time

10 ms

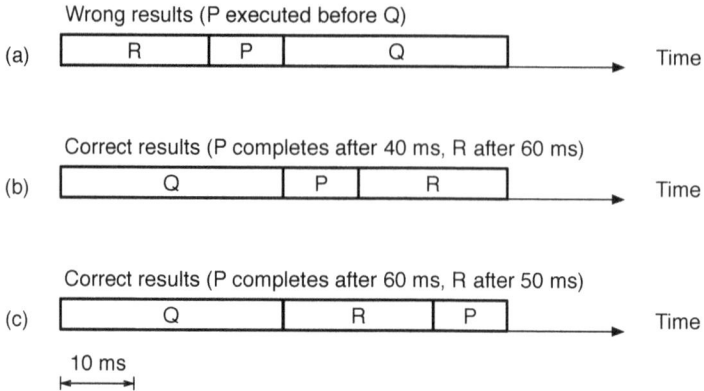

Figure 3.2 Unsuitable process interleavings may produce incorrect results. Process interleaving, even when it is correct, also affects system timing. All processes are ready for execution at $t = 0$.

introducing and enforcing appropriate constraints on the scheduler decisions. They will be discussed in Chapters 5 and 6.

Another aspect of paramount importance for real-time systems—that is, systems in which there are timing constraints on system activities—is that, even if the correct application of concurrent programming techniques guarantees that the results of the concurrent program will be correct, the scheduling decisions made by the operating system may still affect the behavior of the system in undesirable ways, concerning timing.

This is due to the fact that, even when all constraints set forth by the interprocess communication and synchronization primitives are met, there are still many acceptable scheduling sequences, or process *interleaving*. Choosing one or another does not affect the overall result of the computation but may change the timing of the processes involved.

As an example, Figure 3.2 shows three different interleavings of processes P, Q, and R. All of them are ready for execution at $t = 0$, and their execution requires 10, 30, and 20 ms of processor time, respectively. Since Q produces some data used by P, P cannot be executed before Q. For simplicity, it is also assumed that processes are always run to completion once started and that there is a single processor in the system.

Interleaving (a) is unsuitable from the concurrent programming point of view because it does not satisfy the precedence constraint between P and Q stated in the requirements, and will lead P to produce incorrect results. On the other hand, interleavings (b) and (c) are both correct in this respect—the precedence constraint is met in both cases—but they are indeed very different from the system timing point of view. As shown in the figure, the completion time of P and R will be very different. If we are dealing with a real-time system and, for example, process P must conclude within 50 ms, interleaving (b) will satisfy this requirement, but interleaving (c) will not.

In order to address this issue, real-time systems use specially devised scheduling algorithms, to be discussed in Chapters 10 and 11. Those algorithms, complemented by appropriate analysis techniques, guarantee that a concurrent program will not only produce correct results but it will also satisfy its timing constraints for all permitted interleavings. This will be the main subject of Chapters 12 through 15.

3.3 PROCESS STATE

In the previous section, we discussed how the concept of process plays a central role in concurrent programming. Hence, it is very important to clearly define and understand what the "contents" of a process are, that is, what the process state components are. Interestingly enough, a correct definition of the process state is very important from the practical viewpoint, too, because it also represents the information that the operating system must save and restore in order to perform a *transparent* context switch from one process to another.

There are four main process state components, depicted in Figure 3.3:

1. Program code, sometimes called the *text* of the program;
2. Processor state: program counter, general purpose registers, status words, etc.
3. Data memory, comprising the program's global variables, as well as the procedure call stack that, for many programming languages, also holds local variables;

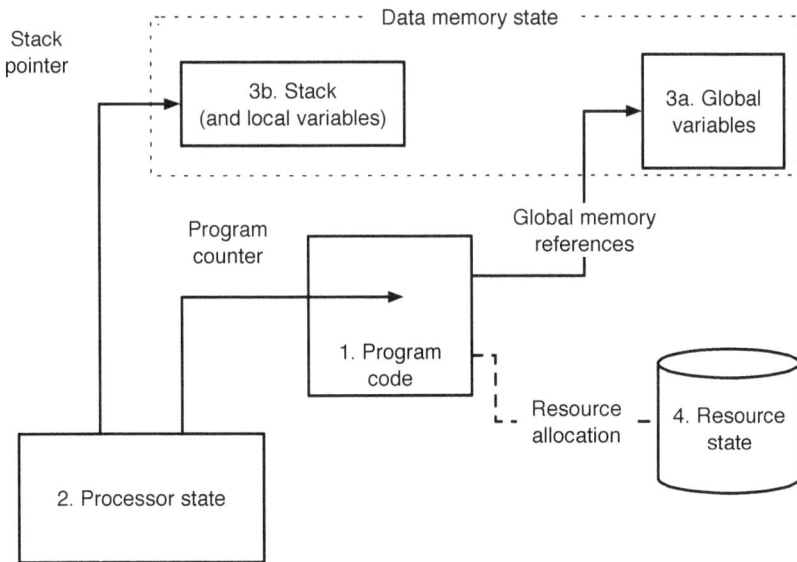

Figure 3.3 Graphical representation of the process state components.

4. The state of all operating system resources currently assigned to, and being used by the process: open files, input–output devices, etc.

Collectively, all memory locations a process can have access to are often called *address space*. The address space therefore includes both the program code and the data memory. See Chapter 2 for more general information about this and other related terms from the application programmer's point of view.

The need of including the program code in the process state is rather obvious because, by intuition, the execution of different programs will certainly give rise to different activities in the computer system. On the other hand, the program code is certainly not enough to characterize a process. For example, even if the program code is the same, different execution activities still come out if we observe the system behavior at different phases of program execution, that is, for different values of the program counter. This observation can be generalized and leads to the inclusion of the whole processor state into the process state.

However, this is still not enough because the same program code, with the same processor state, can still give origin to distinct execution activities depending on the memory state. The same instruction, for example, a division, can in fact correspond to very different activities, depending on the contents of the memory word that holds the divisor. If the divisor is not zero, the division will be carried out normally; if it is zero, most processors will instead take a trap.

The last elements the process state must be concerned with are the operating system resources currently assigned to the process itself. They undoubtedly have an influence on program execution—that is, in the final analysis, on the process—because, for example, the length and contents of an input file may affect the behavior of the program that is reading it.

It should be noted that none of the process state components discussed so far have anything to do with *time*. As a consequence, by design, a context switch operation will be transparent with respect to the results computed by the process but may *not* be transparent for what concerns its timeliness. This is another way to justify why different scheduling decisions—that is, performing a context switch at a certain instant instead of another—will not affect process results but may lead to either an acceptable or an unacceptable timing behavior. It also explains why other techniques are needed to deal with, and satisfy, timing constraints in real-time systems.

The fact that program code is one of the process components but not the only one, also implies that there are some decisive differences between programs and processes, and that those two terms must not be used interchangeably. Similarly, processes and processors are indeed not synonyms. In particular:

- A *program* is a static entity. It basically describes an algorithm, in a formal way, by means of a programming language. The machine is able to understand this description, and execute the instructions it contains, after a suitable translation.
- A *process* is a dynamic concept and captures the notion of program execution. It is the activity carried out by a processor when it is executing a

certain program and, therefore, requires some state information, besides the program code itself, to be characterized correctly.

- A *processor* is a physical or virtual entity that supports program execution and, therefore, makes processes progress. There is not necessarily a one-to-one correspondence between processors and processes, because a single processor can be time-shared among multiple processes through multipro-gramming.

3.4 PROCESS LIFE CYCLE AND PROCESS STATE DIAGRAM

In the previous section, we saw that a process is characterized by a certain amount of state information. For a process to exist, it is therefore necessary to reserve space for this information within the operating system in a data structure often called *Process Control Block* (PCB), and initialize it appropriately.

This initialization, often called process *creation*, ensures that the new process starts its life in a well-known situation and that the system will actually be able to keep track of it during its entire lifetime. At the same time, most operating systems also give to each process a *process identifier* (PID), which is guaranteed to be unique for the whole lifetime of the process in a given system and must be used whenever it is necessary to make a reference to the process.

Depending on the purpose and complexity of the system, it may be possible to precisely determine how many and which processes will be needed right when the system itself is turned on. Process creation becomes simpler because all processes can be *statically* created while the system as a whole is being initialized, and it will not be possible to create new processes afterward.

This is often the case with simple, real-time control systems, but it becomes more and more impractical as the complexity of the system and the variety of functions it must fulfill from time to time grow up. The most extreme case happens in general purpose systems: it would be extremely impractical for users to have to decide which applications they will need during their workday when they turn on their personal computer in the morning and, even worse, being constrained to reconfigure and restart it whenever they want to start an application they did not think about before.

In addition, this approach may be quite inefficient from the point of view of resource usage, too, because all the processes may start consuming system resources a long time before they are actively used. For all these reasons, virtually all general purpose operating systems, as well as many real-time operating systems, also contemplate *dynamic* process creation. The details of this approach vary from one system to another but, in general terms:

- During initialization, the operating system crafts a small number, or even one single process. For historical reasons, this process is often called *init*, from the terminology used by most Unix and Unix-like systems [66].
- All other processes are created (directly or indirectly) by *init*, through the invocation of an appropriate operating system service. All newly created processes can, in turn, create new processes of their own.

This kind of approach also induces a hierarchical relationship among processes, in which the creation of a new process sets up a relation between the existing process, also called the *parent*, and the new one, the *child*. All processes, except *init*, have exactly one parent, and zero or more children. This relationship can be conveniently represented by arranging all processes into a tree, in which

*   Each process corresponds to a node
*   Each parent–child relation corresponds to an arc going from the parent to the child.

Some operating systems keep track and make use of this relationship in order to define the scope of some service requests related to the processes themselves. For example, in most Unix and Unix-like systems only the parent of a process can wait for its termination and get its final termination status. Moreover, the parent–child relation also controls resource inheritance, for example, open files, upon process creation.

During their life, processes can be in one of several different states. They go from one state to another depending on their own behavior, operating system decision, or external events. At any instant, the operating systems has the responsibility of keeping track of the current state of all processes under its control.

A useful and common way to describe in a formal way all the possible process states and the transition rules is to define a directed graph, called *Process State Diagram* (PSD), in which nodes represent states and arcs represent transitions.

A somewhat simplified process state diagram is shown in Figure 3.4. It should be remarked that real-world operating systems tend to have more states and transitions, but in most cases they are related to internal details of that specific operating systems and are therefore not important for a general discussion.

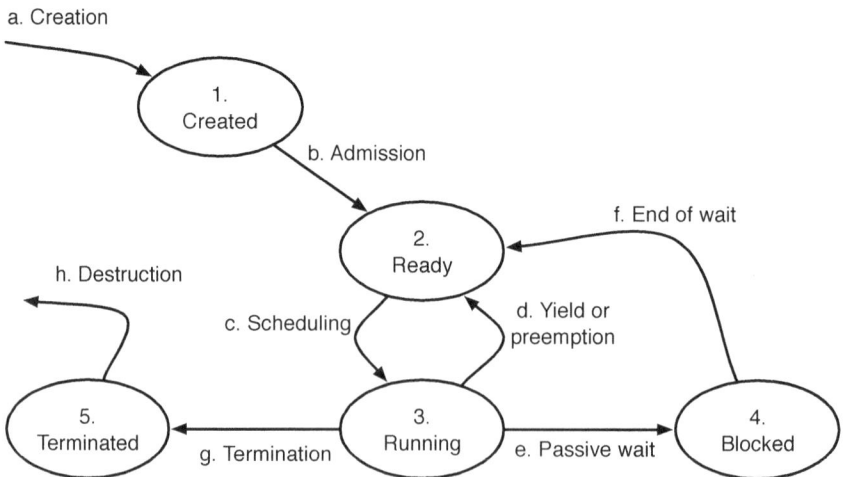

Figure 3.4   An example of process state diagram.

Looking at the diagram, at any instant, a process can be in one of the following states:

1. A process is in the *Created* state immediately after creation. It has a valid PCB associated with it, but it does not yet compete for execution with the other processes present in the system.
2. A process is *Ready* when it is willing to execute, and competes with the other processes to do so, but at the moment there is not any processor available in the system to actually execute it. This happens, for example, when all processors are busy with other processes. As a consequence, processes do not make any progress while they are ready.
3. A process being actively executed by a processor is in the *Running* state. The upper limit to the number of running processes, at any given time, is given by the total number of processors available in the system.
4. Sometimes, a process will have to wait for an external event to occur, for example, the completion of an input–output (I/O) operation. In other cases, discussed in Chapters 5 and 6, it may be necessary to block a process, that is, temporarily stop its execution, in order to correctly synchronize it with other processes and let them communicate in a meaningful way. All those processes are put in the *Blocked* state. A process does not compete for execution as long as it is blocked.
5. Most operating systems do not destroy a process immediately when it terminates, but put it in the *Terminated* state instead. In this state, the process can no longer be executed, but its PCB is still available to other processes, giving them the ability to retrieve and examine the summary information it contains. In this way, it is possible, for example, to determine whether the process terminated spontaneously or due to an error.

The origin of a state transition may be either a voluntary action performed by the process that undergoes it, or the consequence of an operating system decision, or the occurrence of an event triggered by a hardware component or another process. In particular:

a. The initial transition of a newly created process into the Created state occurs when an existing process creates it and after the operating system has correctly initialized its PCB.
   In most cases, during process creation, the operating system also checks that the bare minimum amount of system resources needed by the new process, for example, an adequate amount of memory to hold the program text, are indeed available.
b. The transition from the Created to the Ready state is under the control of an operating system function usually known as *admission control*. For general-purpose operating systems, this transition is usually immediate, and they may even lack the distinction between the Created and Ready states.
   On the contrary, real-time operating systems must be much more careful because, as outlined in Section 3.2, the addition of a new process actively competing for execution can adversely affect the timings of the whole system.

In this case, a new process is admitted into the Ready state only after one of the schedulability and mode transition analysis techniques, to be described in Chapters 12 through 15, reaffirmed that the system will still meet all its timing requirements.

c. The transition from the Ready to the Running state is controlled by the operating system scheduler, according to its scheduling algorithm, and is transparent to the process that experiences it. Scheduling algorithms play a central role in determining the performance of a real-time system, and will be the main topic of Chapters 10 and 11.

d. The opposite transition, from Running to Ready, can be due to two distinct reasons:
   - *Preemption*, decided by the operating system scheduler, in which a process is forced to relinquish the processor even if it is still willing to execute.
   - *Yield*, requested by the process itself to ask the system to reconsider its scheduling decision and possibly hand over the processor to another process.

   The high-level result is the same in both cases, that is, the process goes back to the Ready state both after a preemption and after a successful yield. The most important difference depends on the fact that, from the point of view of the process experiencing it, the transition is involuntary in the first case, and voluntary in the second.

   Hence, a preemption may occur anywhere and at any time during process execution, whereas a yield may occur only at specific locations in the code, and at the time the process requests it.

e. A process transitions from the Running to the Blocked state when it voluntarily hands over the processor, because it is about to start a passive wait. This transition is typically a consequence of a synchronous input–output request or interprocess communication, to be discussed in Chapters 5 and 6. In all cases, the goal of the process is to wait for an external event to occur—for instance, it may be either the completion of the input–output operation or the availability of data from another process—without wasting processor cycles in the meantime.

f. When the event the process is waiting for eventually occurs, the process is awakened and goes back to the Ready state, and hence, it starts competing with the other processes for execution again. The process is not brought back directly to the Running state, because the fact that it has just been awakened does not guarantee that it actually is the most important process in the system at the moment. This decision pertains to the scheduler, and not to the passive wait mechanism. The source of the awakening event may be either another process, for interprocess communication, or a hardware component for synchronous I/O operations. In the latter case, the I/O device usually signals the occurrence of the event to the processor by means of an interrupt request, and the process is awakened as part of the consequent interrupt handling activity.

g. A process may go from the Running to the Terminated state for two distinct reasons:

- When it voluntarily ends its execution because, for example, it is no longer needed in the system.
- When an unrecoverable error occurs, unlike in the previous case, this transition is involuntary.

h. After termination, a process and its PCB are ultimately removed from the system with a final transition out of the Terminated state. After this transition, the operating system can reuse the same process identifier formerly assigned to the process for a new one.

This is crucial for what concerns process identification because, after this transition occurs, all uses of that PID become invalid or, even worse, may refer to the wrong process. It is therefore the responsibility of the programmer to avoid any use of a PID after the corresponding process went out of the Terminated state.

In some operating systems, the removal of the process and its PCB from the system is performed automatically, immediately after the summary information of the terminated process has been retrieved successfully by another process. In other cases, an explicit request is necessary to collect and reuse PCBs related to terminated processes.

From this discussion, it becomes evident that the PCB must contain not only the concrete representation of the main process state components discussed in Section 3.3, but also other information pertaining to the operating system and its process management activities.

This includes the current position of the process within the Process State Diagram and other process attributes that drive scheduling decisions and depend on the scheduling algorithm being used. A relatively simple scheduling algorithm may only support, for example, a numeric attribute that represents the relative process priority, whereas more sophisticated scheduling techniques may require more attributes.

3.5 MULTITHREADING

According to the definition given in Section 3.3, each process can be regarded as the execution of a sequential program on "its own" processor. That is, the process state holds enough state information to fully characterize its address space, the state of the resources associated with it, and *one* single flow of control, the latter being represented by the processor state.

In many applications, there are several distinct activities that are nonetheless related to each other, for example, because they have a common goal. For example, in an interactive media player, it is usually necessary to take care of the user interface while decoding and playing an audio stream, possibly retrieved from the Internet. Other background activities may be needed as well, such as retrieving the album artwork and other information from a remote database.

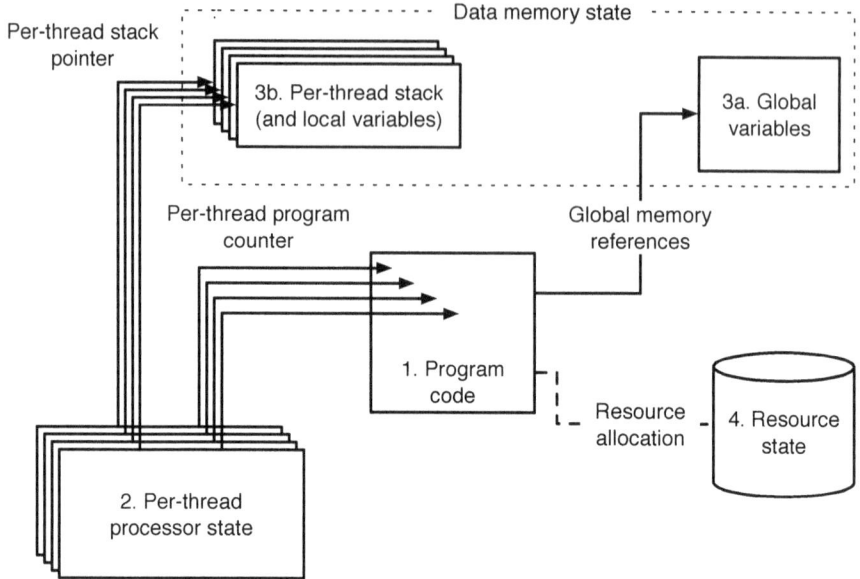

Figure 3.5 Graphical representation of the process state components in a multithreading system.

It may therefore be useful to manage all these activities as a group and share system resources, such as files, devices, and network connections, among them. This can be done conveniently by envisaging *multiple* flows of control, or *threads*, within a single process. As an added bonus, all of them will implicitly refer to the same address space and thus share memory. This is a useful feature because many interprocess communication mechanisms, for instance, those discussed in Chapter 5, are indeed based on shared variables.

Accordingly, many modern operating systems support *multithreading*, that is, they support multiple threads within the same process by splitting the process state into per-process and per-thread components as shown in Figure 3.5. In particular,

- The program code is the same for all threads, so that all of them execute from the same code base.
- Each thread has its own processor state and procedure call stack, in order to make the flows of control independent from each other.
- All threads evolve autonomously for what concerns execution, and hence, each of them has its own position in the PSD and its own scheduling attributes.
- All threads reside in the same address space and implicitly share memory.
- All resources pertaining to the process are shared among its threads.

Another important reason for being aware of multithreading is that a full-fledged implementation of multiple processes must be assisted by hardware, particularly to enforce address space separation and protection. For example, on contemporary Intel processors, this is accomplished by means of a Memory Management Unit (MMU) integral to the processor architecture and other related hardware components, such as the Translation Lookaside Buffer (TLB) [49].

The main disadvantage of MMUs is that they consume a significant amount of silicon area and power. Moreover, since they contribute to chip complexity, they are also likely to increase the cost of the processor. For this reason some processor architectures such as, for example, the ARM architectures v6 [5] and v7-M [6] offer a choice between a full MMU and a simpler hardware component called Memory Protection Unit (MPU), which does not provide address translation but is still able to ensure that the address space of each process is protected from unauthorized access by other processes.

Nonetheless, many processors of common use in embedded systems have neither an MMU nor an MPU. Any operating systems running on those processors are therefore forced to support only *one* process because they are unable to provide address space protection. This is the case for many small real-time operating systems, too. In all these situations, the only way to still support multiprogramming despite the hardware limitations is through multithreading.

For example, some FreeRTOS ports, like the one for the ARM Cortex-M3 [7], support an MPU and can make use of it if it is available. If it is not, the operating system still supports multiple threads, which share the same address space and can freely read and write each other's data.

3.6   SUMMARY

In this chapter, the concept of *process* has been introduced. A process is an abstraction of an executing program and encompasses not only the program itself, which is a static entity, but also the *state* information that fully characterizes execution.

The notion of process as well as the distinction between programs and processes become more and more important when going from *sequential* to *concurrent* programming because it is essential to describe, in a sound and formal way, all the activities going on in parallel within a concurrent system. This is especially important for real-time applications since the vast majority of them are indeed concurrent.

The second main concept presented in this chapter is the PSD. Its main purpose is to define and represent the different states a process may be in during its lifetime. Moreover, it also formalizes the rules that govern the transition of a process from one state to another.

As it will be better explained in the next chapters, the correct definition of process *states* and *transitions* plays a central role in understanding how processes are scheduled for execution, when they outnumber the processors available in the systems, how they exchange information among themselves, and how they interact with the outside world in a meaningful way.

Last, the idea of having more than one execution flow within the same process, called *multithreading*, has been discussed. Besides being popular in modern, general-purpose systems, multithreading is of interest for real-time systems, too. This is because hardware limitations may sometimes prevent real-time operating systems from supporting multiple processes in an effective way. In that case, typical of small embedded systems, multithreading is the only option left to support concurrency anyway.

3.7 EXERCISES

EXERCISE 1

Four tasks A, B, C, and D execute concurrently on a non-preemptive system. B makes use of a value computed by A. D makes use of values computed by A and B. Which of the following execution sequences produce correct results:

1. A, D, B, C.
2. C, A, B, D.
3. B, A, C, D.
4. A, C, B, D.

EXERCISE 2

May a task bring itself out of the *Blocked* state? What about the *Ready* state?

EXERCISE 3

Describe the differences between *processes* and *threads*. Is it easier to share data among distinct processes or among threads belonging to the same process?

4 Deadlock

In any concurrent system, process execution relies on the availability of a number of *resources* such as, for example, disk blocks, memory areas, and input–output (I/O) devices. Resources are often in scarce supply, hence they must be shared among all processes in the system, and processes compete with each other to acquire and use them.

In this chapter we will see that uncontrolled resource sharing may be very dangerous and, in particular, may prevent whole groups of processes from performing their job. Even if the probability of occurrence of this unfortunate phenomenon, known as *deadlock*, may be very low, it must still be dealt with adequately, especially in a real-time system.

4.1 A SIMPLE EXAMPLE

In any multiprogrammed system, many processes usually share a certain number of *resources* and compete for their use. The concept of resource is very broad and includes both physical resources, for example, printers, disk blocks, and memory areas, as well as logical ones, like entries in the operating system's data structures or filesystem tables.

Some kinds of resource, such as a read-only data structure, pose no problems in this respect because many processes can access them concurrently with correct results. However, many other resources can intrinsically be used by only one process at a time. For instance, having multiple processes simultaneously using the same disk block for storage must be avoided because this would lead to incorrect results and loss of data.

To deal with this problem, most operating systems compel the processes under their control to *request* resources before using them and wait if those resources are currently assigned to another process, so that they are not immediately available for use. Processes must also *release* their resources when they no longer need them, in order to make them available to others. In this way, the operating system acts as an arbiter for what concerns resource allocation and can ensure that processes will have exclusive access to them when required.

Unless otherwise specified, in this chapter we will only be concerned with *reusable* resources, a term taken from historical IBM literature [37]. A reusable resource is a resource that, once a process has finished with it, is returned to the system and can be used by the same or another process again and again. In other words, the value of the resource or its functionality do not degrade with use. This is in contrast with the concept of *consumable* resource, for example, a message stored in a FIFO queue, that is created at a certain point and ceases to exist as soon as it is assigned to a process.

DOI: 10.1201/9781003593416-4   **73**

In most cases, processes need more than one resource during their lifetime in order to complete their job, and request them in succession. A process A wanting to print a file may first request a memory buffer in order to read the file contents into it and have a workspace to convert them into the printer-specific page description language. Then, it may request exclusive use of the printer and send the converted data to it. We leave out, for clarity, the possibly complex set of operations A must perform to get access to the file.

If the required amount of memory is not immediately available, it is reasonable for the process to wait until it is, instead of failing immediately because it is likely that some other process will release part of its memory in the immediate future. Likewise, the printer may be assigned to another process at the time of the request and, also in this case, it is reasonable to wait until it is released.

Sadly, this very common situation can easily lead to an anomalous condition, known as *deadlock*, in which a whole set of processes is blocked forever and will no longer make any progress. Not surprisingly, this problem has received a considerable amount of attention in computer science; in fact, one of the first formal definitions of deadlock was given by in 1965 by Dijkstra [25], who called it "deadly embrace."

To illustrate how a deadlock may occur in our running example, let us consider a second process B that runs concurrently with A. It has the same goal as process A, that is, to print a file, but is has been coded in a different way. In particular, process B request the printer *first*, and *then* it tries to get the memory buffer it needs. The nature of this difference is not at all important (it may be due, for example, to the fact that A and B have been written by two different programmers unaware of each other's work), but it is nonetheless important to realize that both approaches are meaningful and there is nothing wrong with either of them.

In this situation, the sequence of events depicted in Figure 4.1 may occur:

1. Process B request the printer, P. Since the printer has not been assigned to any process yet, the request is granted immediately and B continues.
2. Process A requests a certain amount M_A of memory. If we assume that the amount of free memory at the time of the request is greater than M_A, this request is granted immediately, too, and A proceeds.
3. Now, it is the turn of process B to request a certain amount of memory M_B. If the request is sensible, but there is not enough free memory in the system at the moment, the request is not declined immediately. Instead, B is blocked until a sufficient amount of memory becomes available.
   This may happen, for example, when the total amount of memory M in the system is greater than both M_A and M_B, but less than $M_A + M_B$, so that both requests can be satisfied on their own, but not together.
4. When process A requests the printer P, it finds that it has been assigned to B and that it has not been released yet. As a consequence, A is blocked, too.

At this point, both A and B will stay blocked forever because they own a resource, and are waiting for another resource that will never become available since it has been assigned to the other process involved in the deadlock.

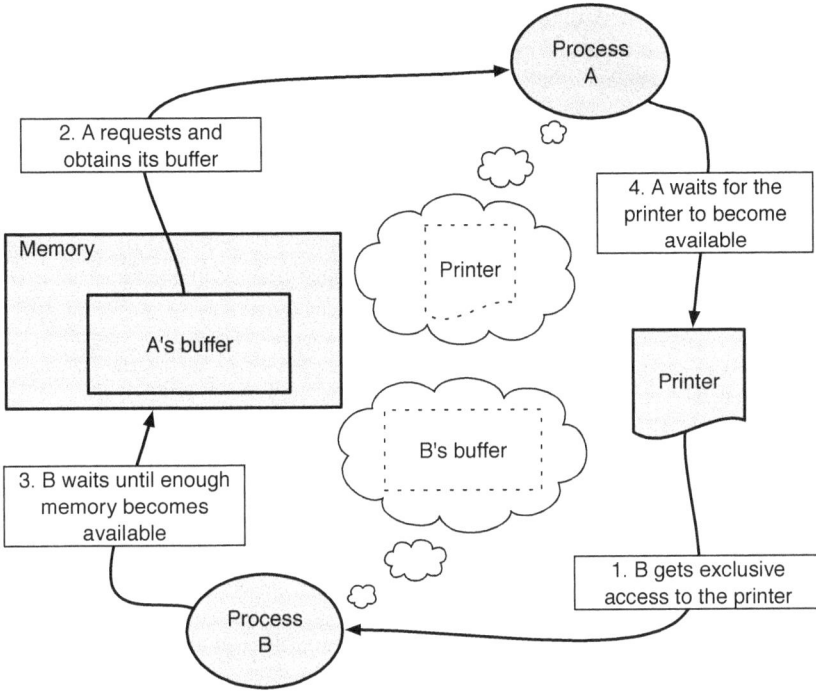

Figure 4.1 A simple example of deadlock involving two processes and two resources.

Even in this very simple example, it is evident that a deadlock is a complex phenomenon with a few noteworthy characteristics:

- It is a *time-dependent* issue. More precisely, the occurrence of a deadlock in a system depends on the relationship among process timings. The chain of events leading to a deadlock may be very complex, hence the probability of actually observing a deadlock during bench testing may be very low. In our example, it can easily be observed that no deadlock occurs if process *A* is run to completion before starting *B* or vice versa.

  Unfortunately, this means that the code will be hard to debug, and even the insertion of a debugger or code instrumentation to better understand what is happening may perturb system timings enough to make the deadlock disappear. This is a compelling reason to address deadlock problems from a theoretical perspective, during system *design*, rather than while testing or debugging it.

- It also depends on a few specific *properties* of the resources involved and on how the operating system manages them. For example, albeit this technique is not widespread in real-time operating systems, some general-purpose operating systems are indeed able to *swap* process images in and out of main memory with the assistance of a mass storage device. Doing this, they are able to accommodate processes whose total memory requirements exceed the available memory.

In this case, the memory request performed by B in our running example does not necessarily lead to an endless wait because the operating system can temporarily take away—or *preempt*—some memory from A in order to satisfy B's request, so that both process will be eventually able to complete their execution. As a consequence, the same processes may or may not be at risk for what concerns deadlock, when they are executed by operating systems employing dissimilar memory management or, more generally, resource management techniques.

4.2   FORMAL DEFINITION OF DEADLOCK

In the most general way, a deadlock can be defined formally as a situation in which a set of processes passively waits for an event that can be triggered only by another process in the same set. More specifically, when dealing with resources, there is a deadlock when all processes in a set are waiting for some resources previously allocated to other processes in the same set. As discussed in the example of Section 4.1, a deadlock has therefore two kinds of adverse consequences:

- The *processes* involved in the deadlock will no longer make any progress in their execution, that is, they will wait forever.
- Any *resource* allocated to them will never be available to other processes in the system again.

Havender [37] and Coffman et al. [21] were able to formulate four conditions that are *individually necessary* and *collectively sufficient* for a deadlock to occur. These conditions are useful, first of all because they define deadlock in a way that abstracts away as much as possible from any irrelevant characteristics of the processes and resources involved.

Second, they can and have been used as the basis for a whole family of deadlock prevention algorithms because, if an appropriate policy is able to prevent (at least) one of them from ever being fulfilled in the system, then no deadlock can possibly occur by definition. The four conditions are

1. *Mutual exclusion*: Each resource can be assigned to, and used by, at most one process at a time. As a consequence, a resource can only be either free or assigned to one particular process. If any process requests a resource currently assigned to another process, it must wait.
2. *Hold and Wait*: For a deadlock to occur, the processes involved in the deadlock must have successfully obtained at least one resource in the past and have not released it yet, so that they *hold* those resources and then *wait* for additional resources.
3. *Nonpreemption*: Any resource involved in a deadlock cannot be taken away from the process it has been assigned to without its consent, that is, unless the process voluntarily releases it.

4. *Circular wait*: The processes and resources involved in a deadlock can be arranged in a circular chain, so that the first process waits for a resource assigned to the second one, the second process waits for a resource assigned to the third one, and so on up to the last process, which is waiting for a resource assigned to the first one.

4.3 REASONING ABOUT DEADLOCK: THE RESOURCE ALLOCATION GRAPH

The *resource allocation graph* is a tool introduced by Holt [43] with the twofold goal of describing in a precise, rigorous way the resource allocation state in a system at a given instant, as well as being able to reason about and detect deadlock conditions. Figure 4.2 shows an example.

In its simplest form, a *resource allocation graph* is a directed graph with two kinds of nodes and two kinds of arcs. The two kinds of nodes represent the processes and resources of interest, respectively. In the most common notation, used, for example, by Tanenbaum and Woodhull [88] and Silbershatz et al. [87]:

1. Processes are shown as *circles*.
2. Resources are shown as *squares*.

For instance, in Figure 4.2, P_2 represents a process and R_1 represents a resource. Of course, the exact geometric shape used to represent processes and resources is not at all important. As a matter of fact, in the Holt's paper [43] the notation was exactly the opposite.

On the other hand, the two kinds of arcs express the request and ownership relations between processes and resources. In particular,

1. An arc directed from a *process* to a *resource* (similar to the arc labeled A in Figure 4.2, going from process P_1 to resource R_1) indicates that the process is *waiting for* the resource.

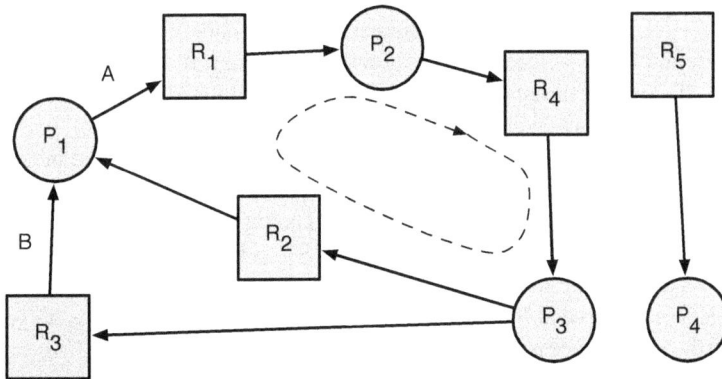

Figure 4.2 A simple resource allocation graph, indicating a deadlock.

2. An arc directed from a *resource* to a *process* (for example, arc B in Figure 4.2, going from resource R$_3$ to process P$_1$) denotes that the resource is currently *assigned to*, or owned by, the process.

Hence, the resource allocation graph shown in Figure 4.2 represents the following situation:

- Process P$_1$ owns resources R$_2$ and R$_3$, and is waiting for R$_1$.
- Process P$_2$ owns resource R$_1$ and is waiting for resource R$_4$.
- Process P$_3$ owns resource R$_4$ and is waiting for two resources to become available: R$_2$ and R$_3$.
- Process P$_4$ owns resource R$_5$ and is not waiting for any other resource.

It should also be noted that arcs connecting either two processes, or two resources, have got no meaning and are therefore not allowed in a resource allocation graph. More formally, the resource allocation graph must be *bipartite* with respect to process and resource nodes.

The same kind of data structure can also be used in an operating system to keep track of the evolving resource request and allocation state. In this case,

- When a process *P* requests a certain resource *R*, the corresponding "request arc," going from *P* to *R*, is added to the resource allocation graph.
- As soon as the request is granted, the request arc is replaced by an "ownership arc," going from *R* to *P*. This may either take place immediately or after a wait. The latter happens, for example, if *R* is busy at the moment. Deadlock avoidance algorithms, discussed in Section 4.6, may compel a process to wait, even if the resource it is requesting is free.
- When a process *P* releases a resource *R* it has previously acquired, the ownership arc going from *R* to *P* is deleted. This arc must necessarily be present in the graph, because it must have been created when *R* has been granted to *P*.

For this kind of resource allocation graph, it has been proved that the presence of a *cycle* in the graph is a *necessary* and *sufficient* condition for a deadlock. It can therefore be used as a tool to check whether a certain sequence of resource requests, allocations, and releases leads to a deadlock. It is enough to keep track of them, by managing the arcs of the resource allocation graph as described earlier, and check whether or not there is a cycle in the graph after each step.

If a cycle is found, then there is a deadlock in the system, and the deadlock involves precisely the set of processes and resources belonging to the cycle. Otherwise the sequence is "safe" from this point of view. The resource allocation graph shown in Figure 4.2 models a deadlock because P$_1 \rightarrow$ R$_1 \rightarrow$ P$_2 \rightarrow$ R$_4 \rightarrow$ P$_3 \rightarrow$ R$_2 \rightarrow$ P$_1$ is a cycle. Processes P$_1$, P$_2$, and P$_3$, as well as resources R$_1$, R$_2$, and R$_4$ are involved in the deadlock. Likewise, P$_1 \rightarrow$ R$_1 \rightarrow$ P$_2 \rightarrow$ R$_4 \rightarrow$ P$_3 \rightarrow$ R$_3 \rightarrow$ P$_1$ is a cycle, too, involving the same processes as before and resources R$_1$, R$_3$, and R$_4$.

As for any directed graph, also in this case arc orientation must be taken into account when assessing the presence of a cycle. Hence, referring again to Figure 4.2,

$P_1 \leftarrow R_2 \leftarrow P_3 \rightarrow R_3 \rightarrow P_1$ is *not* a cycle and does not imply the presence of any deadlock in the system.

The deadlock problem becomes more complex when there are different kinds (or classes) of resources in a system and there is, in general, more than one resource of each kind. All resources of the same kind are *fungible*, that is, they are interchangeable so that any of them can be used to satisfy a resource request for the class they belong to.

This is a common situation in many cases of practical interest: if we do not consider data access time optimization, disk blocks are fungible resources because, when any process requests a disk block to store some data, any free block will do. Other examples of fungible resources include memory page frames and entries in most operating system tables.

The definition of resource allocation graph can be extended to handle multiple resource instances belonging to the same class, by using one rectangle for each resource class, and representing each instance by means of a dot drawn in the corresponding rectangle. In Reference [43], this is called a *general* resource graph.

However, in this case, the theorem that relates cycles to deadlocks becomes weaker. It can be proved [43] that the presence of a cycle is still a necessary condition for a deadlock to take place, but it is no longer sufficient. The theorem can hence be used only to deny the presence of a deadlock, that is, to state that if there is *not* any cycle in an extended resource allocation graph, then there is *not* any deadlock in the system.

4.4   LIVING WITH DEADLOCK

Deadlock is a time-dependent problem, and the probability of actually encountering a deadlock during operation is often quite small with respect to other issues such as, for example, power interruptions or other software bugs. On the other hand, deadlock may also be a very complex problem to deal with, and its handling may consume a considerable amount of system resources.

It is therefore not a surprise if some operating system designers deliberately decided to completely neglect the problem, at least in some cases, and adopt a strategy sometimes called "the Ostrich algorithm" [88]. In this sense, they are trading off a small probability of being left with a (partially) deadlocked system for a general improvement of system performance.

This reasoning usually makes sense for general-purpose operating systems. For example, Bach [10] highlights the various situations in which the Unix operating system is, or (more commonly) is *not*, protected against deadlock. Most other operating systems derived from Unix, for example, Linux, suffer from the same problems.

On the contrary, in many cases, real-time applications cannot tolerate any latent deadlock, regardless of its probability of occurrence, for instance, due to safety concerns. Once it has been decided to actually "do something" about deadlock, the algorithms being used can be divided into three main families:

1. Try to *prevent* deadlock, by imposing some constraints during system design and implementation.

2. Check all resource requests made by processes as they come by and *avoid* deadlock by delaying resource allocation when appropriate.
3. Let the system possibly go into a deadlock, but then *detect* this situation and put into effect a *recovery* action.

The main trade-offs between these techniques have to do with several different areas of application development and execution, and all of them should be considered to choose the most fruitful technique for any given case. In particular, the various families differ regarding

- When deadlock handling takes place. Some techniques must be applied early in system design, other ones take action at run time.
- How much influence they have in the way designers and programmers develop the application.
- How much and what kind of information about processes behavior they need, in order to work correctly.
- The amount of run time overhead they inflict on the system.

4.5   DEADLOCK PREVENTION

The general idea behind all deadlock prevention techniques is to prevent deadlocks from occurring by making sure that (at least) one of the individually necessary conditions presented in Section 4.2 can never be satisfied in the system. In turn, this property is ensured by putting into effect and enforcing appropriate design or implementation rules, or constraints.

Since there are four necessary conditions, four different deadlock prevention strategies are possible, at least in principle. Most of the techniques to be presented in this section have been proposed by Havender [37].

1. The *mutual exclusion* condition can be attacked by allowing multiple processes to use the same resource concurrently, without waiting when the resource has been assigned to another process.
   This goal cannot usually be achieved by working directly on the resource involved because the need for mutual exclusion often stems from some inescapable hardware characteristics of the resource itself. For example, there is no way to modify a CD burner and allow two processes to write on a CD concurrently.
   On the contrary, an equivalent result can sometimes be obtained by interposing an additional software layer between the resource and the processes competing for it: for printers, it is common to have *spoolers* devoted to this purpose. In its simplest form, a spooler is a system process that has permanent ownership of a certain printer. Its role is to collect print requests from all the other processes in the system, and carry them out one at a time.
   Even if the documents to be printed are still sent to the printer sequentially, in order to satisfy the printer's mutual exclusion requirements, the spooler

will indeed accept multiple, concurrent print requests because it will use another kind of media, for example, a magnetic disk, to collect and temporarily store the documents to be printed.

Hence, from the point of view of the requesting processes, resource access works "as if" the mutual exclusion constraint had been lifted, and deadlock cannot occur, at least as long as the disk space available for spooling is not so scarce to force processes to wait for disk space while they are producing their output. The latter condition may intuitively lead to other kinds of deadlock, because we are merely "shifting" the deadlock problem from one resource (the printer) to another (the spooling disk space).

Nevertheless, the main problem of spooling techniques is their limited applicability: many kinds of resource are simply not amenable to be spooled. To make a very simple example, it is totally unclear how it would be possible to spool an operating system data structure.

2. The *hold and wait* condition is actually made of two parts that can be considered separately. The *wait* part can be invalidated by making sure that no processes will ever wait for a resource. One particularly simple way of achieving this is to force processes to request all the resources they may possibly need during execution all together and right at the beginning of their execution.

Alternatively, processes can be constrained to release all the resources they own before requesting new ones, thus invalidating the *hold* part of the condition. The new set of resources being requested can include, of course, some of the old ones if they are still needed, but the process must accept a temporary loss of resource ownership anyway. If a stateful resource, like for instance a printer, is lost and then reacquired, the resource state will be lost, too.

In a general purpose system, it may be difficult, or even impossible, to know in advance what resources any given application will need during its execution. For example, the amount of memory or disk space needed by a word processor is highly dependent on what the user is doing with it and is hard to predict in advance.

Even when it is possible to agree upon a reasonable set of resources to be requested at startup, the efficiency of resource utilization will usually be low with this method, because resources are requested on the basis of *potential*, rather than *actual* necessity. As a result, resources will usually be requested a long time before they are really used. When following this approach, the word processor would immediately request a printer right when it starts—this is reasonable because it may likely need it in the future—and retain exclusive access rights to it, even if the user does not actually print anything for hours.

When considering simple real-time systems, however, the disadvantages of this method, namely, early resource allocation and low resource utilization, are no longer a limiting factor, for two reasons. First of all, low resource utilization is often not a concern in those systems. For example, allocating an analog-to-digital converter (ADC) to the process that will handle the data it produces well before it is actively used may be perfectly acceptable. This is due to the fact that the resource is somewhat hardwired to the process, and no other processes in the system would be capable of using the same device anyway even if it were available.

Second, performing on-demand resource allocation to avoid allocating resources too early entails accepting the *wait* part of this condition. As a consequence, processes must be prepared to wait for resources during their execution. In order to have a provably working hard real-time system, it must be possible to derive an upper bound of the waiting time, and this may be a difficult task in the case of on-demand resource allocation.

3. The *nonpreemption* condition can be attacked by making provision for a resource preemption mechanism, that is, a way of forcibly take away a resource from a process against its will. Like for the mutual exclusion condition, the difficulty of doing this heavily depends on the kind of resource to be handled. One the one hand, preempting a resource such as a processor is quite easy, and is usually done (by means of a context switch) by most operating systems.

On the other hand, preempting a print operation already in progress in favor of another one entails losing a certain amount of work (the pages printed so far) even if the print operation is later resumed at the right place, unless a patient operator is willing to put the pieces together. As awkward as it may seem today, this technique was actually used in the past: in fact, the THE operating system might preempt the printer on a page-by-page basis [28, 26].

4. The *circular wait* condition can be invalidated by imposing a total ordering on all resource classes and imposing that, by design, all processes follow that order when they request resources. In other words, an integer-valued function $f(R_i)$ is defined on all resource classes R_i and it has a unique value for each class. Then, if a process already owns a certain resource R_j, it can request an additional resource belonging to class R_k if, and only if, $f(R_k) > f(R_j)$.

It can easily be proved that, if all processes obey this rule, no circular wait may occur in the system [87]. The proof proceeds by *reductio ad absurdum*. Assume that, although all processes followed the rule, there is indeed a circular wait in the system. Without loss of generality, let us assume that the circular wait involves processes $P_1, \ldots, P_m$ and resource classes $R_1, \ldots, R_m$, so that process P_1 owns a resource of class R_1 and is waiting for a resource of class R_2, process P_2 owns a resource of class R_2 and is waiting for a resource of class R_3, ... process P_m owns a resource of class R_m and is waiting for a resource of class R_1, thus closing the circular wait. If process P_1 followed the resource request rule, then it must be

$$f(R_2) > f(R_1) \tag{4.1}$$

because P_1 is requesting a resource of class R_2 after it already obtained a resource of class R_1.

The same reasoning can be repeated for all processes, to derive the following set of inequalities:

$$
\begin{cases}
f(R_2) & > & f(R_1) & \text{for } P_1 \\
f(R_3) & > & f(R_2) & \text{for } P_2 \\
& \cdots & & \\
f(R_m) & > & f(R_{m-1}) & \text{for } P_{m-1} \\
f(R_1) & > & f(R_m) & \text{for } P_m
\end{cases} \tag{4.2}
$$

Due to the transitive property of inequalities, we come to the absurd:

$$
f(R_1) > f(R_1) \tag{4.3}
$$

and are able to disprove the presence of a circular wait.

In large systems, the main issue of this method is the difficulty of actually enforcing the design rules and check whether they have been followed or not. For example, the multithreaded FreeBSD operating system kernel uses this approach and orders its internal locks to prevent deadlocks. However, even after many years of improvements, a fairly big number of "lock order reversals"—that is, situations in which locks are actually requested in the wrong order—are still present in the kernel code. In fact, a special tool called witness was even specifically designed to help programmers detect them [12].

4.6 DEADLOCK AVOIDANCE

Unlike deadlock prevention algorithms, discussed in Section 4.5, deadlock avoidance algorithms take action later, while the system is *running*, rather than during system *design*. As in many other cases, this choice involves a trade-off: on the one hand, it makes programmers happier and more productive because they are no longer constrained to obey any deadlock-related design rule. On the other hand, it entails a certain amount of overhead.

The general idea of any deadlock avoidance algorithm is to check resource allocation requests, as they come from processes, and determine whether they are safe or unsafe for what concerns deadlock. If a request is deemed to be unsafe, it is postponed, even if the resources being requested are free. The postponed request will be reconsidered in the future, and eventually granted if and when its safety can indeed be proved. Usually, deadlock avoidance algorithms also need a certain amount of preliminary information about process behavior to work properly.

Among all the deadlock avoidance methods, we will discuss in detail the *banker's algorithm*. The original version of the algorithm, designed for a single resource class, is due to Dijkstra [25]. It was later extended to multiple resource classes by Habermann [36].

In the following, we will sometimes refer to the j-th column of a certain matrix M as $\mathbf{m_j}$ and treat it as a (column) vector. To simplify the notation, we also introduce a weak ordering relation between vectors. In particular, we will state that

$$\mathbf{v} \leq \mathbf{w} \text{ if and only if } v_i \leq w_i \ \forall i \tag{4.4}$$

Informally speaking, a vector $\mathbf{v}$ is less than or equal to another vector $\mathbf{w}$, if and only if all its elements are less than or equal to the corresponding elements of the other one. Analogously, the strict inequality is defined as

$$\mathbf{v} < \mathbf{w} \text{ if and only if } \mathbf{v} \leq \mathbf{w} \wedge \mathbf{v} \neq \mathbf{w} \tag{4.5}$$

If we use n to denote the total number of processes in the system and m to denote the total number of resource classes, the banker's algorithm needs and manages the following data structures:

- A (column) vector $\mathbf{t}$, representing the total number of resources of each class initially available in the system:

$$\mathbf{t} = \begin{pmatrix} t_1 \\ \vdots \\ t_m \end{pmatrix} \tag{4.6}$$

  Accordingly, t_i indicates the number of resources belonging to the i-th class initially available in the system. It is assumed that $\mathbf{t}$ does not change with time, that is, resources never break up or become unavailable for use, either temporarily or permanently, for any other reason.

- A matrix C, with m rows and n columns, that is, a column for each process and a row for each resource class, which holds the current resource allocation state:

$$C = \begin{pmatrix} c_{11} & c_{12} & \cdots & c_{1n} \\ c_{21} & & & \cdots \\ & & \cdots & \\ c_{m1} & & \cdots & c_{mn} \end{pmatrix} \tag{4.7}$$

  The value of each individual element of this matrix, c_{ij}, represents how many resources of class i have been allocated to the j-th process, and $\mathbf{c_j}$ is a vector that specifies how many resources of each class are currently allocated to the j-th process.

  Therefore, unlike $\mathbf{t}$, the value of C varies as the system evolves. Initially, $c_{ij} = 0 \ \forall i, j$, because no resources have been allocated yet.

- A matrix X, also with m rows and n columns, containing information about the maximum number of resources that each process may possibly require, for each resource class, during its whole life:

$$X = \begin{pmatrix} x_{11} & x_{12} & \cdots & x_{1n} \\ x_{21} & & & \cdots \\ & & \cdots & \\ x_{m1} & & \cdots & x_{mn} \end{pmatrix} \tag{4.8}$$

This matrix indeed represents an example of the auxiliary information about process behavior needed by this kind of algorithm, as it has just been mentioned earlier. That is, it is assumed that each process P_j will be able to declare in advance its worst-case resource needs by means of a vector $\mathbf{x_j}$:

$$\mathbf{x_j} = \begin{pmatrix} x_{1j} \\ \vdots \\ x_{mj} \end{pmatrix} \tag{4.9}$$

so that, informally speaking, the matrix X can be composed by placing all the vectors $\mathbf{x_j}, \forall j = 1, \ldots, n$ side by side.

It must clearly be $\mathbf{x_j} \leq \mathbf{t} \;\; \forall j$; otherwise, the j-th process could never be able to conclude its work due to lack of resources even if it is executed alone. It should also be noted that processes cannot "change their mind" and ask for matrix X to be updated at a later time unless they have no resources allocated to them.

- An auxiliary matrix N, representing the worst-case future resource needs of the processes. It can readily be calculated as

$$N = X - C \tag{4.10}$$

and has the same shape as C and X. Since C changes with time, N also does.

- A vector $\mathbf{r}$, representing the resources remaining in the system at any given time:

$$\mathbf{r} = \begin{pmatrix} r_1 \\ \vdots \\ r_m \end{pmatrix} . \tag{4.11}$$

The individual elements of $\mathbf{r}$ are easy to calculate for a given C, as follows:

$$r_i = t_i - \sum_{j=1}^{n} C_{ij} \;\; \forall i = 1, \ldots, n \;\; . \tag{4.12}$$

In informal language, this equation means that r_i, representing the number of remaining resources in class i, is given by the total number of resources belonging to that class t_i, minus the resources of that class currently allocated to any process, which is exactly the information held in the i-th row of matrix C. Hence the summation of the elements belonging to that row must be subtracted from t_i to get the value we are interested in.

Finally, a resource request coming from the j-th process will be denoted by the vector $\mathbf{q_j}$:

$$\mathbf{q_j} = \begin{pmatrix} q_{1j} \\ \vdots \\ q_{mj} \end{pmatrix} , \tag{4.13}$$

where the i-th element of the vector, q_{ij}, indicates how many resources of the i-th class the j-th process is requesting. Of course, if the process does not want to request any resource of a certain class, it is free to set the corresponding q_{ij} to 0.

Whenever it receives a new request $\mathbf{q_j}$ from the j-th process, the banker executes the following algorithm:

1. It checks whether the request is *legitimate* or not. In other words, it checks if, by submitting the request being analyzed, the j-th process is trying to exceed the maximum number of resources it declared to need beforehand, $\mathbf{x_j}$.

   Since the j-th column of N represents the worst-case future resource needs of the j-th process, given the current allocation state C, this test can be written as

   $$\mathbf{q_j} \leq \mathbf{n_j} \qquad (4.14)$$

   If the test is satisfied, the banker proceeds with the next step of the algorithm. Otherwise, the request is refused immediately and an *error* is reported back to the requesting process. It should be noted that this error indication is not related to deadlock but to the detection of an illegitimate behavior of the process.

2. It checks whether the request could, in principle, be granted immediately or not, depending on current resource *availability*.

   Since $\mathbf{r}$ represents the resources that currently remain available in the system, the test can be written as

   $$\mathbf{q_j} \leq \mathbf{r} \qquad (4.15)$$

   If the test is satisfied, there are enough available resources in the system to grant the request and the banker proceeds with the next step of the algorithm. Otherwise, regardless of any deadlock-related reasoning, the request cannot be granted immediately, due to lack of resources, and the requesting process has to *wait*.

3. If the request passed both the preliminary checks described earlier, the banker *simulates* the allocation and generates a new state that reflects the effect of granting the request on resource allocation ($\mathbf{c_j'}$), future needs ($\mathbf{n_j'}$), and availability ($\mathbf{r'}$), as follows:

   $$\begin{cases} \mathbf{c_j'} &= \mathbf{c_j} + \mathbf{q_j} \\ \mathbf{n_j'} &= \mathbf{n_j} - \mathbf{q_j} \\ \mathbf{r'} &= \mathbf{r} - \mathbf{q_j} \end{cases} \qquad (4.16)$$

   Then, the new state is analyzed to determine whether it is *safe* or not for what concerns deadlock. If the new state is safe, then the request is granted and the simulated state becomes the new, actual state of the system:

   $$\begin{cases} \mathbf{c_j} &:= \mathbf{c_j'} \\ \mathbf{n_j} &:= \mathbf{n_j'} \\ \mathbf{r} &:= \mathbf{r'} \end{cases} \qquad (4.17)$$

Otherwise, the simulated state is discarded, the request is not granted immediately even if enough resources are available, and the requesting process has to *wait*.

To assess the safety of a resource allocation state, during step 3 of the preceding algorithm, the banker uses a conservative approach. It tries to compute at least one sequence of processes—called a *safe* sequence—comprising all the n processes in the system and that, when followed, allows each process in turn to attain the *worst-case* resource need it declared, and thus successfully conclude its work. The safety assessment algorithm uses two auxiliary data structures:

- A (column) vector $\mathbf{w}$ that is initially set to the currently available resources (i.e., $\mathbf{w} = \mathbf{r}'$ initially) and tracks the evolution of the available resources as the safe sequence is being constructed.
- A (row) vector $\mathbf{f}$, of n elements. The j-th element of the vector, f_j, corresponds to the j-th process: $f_j = 0$ if the j-th process has not yet been inserted into the safe sequence, $f_j = 1$ otherwise. The initial value of $\mathbf{f}$ is zero, because the safe sequence is initially empty.

The algorithm can be described as follows:

1. Try to find a new, suitable candidate to be appended to the safe sequence being constructed. In order to be a suitable candidate, a certain process P_j must not already be part of the sequence and it must be able to reach its worst-case resource need, given the current resource availability state. In formulas, it must be

$$
\begin{aligned}
f_j &= 0 &&(P_j \text{ is not in the safe sequence yet}) \\
&\wedge \\
\mathbf{n_j}' &\leq \mathbf{w} &&(\text{there are enough resources to satisfy } \mathbf{n_j}')
\end{aligned}
\tag{4.18}
$$

   If no suitable candidates can be found, the algorithm ends.

2. After discovering a candidate, it must be appended to the safe sequence. At this point, we can be sure that it will eventually conclude its work (because we are able to grant it all the resources it needs) and will release all the resources it holds. Hence, we shall update our notion of available resources accordingly:

$$
\begin{aligned}
f_j &:= 1 &&(P_j \text{ belongs to the safe sequence now}) \\
\mathbf{w} &:= \mathbf{w} + \mathbf{c_j}' &&(\text{it releases its resources upon termination})
\end{aligned}
\tag{4.19}
$$

   Then, the algorithm goes back to step 1, to extend the sequence with additional processes as much as possible.

At end, if $f_j = 1 \ \forall j$, then all processes belong to the safe sequence and the simulated state is certainly safe for what concerns deadlock. On the contrary, being unable to find a safe sequence of length n does not necessarily imply that a deadlock will *definitely* occur because the banker's algorithm is considering the worst-case resource requirements of each process, and it is therefore being conservative.

Even if a state is unsafe, all processes could still be able to conclude their work without deadlock if, for example, they never actually request the maximum number of resources they declared.

It should also be remarked that the preceding algorithm does not need to backtrack when it picks up a sequence that does not ensure the successful termination of all processes. A theorem proved in Reference [36] guarantees that, in this case, no safe sequences exist at all. As a side effect, this property greatly reduces the computational complexity of the algorithm.

Going back to the overall banker's algorithm, we still have to discuss the fate of the processes which had their requests postponed and were forced to *wait*. This can happen for two distinct reasons:

- Not enough resources are available to satisfy the request
- Granting the request would bring the system into an unsafe state

In both cases, if the banker later grants other resource allocation requests made by other processes, by intuition the state of affairs gets even worse from the point of view of the waiting processes. Given that their requests were postponed when more resources were available, it seems even more reasonable to further postpone them without reconsideration when further resources have been allocated to others.

On the other hand, when a process P_j releases some of the resources it owns, it presents to the banker a release vector, $\mathbf{l_j}$. Similar to the request vector $\mathbf{q_j}$, it contains one element for each resource class but, in this case, the i-th element of the vector, l_{ij}, indicates how many resources of the i-th class the j-th process is releasing. As for resource requests, if a process does not want to release any resource of a given kind, it can leave the corresponding element of $\mathbf{l_j}$ at zero. Upon receiving such a request, the banker updates its state variables as follows:

$$\begin{cases} \mathbf{c_j} & := & \mathbf{c_j} - \mathbf{l_j} \\ \mathbf{n_j} & := & \mathbf{n_j} + \mathbf{l_j} \\ \mathbf{r} & := & \mathbf{r} + \mathbf{l_j} \end{cases} \tag{4.20}$$

As expected, the state update performed on release (4.20) is almost exactly the opposite of the update performed upon resource request (4.16), except for the fact that, in this case, it is not necessary to check the new state for safety and the update can therefore be made directly, without simulating it first.

Since the resource allocation situation does improve in this case, this is the right time to reconsider the requests submitted by the waiting processes because some of them might now be granted safely. In order to do this, the banker follows the same algorithm already described for newly arrived requests.

The complexity of the banker's algorithm is $\mathscr{O}(mn^2)$, where m is the number of resource classes in the system, and n is the number of processes. This overhead is incurred on every resource allocation *and* release due to the fact that, in the latter case, any waiting requests shall be reconsidered.

The overall complexity is dominated by the safety assessment algorithm because all the other steps of the banker's algorithm (4.14)–(4.17) are composed of a constant number of vector operations on vectors of length m, each having a complexity of $\mathcal{O}(m)$.

In the safety assessment algorithm, we build the safe sequence one step at a time. In order to do this, we must inspect at most n candidate processes in the first step, then $n-1$ in the second step, and so on. When the algorithm is able to build a safe sequence of length n, the worst case for what concerns complexity, the total number of inspections is therefore

$$n + (n-1) + \ldots + 1 = \frac{n(n+1)}{2} \tag{4.21}$$

Each individual inspection (4.18) is made of a scalar comparison, and then of a vector comparison between vectors of length m, leading to a complexity of $\mathcal{O}(m)$ for each inspection and to a total complexity of $\mathcal{O}(mn^2)$ for the whole inspection process.

The insertion of each candidate into the safe sequence (4.19), an operation performed at most n times, does not make the complexity any larger because the complexity of each insertion is $\mathcal{O}(m)$, giving a complexity of $\mathcal{O}(mn)$ for them all.

As discussed in Chapter 3, Section 3.4, many operating systems support dynamic process creation and termination. The creation of a new process P_{n+1} entails the extension of matrices C, X, and N with an additional column, let it be the rightmost one. The additional column of C must be initialized to zero because, at the very beginning of its execution, the new process does not own any resource.

On the other hand, as for all other processes, the additional column of X must hold the maximum number of resources the new process will need during its lifetime for each resource class, represented by $\mathbf{x_{n+1}}$. The initial value of the column being added to N must be $\mathbf{x_{n+1}}$, according to how this matrix has been defined in Equation (4.10).

Similarly, when process P_j terminates, the corresponding j-th column of matrices C, X, and N must be suppressed, perhaps after checking that the current value of $\mathbf{c_j}$ is zero. Finding any non-zero value in this vector means that the process concluded its execution without releasing all the resources that have been allocated to it. In this case, the residual resources shall be released forcibly, to allow them to be used again by other processes in the future.

4.7 DEADLOCK DETECTION AND RECOVERY

The deadlock prevention approach described in Section 4.5 poses significant restrictions on system design, whereas the banker's algorithm presented in Section 4.6 requires information that could not be readily available and has a significant run-time overhead. These qualities are typical of any other deadlock avoidance algorithm.

To address these issues, a third family of methods acts even later than deadlock avoidance algorithms. That is, these methods allow the system to enter a deadlock condition but are able to *detect* this fact and react accordingly with an appropriate *recovery* action. For this reason, they are collectively known as deadlock *detection and recovery* algorithms.

If there is only one resource for each resource class in the system, a straightforward way to detect a deadlock condition is to maintain a resource allocation graph, updating it whenever a resource is requested, allocated, and eventually released. Since this maintenance only involves adding and removing arcs from the graph, it is not computationally expensive and, with a good supporting data structure, can be performed in constant time.

Then, the resource allocation graph is examined at regular intervals, looking for cycles. Due to the theorem discussed in Section 4.3, the presence of a cycle is a necessary and sufficient indication that there is a deadlock in the system. Actually, it gives even more information, because the processes and resources belonging to the cycle are exactly those suffering from the deadlock. We will see that this insight turns out to be useful in the subsequent deadlock recovery phase.

If there are multiple resource instances belonging to the same resource class, this method cannot be applied. On its place we can, for instance, use another algorithm, due to Shoshani and Coffman [86], and reprinted in Reference [21]. Similar to the banker's algorithm, it maintains the following data structures:

- A matrix C that represents the current resource allocation state
- A vector $\mathbf{r}$, indicating how many resources are currently available in the system

Furthermore, for each process P_j in the system, the vector

$$\mathbf{s_j} = \begin{pmatrix} s_{1j} \\ \vdots \\ s_{mj} \end{pmatrix}$$

indicates how many resources of each kind process P_j is currently requesting and waiting for, in addition to the resources $\mathbf{c_j}$ it already owns. If process P_j is not requesting any additional resources at a given time, the elements of its $\mathbf{s_j}$ vector will all be zero at that time.

As for the graph-based method, all these data structures evolve with time and must be updated whenever a process requests, receives, and relinquishes resources. However, again, all of them can be maintained in constant time. Deadlock detection is then based on the following algorithm:

1. Start with the auxiliary (column) vector $\mathbf{w}$ set to the currently available resources (i.e., $\mathbf{w} = \mathbf{r}$ initially) and the (row) vector $\mathbf{f}$ set to zero. Vector $\mathbf{w}$ has one element for each resource class, whereas $\mathbf{f}$ has one element for each process.
2. Try to find a process P_j that has not already been marked and whose resource request can be satisfied, that is,

$$\begin{aligned} f_j &= 0 \qquad\qquad (P_j \text{ has not been marked yet}) \\ &\wedge \\ \mathbf{s_j} &\le \mathbf{w} \quad \text{(there are enough resources to satisfy its request)} \end{aligned} \qquad (4.22)$$

If no suitable process exists, the algorithm ends.

3. Mark P_j and return the resources it holds to the pool of available resources:

$$
\begin{aligned}
f_j &:= 1 &&\text{(mark } P_j) \\
\mathbf{w} &:= \mathbf{w} + \mathbf{c_j} &&\text{(releases its resources)}
\end{aligned}
\tag{4.23}
$$

Then, the algorithm goes back to step 2 to look for additional processes.

It can be proved that a deadlock exists if, and only if, there are unmarked processes—in other word, at least one element of $\mathbf{f}$ is still zero—at the end of the algorithm. Rather obviously, this algorithm bears a strong resemblance to the state safety assessment part of the banker's algorithm. Unsurprisingly, they also have the same computational complexity.

From the conceptual point of view, the main difference is that the latter algorithm works on the *actual* resource requests performed by processes as they execute (and represented by the vectors $\mathbf{s_j}$), whereas the banker's algorithm is based on the *worst-case* resource needs forecast (or guessed) by each process (represented by $\mathbf{x_j}$).

As a consequence, the banker's algorithm results are conservative, and a state can pessimistically be marked as unsafe, even if a deadlock will not necessarily ensue. On the contrary, the last algorithm provides exact indications.

It can be argued that, in general, since deadlock detection algorithms have a computational complexity comparable to the banker's algorithms, there is apparently nothing to be gained from them, at least from this point of view. However, the crucial difference is that the banker's algorithm *must* necessarily be invoked on every resource request and release, whereas the frequency of execution of the deadlock detection algorithm is a parameter and can be chosen freely.

Therefore, it can be adjusted to obtain the best trade-off between contrasting system properties such as, for example, the maximum computational overhead that can tolerably be imposed on the system and the "reactivity" to deadlocks of the system itself, that is, the maximum time that may elapse between a deadlock and the subsequent recovery.

The last point to be discussed, that is, deciding how to recover from a deadlock, is a major problem indeed. A crude recovery principle, suggested in Reference [21], consists of aborting each of the deadlocked processes or, more conservatively, abort them one at a time, until the additional resources made available by the aborted processes remove the deadlock from the remaining ones.

More sophisticated algorithms, one example of which is also given in References [21, 86], involve the forced removal, or preemption, of one or more resources from deadlocked processes on the basis of a cost function. Unfortunately, as in the case of choosing which processes must be aborted, assigning a cost to the removal of a certain resource from a given process may be a difficult job because it depends on several factors such as, for example:

- the importance, or priority of the victim process
- the possibility for the process to recover from the preemption

Symmetrically, other recovery techniques act on resource *requests*, rather than *allocation*. In order to recover from a deadlock, they deny one or more pending resource requests and give an error indication to the corresponding processes. In this way, they force some of the s_j vectors to become zero and bring the system in a more favorable state with respect to deadlock. The choice of the "right" requests to deny is still subject to cost considerations similar to those already discussed.

The last, but not the less important, aspect is that any deadlock recovery technique—which involves either aborting a process, preempting some of the resources it needs to perform its job, or denying some of its resource requests—will certainly have adverse effects on the timeliness of the affected processes and may force them to violate their timing requirements.

In any case, if one wants to use this technique, processes must be prepared beforehand to the deadlock recovery action and be able to react in a meaningful way. This is not always easy to do in a real-time system where, for example, the idea of aborting a process at an arbitrary point of its execution, is per se totally unacceptable in most cases.

4.8   SUMMARY

Starting with a very simple example involving only two processes, this chapter introduced the concept of *deadlock*, an issue that may arise whenever processes compete with each other to acquire and use some resources. Deadlock is especially threatening in a real-time system because its occurrence blocks one or more processes forever and therefore jeopardizes their timing.

Fortunately, it is possible to define formally what a deadlock is, and when it takes place, by introducing four conditions that are individually necessary and collectively sufficient for a deadlock to occur. Starting from these conditions, it is possible to define a whole family of *deadlock avoidance* algorithms. Their underlying idea is to ensure, by design, that at least one of the four conditions cannot be satisfied in the system being considered so that a deadlock cannot occur. This is done by imposing various rules and constraints to be followed during system design and implementation.

When design-time constraints are unacceptable, other algorithms can be used as well, at the expense of a certain amount of run-time overhead. They operate during system *execution*, rather than design, and are able to *prevent* deadlock by checking all resource allocation requests. They make sure that the system never enters a risky state for what concerns deadlock by postponing some requests on purpose, even if the requested resources are free.

To reduce the overhead, it is also possible to deal with deadlock even later by using a deadlock *detection and recovery* algorithm. Algorithms of this kind let the system enter a deadlock state but are able to detect deadlock and recover from it by aborting some processes or denying resource requests and grants forcibly.

For the sake of completeness, it should also be noted that deadlock is only one aspect of a more general group of phenomenons, known as *indefinite wait*, *indefinite postponement*, or *starvation*. A full treatise of indefinite wait is very complex and well beyond the scope of this book, but an example taken from the material presented in this chapter may still be useful to grasp the full extent of this issue. A good starting point for readers interested in a more thorough discussion is, for instance, the work of Owicki and Lamport [74].

In the banker's algorithm discussed in Section 4.6, when more than one request can be granted safely but not all of them, a crucial point is how to pick the "right" request, so that no process is forced to wait indefinitely in favor of others.

Even if, strictly speaking, there is not any deadlock under these circumstances, if the choice is not right, there may still be some processes that are blocked for an indefinite amount of time because their resource requests are always postponed. Similarly, Reference [42] pointed out that, even granting safe requests as soon as they arrive, without reconsidering postponed requests first, may lead to other forms of indefinite wait.

4.9 EXERCISES

EXERCISE 1

Consider the following resource allocation graph:

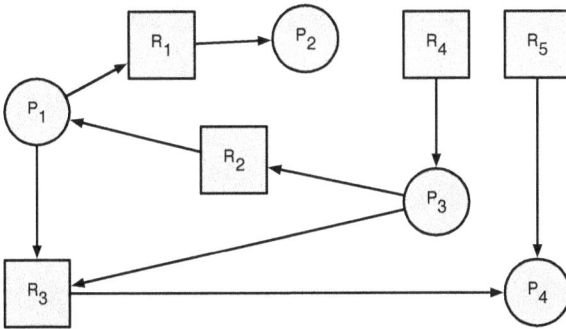

Is there a deadlock in the system? How can the deadlock be resolved?

EXERCISE 2

Same questions as in Exercise 1, but with the following resource allocation graph:

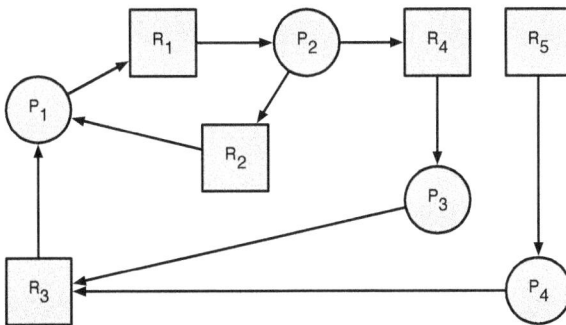

EXERCISE 3

In a system, process P needs to acquire mutually exclusive, reusable resources A and B together for part of its execution, it requests A and then B. Process Q needs resources B and C, it requests B and then C. Process R needs resources A and C, it requests C and then A.

- Is the system susceptible to deadlock? If so, make an example of how a deadlock may occur.
- Introduce a suitable total resource ordering and discuss how the processes must be modified to follow it.

EXERCISE 4

Given the following resource availability and allocation state, determine whether the state is safe or not using the banker's algorithm.

$$C = \begin{pmatrix} 3 & 3 & 4 \\ 1 & 0 & 0 \\ 2 & 2 & 3 \end{pmatrix}, \quad X = \begin{pmatrix} 4 & 4 & 4 \\ 3 & 0 & 1 \\ 3 & 3 & 3 \end{pmatrix}, \quad \mathbf{t} = \begin{pmatrix} 10 \\ 3 \\ 7 \end{pmatrix}.$$

EXERCISE 5

Starting from the resource availability and allocation state given in the previous exercise, discuss the following resource allocation requests:

1.

$$\mathbf{q_2} = \begin{pmatrix} 2 \\ 2 \\ 2 \end{pmatrix} \quad \text{(from } P_2\text{)}$$

2.

$$\mathbf{q_1} = \begin{pmatrix} 1 \\ 1 \\ 1 \end{pmatrix} \quad \text{(from } P_1\text{)}$$

3.

$$\mathbf{q_3} = \begin{pmatrix} 0 \\ 1 \\ 0 \end{pmatrix} \quad \text{(from } P_3\text{)}$$

4.

$$\mathbf{q_1} = \begin{pmatrix} 0 \\ 2 \\ 0 \end{pmatrix} \quad \text{(from } P_1\text{)}$$

5 Interprocess Communication Based on Shared Variables

More often than not, in both general-purpose and real-time systems, processes do not live by themselves and are not independent of each other. Rather, several processes are brought together to form the application software, and they must therefore *cooperate* to solve the problem at hand.

Processes must therefore be able to *communicate*, that is, exchange information, in a meaningful way. As discussed in Chapter 2, it is quite possible to share some memory among processes in a controlled way by making part of their address space refer to the same physical memory region.

However, this is only part of the story. In order to implement a correct and meaningful data exchange, processes must also *synchronize* their actions in some ways. For instance, they must not try to use a certain data item if it has not been set up properly yet. Another purpose of synchronization, presented in Chapter 4, is to regulate process access to shared resources.

This chapter addresses the topic, explaining how shared variables can be used for communication and introducing various kinds of hardware- and software-based synchronization approaches.

5.1 RACE CONDITIONS AND CRITICAL REGIONS

At first sight, using a set of shared variables for interprocess communication may seem a rather straightforward extension of what is usually done in sequential programming. In a sequential program written in a high-level language, it is quite common to have a set of functions, or procedures, that together form the application and exchange data exactly in the same way: there is a set of *global* variables defined in the program, all functions have access to them (within the limits set forth by the scoping rules of the programming language), and they can get and set their value as required by the specific algorithm they implement.

A similar thing also happens at the function call level, in which the caller prepares the function arguments and stores them in a well-known area of memory, often allocated on the stack. The called function then reads its arguments from there and uses them as needed. The value returned by the function is handled in a similar way. When possible, for instance, when the function arguments and return value are small enough, the whole process may be optimized by the language compiler to use some processor registers instead of memory, but the general idea is still the same.

DOI: 10.1201/9781003593416-5

Unfortunately, when trying to apply this idea to a concurrent system, one immediately runs into several, deep problems, even in trivial cases. If we want, for instance, to count how many events of a certain kind happened in a sequential programming framework, it is quite intuitive to define a memory-resident, global variable (the definition will be somewhat like int k if we use the C programming language) and then a very simple function void inck(void) that only contains the statement k = k+1.

It should be pointed out that, as depicted in Figure 5.1, no real-world CPU is actually able to increment k in a single, indivisible step, at least when the code is compiled into ordinary assembly instructions. Indeed, even a strongly simplified computer based on the *von Neumann* architecture [34, 89] will perform a sequence of three distinct steps:

1. Load the value of k from memory into an internal processor register; on a simple processor, this register would be the accumulator. From the processor's point of view, this is an external operation because it involves both the processor itself and memory, two distinct units that communicate through the memory bus. The load operation is not destructive, that is, k retains its current value after it has been performed.
2. Increment the value loaded from memory by one. Unlike the previous one, this operation is internal to the processor. It cannot be observed from the outside, also because it does not require any memory bus cycle to be performed. On a simple processor, the result is stored back into the accumulator.
3. Store the new value of k into memory with an external operation involving a memory bus transaction like the first one. It is important to notice that the new value of k can be observed from outside the processor only at this point, not before. In other words, if we look at memory, k retains its original value until this final step has been completed.

Even if real-world architectures are actually much more sophisticated than what is shown in Figure 5.1—and a much more intimate knowledge of their intricacies is necessary to precisely analyze their behavior—the basic reasoning is still the same: most operations that are believed to be indivisible at the programming language level (even simple, short statements like k = k+1), eventually correspond to a sequence of steps when examined at the instruction execution level.

However, this conceptual detail is often overlooked by many programmers because it has no practical consequences as long as the code being considered is executed in a strictly *sequential* fashion. It becomes, instead, of paramount importance as soon as we add *concurrency* to the recipe.

Let us imagine, for example, a situation in which not one but two different processes want to concurrently update the variable k because they are both counting events belonging to the same class but coming from different sources. Both of them use the same code, that is, function inck(), to perform the update.

For simplicity, we will assume that each of those two processes resides on its own physical CPU, and the two CPUs share a single-port memory by means of a common

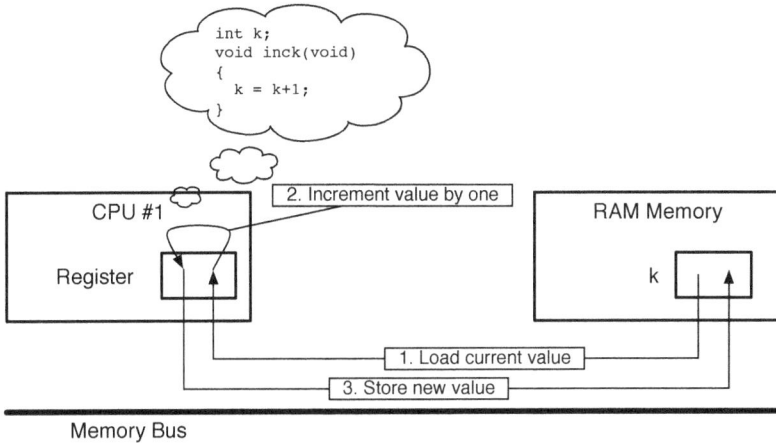

Figure 5.1 Simplified representation of how the CPU increments a memory-resident variable k.

memory bus, as is shown in Figure 5.2. However, the same argument still holds, even if the processes share the same physical processor, by using the multiprogramming techniques discussed in Chapter 3.

Under those conditions, if we let the initial value of k be 0, the following sequence of events *may* occur:

1.1. CPU #1 loads the value of k from memory and stores it into one of its registers, R_1. Since k currently contains 0, R_1 will also contain 0.
1.2. CPU #1 increments its register R_1. The new value of R_1 is therefore 1.

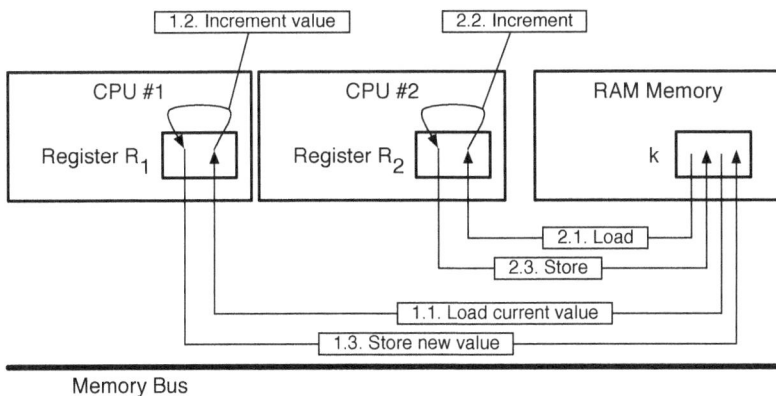

Figure 5.2 Concurrently incrementing a shared variable in a careless way leads to a race condition.

2.1. Now CPU #2 takes over, loads the value of k from memory, and stores it into one of its registers, R_2. Since CPU #1 has not stored the updated value of R_1 back to memory yet, CPU #2 still gets the value 0 from k, and R_2 will also contain 0.

2.2. CPU #2 increments its register R_2. The new value of R_2 is therefore 1.

2.3. CPU #2 stores the contents of R_2 back to memory in order to update k, that is, it stores 1 into k.

1.3. CPU #1 does the same: it stores the contents of R_1 back to memory, that is, it stores 1 into k.

Looking closely at the sequence of events just described, it is easy to notice that taking an (obviously correct) piece of sequential code and using it for concurrent programming in a careless way did not work as expected. That is, two distinct kinds of problems arose:

1. In the particular example just made, the final value of k is clearly incorrect: the initial value of k was 0, two distinct processes incremented it by one, but its final value is 1 instead of 2, as it should have been.
2. The second problem is perhaps even worse: the result of the concurrent update is not only incorrect but it is incorrect only *sometimes*. For example, the sequence **1.1**, **1.2**, **1.3**, **2.1**, **2.2**, **2.3** leads to a correct result.

In other words, the value and correctness of the result depend on how the elemental steps of the update performed by one processor *interleave* with the steps performed by the other. In turn, this depends on the precise timing relationship between the processors, down to the instruction execution level. This is not only hard to determine but will likely change from one execution to another, or if the same code is executed on a different machine.

Informally speaking, this kind of *time-dependent* errors may be hard to find and fix. Typically, they take place with very low probability and may therefore be very difficult to reproduce and analyze. Moreover, they may occur when a certain piece of machinery is working in the field and no longer occur during bench testing because the small, but unavoidable, differences between actual operation and testing slightly disturbed system timings.

Even the addition of software-based instrumentation or debugging code to a concurrent application may subtly change process interleaving and make a time-dependent error disappear. This is also the reason why software development techniques based on concepts like "write a piece of code and check whether it works or not; tweak it until it works," which are anyway questionable even for sequential programming, easily turn into a nightmare when concurrent programming is involved.

These observations also lead to the general definition of a pathological condition, known as *race condition*, that may affect a concurrent system: whenever a set of processes reads and/or writes some shared data to carry out a computation, and the results of this computation depend on the exact way the processes interleaved, there is a race condition.

In this statement, the term "shared data" must be construed in a broad sense: in the simplest case, it refers to a shared variable residing in memory, as in the previous examples, but the definition actually applies to any other kind of shared object, such as files and devices. Since race conditions undermine the correctness of any concurrent system, one of the main goals of concurrent programming will be to eliminate them altogether.

Fortunately, the following consideration is of great help to better focus this effort and concentrate only on a (hopefully small) part of the processes' code. The original concept is due to Hoare [40] and Brinch Hansen [15]:

1. A process spends part of its execution doing *internal* operations, that is, executing pieces of code that do not require or make access to any shared data. By definition, all these operations cannot lead to any race condition, and the corresponding pieces of code can be safely disregarded when the code is analyzed from the concurrent programming point of view.
2. Sometimes a process executes a region of code that makes access to shared data. Those regions of code must be looked at more carefully because they can indeed lead to a race condition. For this reason, they are called *critical regions* or *critical sections*.

With this definition in mind, and going back to the race condition depicted in Figure 5.2, we notice that both processes have a critical region, and it is the body of function inck(). In fact, that fragment of code increments the shared variable k. Even if the critical region code is correct when executed by *one* single process, the race condition stems from the fact that we allowed *two* distinct processes to be in their critical region simultaneously.

We may therefore imagine solving the problem by allowing only one process to be in a critical region at any given time, that is, forcing the *mutual exclusion* between critical regions pertaining to the same set of shared data. For the sake of simplicity, in this book, mutual exclusion will be discussed in rather informal and intuitive terms. See, for example, the works of Lamport [60, 61] for a more formal and general treatment of this topic.

In simple cases, mutual exclusion can be ensured by resorting to special machine instructions that many contemporary processor architectures support. For example, on the Intel® 64 and IA-32 architecture [49], the INC instruction increments a memory-resident integer variable by one.

When executed, the instruction loads the operand from memory, increments it internally to the CPU, and finally stores back the result; it is therefore subject to exactly the same race condition depicted in Figure 5.2. However, it can be accompanied by the LOCK prefix so that the whole sequence is executed atomically, even in a multiprocessor or multicore environment.

Unfortunately, these ad-hoc solutions, which coerce a single instruction to be executed atomically, cannot readily be applied to more general cases, as it will be shown in the following example. Figure 5.3 shows a classic way of solving the so-called *producers–consumers* problem. In this problem, a group of processes $P_1, \ldots, P_n$, called producers, generate data items and make them available to the consumers by means of the prod() function. On the other hand, another group of processes $C_1, \ldots, C_m$, the consumers, use the cons() function to get hold of data items.

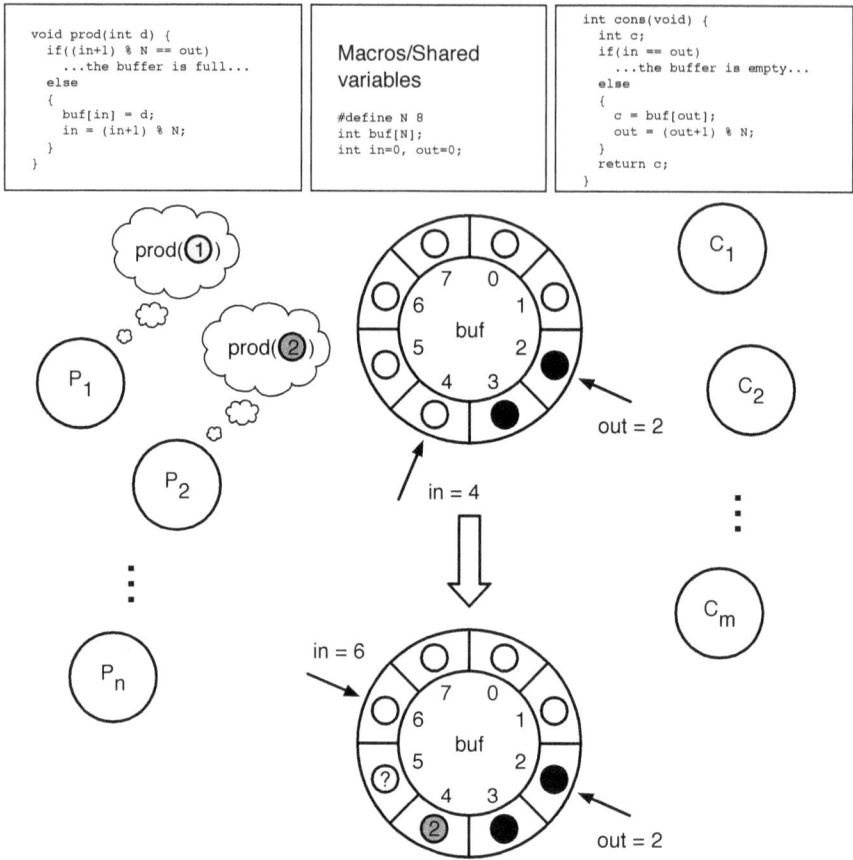

Figure 5.3 Not all race conditions can be avoided by forcing a single instruction to be executed atomically.

To keep the code as simple as possible, data items are assumed to be integer values, held in `int`-typed variables. For the same reason, the error-handling code (which should detect and handle any attempt of putting a data item into a full buffer or, symmetrically, getting an item from an empty buffer) is not shown.

With this approach, producers and consumers exchange data items through a circular buffer with N elements, implemented as a shared, statically allocated array `int buf[N]`. A couple of shared indices, `int in` and `int out`, keep track of the first free element of the buffer and the oldest full element, respectively. Both of them start at zero and are incremented modulus N to circularize the buffer. In particular,

- Assuming that the buffer is not completely full, the function `prod()` first stores `d`, the data item provided by the calling producer, into `buf[in]`, and then increments `in`. In this way, `in` now points to the next free element of the buffer.

- Assuming that the buffer is not completely empty, the function cons()
  takes the oldest data item residing in the buffer from buf[out], stores it
  into the local variable c, and then increments out so that the next consumer
  will get a fresh data item. Last, it returns the value of c to the calling consumer.

If should be noted that, since the condition in == out would be true not only for
a buffer that is completely empty but also for a buffer containing N full elements,
the buffer is never filled completely in order to avoid this ambiguity. In other words,
we must consider the buffer to be full even if one free element—often called the
guard element—is still available. The corresponding predicate is therefore (in+1)
% N == out. As a side effect of this choice, of the N buffer elements, only up to
$N - 1$ can be filled with data.

Taking for granted the following two, quite realistic, hypotheses:

1. any integer variable can be loaded from, or stored to, memory with a single,
   atomic operation;
2. neither the processor nor the memory access subsystem reorders memory
   accesses;

it is easy to show that the code shown in Figure 5.3 works correctly for up to *one*
producer and *one* consumer running concurrently.

For processors that reorder memory accesses—as most modern, high-performance
processors do—the intended sequence can be enforced by using dedicated machine
instructions, often called *fences* or *barriers*. For example, the SFENCE, LFENCE, and
MFENCE instructions of the Intel® 64 and IA-32 architecture [49] provide different
degrees of memory ordering.

The SFENCE instruction is a store barrier; it guarantees that all store operations
that come before it in the instruction stream have been committed to memory and are
visible to all other processors in the system before any of the following store operations becomes visible as well. The LFENCE instruction does the same for memory
load operations, and MFENCE does it for both memory load and store operations.

On the contrary, and quite surprisingly, the code no longer works as it should as
soon as we add a *second* producer to the set of processes being considered. One
obvious issue is with the increment of in (modulus N), but this is the same issue
already considered in Figure 5.2 and, as discussed before, it can be addresses with
some hardware assistance. However, there is another, subtler issue besides this one.

Let us consider two producers, P_1 and P_2: they both concurrently invoke the function prod() to store an element into the shared buffer. For this example, let us assume, as shown in Figure 5.3, that P_1 and P_2 want to put the values 1 and 2 into the
buffer, respectively, although the issue does not depend on these values at all.

It is also assumed that the shared buffer has a total of 8 elements and initially
contains two data items, represented as black dots, whereas white dots represent
empty elements. Moreover, we assume that the initial values of in and out are 4 and
2, respectively. This is the situation shown in the middle of the figure. The following
interleaving could take place:

1. Process P_1 begins executing prod(1) first. Since in is 4, it stores 1 into buf[4].
2. Before P_1 makes further progress, P_2 starts executing prod(2). The value of in is still 4, hence P_2 stores 2 into buf[4] and overwrites the data item just written there by P_1.
3. At this point, both P_1 and P_2 increment in. Assuming that the race condition issue affecting these operations has been addressed, the final value of in will be 6, as it should.

It can easily be seen that the final state of the shared variables after these operations, shown in the lower part of Figure 5.3, is severely corrupted because:

- the data item produced by P_1 is nowhere to be found in the shared buffer: it should have been in buf[4], but it has been *overwritten* by P_2;
- the data item buf[5] is marked as full because in is 6, but it contains an undefined value (denoted as "?" in the figure) since no data items have actually been written into it.

From the consumer's point of view, this means that, on the one hand, the data item produced by P_1 has been lost and the consumer will never get it. On the other hand, the consumer will get and try to use a data item with an undefined value, with dramatic consequences.

With the same reasoning, it is also possible to conclude that a very similar issue also occurs if there is more than one consumer in the system. In this case, multiple consumers could get the same data item, whereas other data items are never retrieved from the buffer. From this example, two important conclusions can be drawn:

1. The correctness of a concurrent program does not merely depend on the presence or absence of concurrency, as it happens in very simple cases, such as that shown in Figure 5.2, but it may also depend on *how many* processes there are, the so-called *degree of concurrency*. In the last example, the program works as long as there are only two concurrent processes in the system, namely, one producer and one consumer. It no longer works as soon as additional producers (or consumers) are introduced.
2. Not all race conditions can be avoided by forcing a *single instruction* to be executed atomically. Instead, we need a way to force mutual exclusion at the *critical region* level, and critical regions may comprise a sequence of many instructions. Going back to the example of Figure 5.3 and considering how critical regions have been defined, it is possible to conclude that the producers' critical region consists of the whole body of prod(), while the consumers' critical region is the whole body of cons(), except for the return statement, which only operates on local variables.

The traditional way of ensuring the mutual exclusion among critical region entails the adoption of a *lock-based* synchronization protocol. A process that wants to access a shared object, by means of a certain critical region, must perform the following sequence of steps:

1. Acquire some sort of *lock* associated with the shared object and wait if it is not immediately available.
2. Use the shared object.
3. Release the lock, so that other processes can acquire it and be able to access the same object in the future.

In the above sequence, step 2 is performed by the code within the critical region, whereas steps 1 and 3 are a duty of two fragments of code known as the critical region *entry* and *exit* code. In this approach, these fragments of code must compulsorily surround the critical region itself. If they are relatively short, they can be incorporated directly by copying them immediately before and after the critical region code, respectively.

If they are longer, it may be more convenient to execute them indirectly by means of appropriate function calls in order to reduce the code size, with the same effect. In the examples presented in this book, we will always follow the latter approach. This also highlights the fact that the concept of critical region is related to code *execution*, and not to the mere presence of some code between the critical region entry and exit code. Hence, for example, if a function call is performed between the critical region entry and exit code, the body of the called function must be considered as part of the critical region itself.

In any case, four conditions must be satisfied in order to have an acceptable solution [88]:

1. It must really work, that is, it must prevent any two processes from simultaneously executing code within critical regions pertaining to the same shared object.
2. Any process that is busy doing internal operations, that is, is not currently executing within a critical region, must not prevent other processes from entering their critical regions, if they so decide.
3. If a process wants to enter a critical region, it must not have to wait forever to do so. This condition guarantees that the process will eventually make progress in its execution.
4. It must work regardless of any low-level details about the hardware or software architecture. For example, the correctness of the solution must not depend on the number of processes in the system, the number of physical processors, or their relative speed.

In the following sections, a set of lock-based methods will be discussed and confronted with the correctness conditions just presented. It should, however, be noted that lock-based synchronization is by far the most common, but it is not the only way to solve the race condition problem. For instance, interested readers are encouraged to look at References [2, 3, 4, 38, 39, 56] to get familiar with some lock-free and wait-free inter-process communication methods of historical and practical interest, which operate *without* using any kind of lock.

5.2 HARDWARE-ASSISTED LOCK VARIABLES

A very simple way of ensuring the mutual exclusion among multiple processes want-ing to access a set of shared variables is to uniquely associate a *lock variable* with them, as shown in Figure 5.4. If there are multiple, independent sets of shared vari-ables, there will be one lock variable for each set.

The underlying idea is to use the lock variable as a flag that will assume two distinct values, depending on how many processes are within their critical regions. In our example, the lock variable is implemented as an integer variable (unsurprisingly) called lock, which can assume either the value 0 or 1:

- 0 means that no processes are currently accessing the set of shared variables associated with the lock;
- 1 means that one process is currently accessing the set of shared variables associated with the lock.

Since it is assumed that no processes are within a critical region when the system is initialized, the initial value of lock is 0. As explained in Section 5.1, each pro-cess must surround its critical regions with the critical region *entry* and *exit* code.

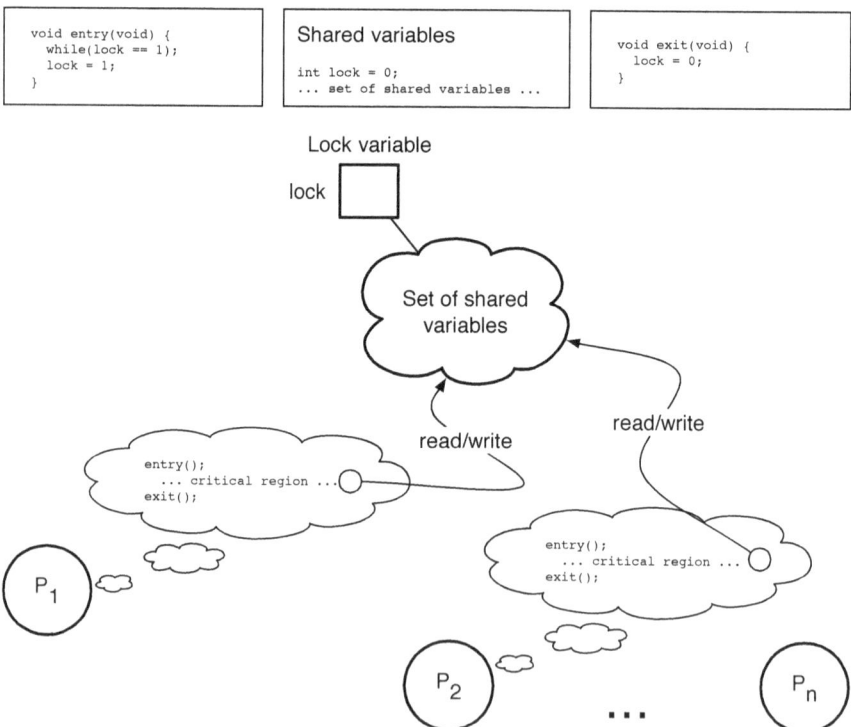

Figure 5.4   Lock variables do not necessarily work unless they are handled correctly.

In Figure 5.4, this is accomplished by calling the `entry()` and `exit()` functions, respectively.

Given how the lock variable has been defined, the contents of these functions are rather intuitive. That is, before entering a critical section, a process P must

1. Check whether the lock variable is 1. If this is the case, another process is currently accessing the set of shared variables protected by the lock. Therefore, P must wait and perform the check again later. This is done by the `while` loop shown in the figure.
2. When P eventually finds that the lock variable is 0, it breaks the `while` loop and is allowed to enter its critical region. Before doing this, it must set the lock variable to 1 to prevent other processes from entering, too.

The exit code is even simpler: whenever P is abandoning a critical region, it must reset the lock variable to 0. This may have two possible effects:

1. If one or more processes are already waiting to enter, one of them will find the lock variable at 0, will set it to 1 again, and will be allowed to enter its critical region.
2. If no processes are waiting to enter, the lock variable will stay at 0 for the time being until the critical region entry code is executed again.

Unfortunately, this naive approach to the problem does not work even if we consider only two processes P_1 and P_2. This is due to the fact that, as described above, the critical region entry code is composed of a *sequence* of two steps that are not executed atomically. The following interleaving is therefore possible:

1.1. P_1 executes the entry code and checks the value of `lock`, finds that `lock` is 0, and immediately escapes from the `while` loop.
2.1. Before P_1 had the possibility of setting `lock` to 1, P_2 executes the entry code, too. Since the value of `lock` is still 0, it exits from the `while` loop as well.
2.2. P_2 sets `lock` to 1.
1.2. P_1 sets `lock` to 1.

At this point, both P_1 and P_2 execute their critical code and violate the mutual exclusion constraint. An attentive reader would have certainly noticed that, using lock variables in this way, the mutual exclusion problem has merely been shifted from one "place" to another. Previously the problem was how to ensure mutual exclusion when accessing the set of shared variables, but now the problem is how to ensure mutual exclusion when accessing the lock variable itself.

Given the clear similarities between the scenarios depicted in Figures 5.2 and 5.4, it is not surprising that the problem has not been solved at all. However, this is one of the cases in which hardware assistance is very effective. In the simplest case, we can assume that the processor provides a *test and set* instruction. This instruction has the address p of a memory word as argument and *atomically* performs the following three steps:

```
void entry(void) {
  while(
    test_and_set(&lock) == 1);
}
```

```
Shared variables

int lock = 0;
... set of shared variables ...
```

```
void exit(void) {
  lock = 0;
}
```

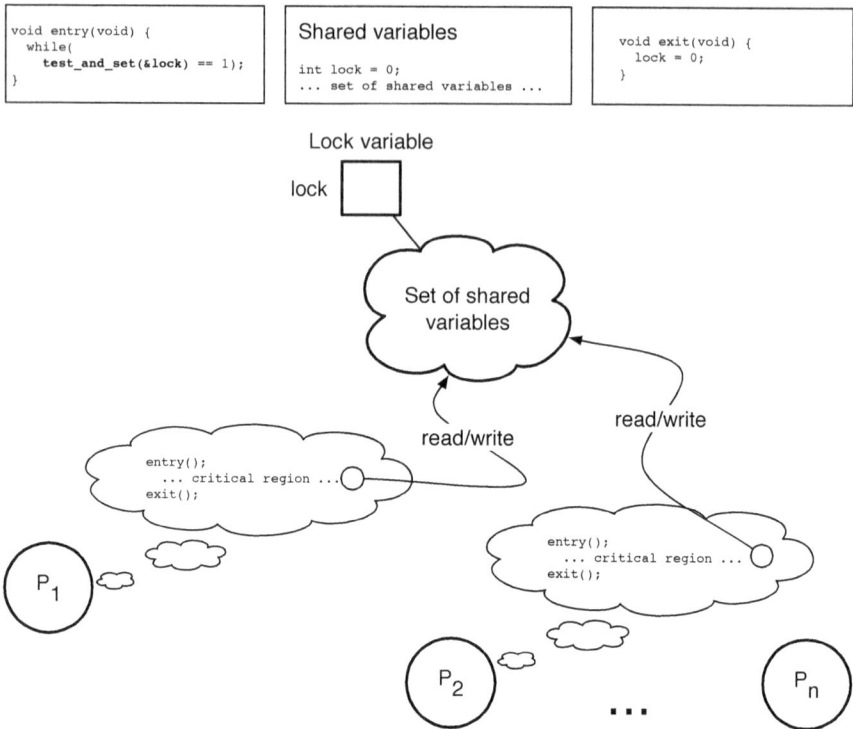

Figure 5.5 Hardware-assisted lock variables work correctly.

1. It reads the value v of the memory word pointed by p.
2. It stores 1 into the memory word pointed by p.
3. It puts v into a register.

As shown in Figure 5.5, this instruction can be used in the critical region entry code to avoid the race condition discussed before because it forces the test of `lock` to be atomic with respect to its update. The rest of the code stays the same. For convenience, the test and set instruction has been denoted as a C function `int test_and_set(int *p)`, assuming that the `int` type indeed represents a memory word.

For what concerns the practical implementation of this technique, on the Intel® 64 and IA-32 architecture [49], the BTS instruction tests and sets a single bit in either a register or a memory location. It also accepts the LOCK prefix so that the whole instruction is executed atomically.

Another, even simpler, instruction is XCHG, which exchanges the contents of a register with the contents of a memory word. In this case, the bus-locking protocol is activated automatically regardless of the presence of the LOCK prefix. The result is the same as the test and set instruction if the value of the register before executing the instruction is 1. Many other processor architectures provide similar instructions.

It can be shown that, considering the correctness conditions stated at the end of Section 5.1, the approach just described is correct with respect to conditions 1 and 2 but does not fully satisfy conditions 3 and 4:

- By intuition, if one of the processes is noticeably slower that the others—because, for example, it is executed on a slower processor in a multiprocessor system—it is placed at a disadvantage when it executes the critical region entry code. In fact, it checks the lock variable less frequently than the others, and this lowers its probability of finding lock at 0. This partially violates condition 4.
- In extreme cases, the execution of the while loop may be so slow that other processes may succeed in taking turns entering and exiting their critical regions, so that the "slow" process never finds lock at 0 and is never allowed to enter its own critical region. This is in contrast with condition 3.

From the practical standpoint, this may or may not be a real issue depending on the kind of hardware being used. For example, using this method for mutual exclusion among cores in a multicore system, assuming that all cores execute at the same speed (or with negligible speed differences), is quite safe.

5.3 SOFTWARE-BASED MUTUAL EXCLUSION

The first correct solution to the mutual exclusion problem between two processes, which does not use any form of hardware assistance, was designed by Dekker [25]. A more compact and elegant solution was then proposed by Peterson [75].

Even if Peterson's algorithm can be generalized to work with an arbitrary (but fixed) number of processes, for simplicity we will only consider the simplest scenario, involving only two processes P_0 and P_1 as shown in Figure 5.6. It is also assumed that the memory access atomicity and ordering constraints set forth in Section 5.1 are either satisfied or can be enforced.

Unlike the other methods discussed so far, the critical region entry and exit functions take a parameter pid that uniquely identifies the invoking process and will be either 0 (for P_0) or 1 (for P_1). The set of shared, access-control variables becomes slightly more complicated, too. In particular,

- There is now one flag for each process, implemented as an array flag[2] of two flags. Each flag will be 1 if the corresponding process wants to, or succeeded in entering its critical section, and 0 otherwise. Since it is assumed that processes neither want to enter, nor already are within their critical region at the beginning of their execution, the initial value of both flags is 0.
- The variable turn is used to enforce the two processes to take turns if both want to enter their critical region concurrently. Its value is a process identifier and can therefore be either 0 or 1. It is initially set to 0 to make sure it has a legitimate value even if, as we will see, this initial value is irrelevant to the algorithm.

```
void entry(int pid) {
  int other = 1-pid;
  flag[pid] = 1;
  turn = pid;
  while(
    flag[other] == 1
    && turn == pid);
}
```

Shared variables

```
int flag[2] = {0, 0};
int turn=0;
... set of shared variables ...
```

```
void exit(int pid) {
  flag[pid] = 0;
}
```

Access-control variables

turn

flag[]

Set of shared variables

read/write

read/write

```
entry(0);
... critical region ...
exit(0);
```

P_0

```
entry(1);
... critical region ...
exit(1);
```

P_1

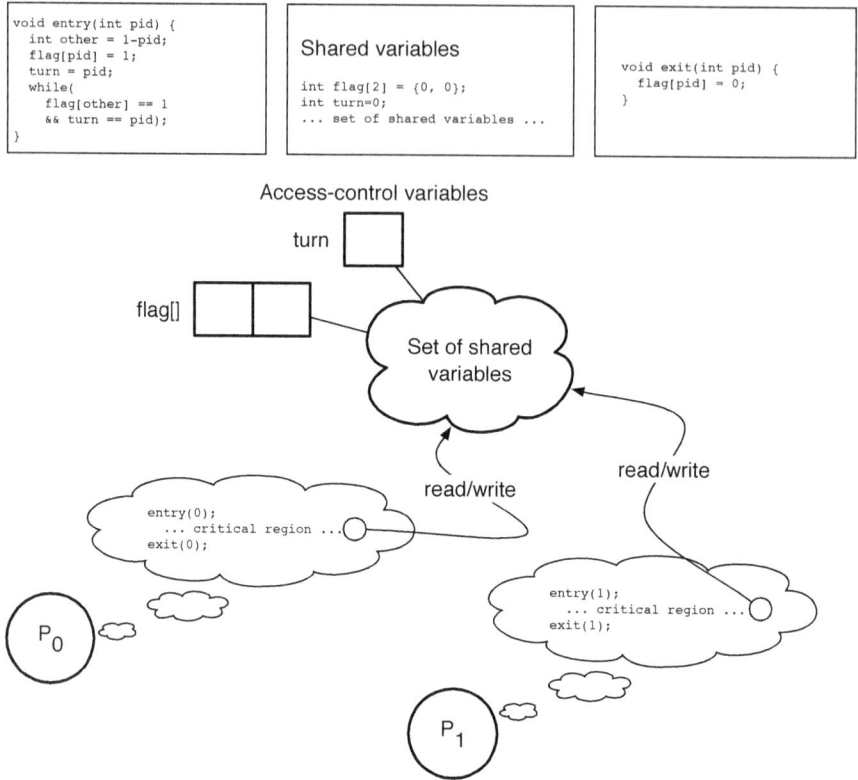

Figure 5.6 Peterson's software-based mutual exclusion for two processes.

A formal proof of the correctness of the algorithm is beyond the scope of this book, but it is anyway useful to gain an informal understanding of how it works and why it behaves correctly. As a side note, the same technique is also useful in gaining a better understanding of how other concurrent programming algorithms are designed and built.

The simplest and easiest-to-understand case for the algorithm happens when the two processes do *not* execute the critical region entry code concurrently but sequentially. Without loss of generality, let us assume that P_0 executes this code first. It will perform the following operations:

1. It sets its own flag, flag[0], to 1. It should be noted that P_0 works on flag[0] because pid is 0 in the instance of enter() it invokes.
2. It sets turn to its own process identifier pid, that is, 0.
3. It evaluates the predicate of the while loop. In this case, the right-hand part of the predicate is true (because turn has just been set to pid), but the left-hand part is false because the other process is not currently trying to enter its critical section.

As a consequence, P_0 immediately abandons the while loop and is granted access to its critical section. The final state of the shared variables after the execution of enter(0) is

- flag[0] == 1 (it has just been set by P_0);
- flag[1] == 0 (P_1 is not trying to enter its critical region);
- turn == 0 (P_0 has been the last process to start executing enter()).

If, at this point, P_1 tries to enter its critical region by executing enter(1), the following sequence of events takes place:

1. It sets its own flag, flag[1], to 1.
2. It sets turn to its own process identifier, 1.
3. It evaluates the predicate of the while loop. Both parts of the predicate are true because:
   - The assignment other = 1-pid implies that the value of other represents the process identifier of the "other" process. Therefore, flag[other] refers to flag[0], the flag of P_0, and this flag is currently set to 1.
   - The right-hand part of the predicate, turn == pid, is also true because turn has just been set this way by P_1, and P_0 does not modify it in any way.

P_1 is therefore trapped in the while loop and will stay there until P_0 exits its critical region and invokes exit(0), setting flag[0] back to 0. In turn, this will make the left-hand part of the predicate being evaluated by P_1 false, break its busy waiting loop, and allow P_1 to execute its critical region. After that, P_0 cannot enter its critical region again because it will be trapped in enter(0).

When thinking about the actions performed by P_0 and P_1 when they execute enter(), just discussed above, it should be remarked that, even if the first *program* statement in the body of enter(), that is, flag[pid] = 1, is the same for both of them, the two *processes* are actually performing very different actions when they execute it.

That is, they are operating on different flags because they have got different values for the variable pid, which belongs to the process state. This further highlights the crucial importance of distinguishing programs from processes when dealing with concurrent programming because, as it happens in this case, the same program fragment produces very different results depending on the executing process state.

It has just been shown that the mutual exclusion algorithm works satisfactorily when P_0 and P_1 execute enter() sequentially, but this, of course, does not cover all possible cases. Now, we must convince ourselves that, *for every possible interleaving* of the two, concurrent executions of enter(), the algorithm still works as intended. To facilitate the discussion, the code executed by the two processes has been listed again in Figure 5.7, replacing the local variables pid and other by their value. As already recalled, the value of these variables depends on the process being considered.

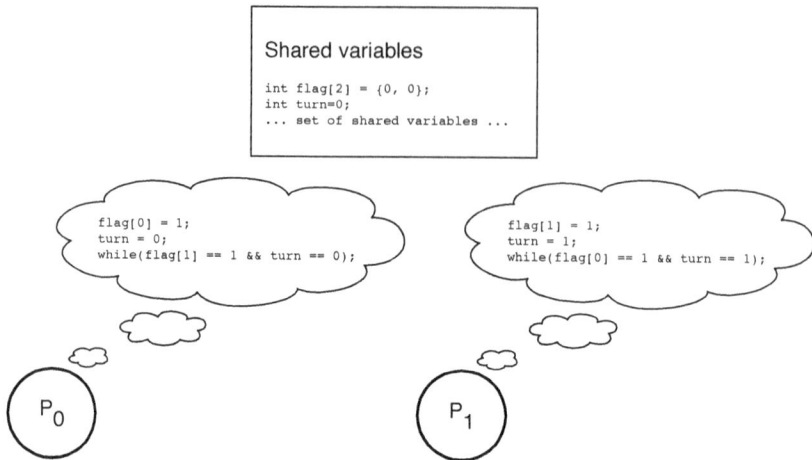

Figure 5.7 Code being executed concurrently by the two processes involved in Peterson's critical region enter code.

For what concerns the first two statements executed by P_0 and P_1, that is, `flag[0] = 1` and `flag[1] = 1`, it is easy to see that the result does not depend on the execution order because they operate on two distinct variables. In any case, after both statements have been executed, both flags will be set to 1.

On the contrary, the second pair of statements, `turn = 0` and `turn = 1`, respectively, work on the same variable `turn`. The result will therefore depend on the execution order but, thanks to the memory access atomicity taken for granted at the single-variable level, the final value of `turn` will be either 0 or 1, and not anything else, even if both processes are modifying it concurrently. More precisely, the final value of `turn` only depends on which process executed its assignment *last*, and represents the identifier of that process.

Let us now consider what happens when both processes evaluate the predicate of their `while` loop:

- The left-hand part of the predicate has no effect on the overall outcome of the algorithm because both `flag[0]` and `flag[1]` have been set to one.
- The right-hand part of the predicate will be true for *one* and *only one* process. It will be true for at least one process because `turn` will be either 0 or 1. In addition, it cannot be true for both processes because `turn` cannot assume two different values at once and no processes can further modify it.

In summary, either P_0 or P_1 (but not both) will be trapped in the `while` loop, whereas the other process will be allowed to enter into its critical region. Due to our considerations about the value of `turn`, we can also conclude that the process that set `turn` last will be trapped, whereas the other will proceed. As before, the `while` loop executed by the trapped process will be broken when the other process resets its flag to 0 by invoking its critical region exit code.

Going back to the correctness conditions outlined in Section 5.1, this algorithm is clearly correct with respect to conditions 1 and 2. For what concerns conditions 3 and 4, it also works better than the hardware-assisted lock variables discussed in Section 5.2. In particular,

- The slower process is no longer systematically put at a disadvantage. Assuming that P_0 is slower than P_1, it is still true that P_1 may initially overcome P_0 if both processes execute the critical region entry code concurrently. However, when P_1 exits from its critical region and then tries to immediately reenter it, it can no longer overcome P_0.

  When P_1 is about to evaluate its `while` loop predicate for the second time, the value of `turn` will, in fact, be 1 (because P_1 set it last), and both flags will be set. Under these conditions, the predicate will be true, and P_1 will be trapped in the loop. At the same time, as soon as `turn` has been set to 1 by P_1, P_0 will be allowed to proceed regardless of its speed because its `while` loop predicate becomes false and stays this way.

- For the same reason, and due to the symmetry of the code, if both processes repeatedly contend against each other for access to their critical region, they will take turns at entering them, so that no process can systematically overcome the other. This property also implies that, if a process wants to enter its critical region, it will succeed within a finite amount of time, that is, at the most the time the other process spends in its critical region.

For the sake of completeness, it should be noted that the algorithm just described, albeit quite important from the historical point of view, is definitely not the only one of this kind. For instance, interested readers may want to look at the famous Lamport's *bakery algorithm* [59]. One of the most interesting properties of this algorithm is that it still works even if memory read and writes are *not* performed in an atomic way by the underlying processor. In this way, it completely solves the mutual exclusion problem without any kind of hardware assistance.

5.4   FROM ACTIVE TO PASSIVE WAIT

All the methods presented in Sections 5.2 and 5.3 are based on *active* or *busy* wait loops: a process that cannot immediately enter into its critical region is delayed by trapping it into a loop. Within the loop, it repeatedly evaluates a Boolean predicate that indicates whether it must keep waiting or can now proceed.

One unfortunate side effect of busy loops is quite obvious: whenever a process is engaged in one of those loops, it does not do anything useful from the application point of view—as a matter of fact, the intent of the loop is to prevent it from proceeding further at the moment—but it wastes processor cycles anyway. Since in many embedded systems processor power is at a premium, due to cost and power consumption factors, it would be a good idea to put these wasted cycles to better use.

Another side effect is subtler but not less dangerous, at least when dealing with real-time systems, and concerns an adverse interaction of busy wait with the concept

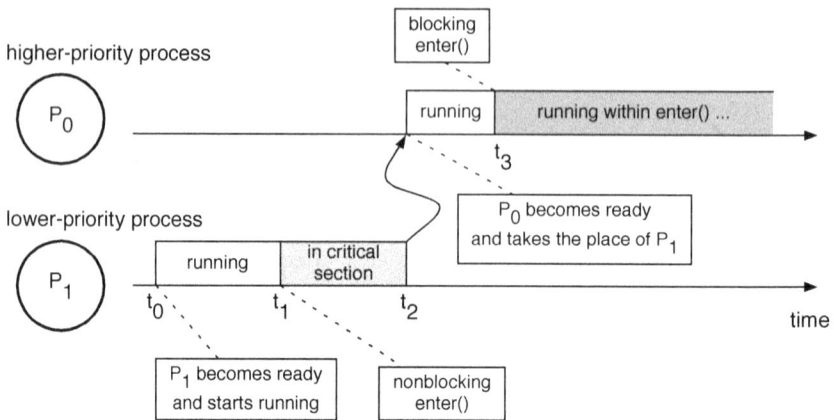

Figure 5.8 Busy wait and fixed-priority assignment may interfere with each other, leading to an unbounded priority inversion.

of process *priority* and the way this concept is often put into practice by a real-time scheduler.

In the previous chapters it has already been highlighted that not all processes in a real-time system have the same "importance" (even if the concept of importance has not been formally defined yet) so that some of them must be somewhat preferred for execution with respect to the others. It is therefore intuitively sound to associate a fixed priority value to each process in the system according to its importance and design the scheduler so that it systematically prefers higher-priority processes when it is looking for a ready process to run.

The intuition is not at all far from reality because several popular real-time scheduling algorithms, to be discussed in Chapter 11, are designed exactly in this way. Moreover, Chapters 12–15 will also make clearer that assigning the right priority to all the processes, and strictly obeying them at run-time, also plays a central role to ensure that the system meets its timing requirements and constraints.

A very simple example of the kind of problems that may occur is given in Figure 5.8. The figure shows two processes, P_0 and P_1, being executed on a single physical processor under the control of a scheduler that systematically prefers P_0 to P_1 because P_0's priority is higher. It is also assumed that these two processes share some data and—being written by a proficient programmer who just read this chapter—therefore contain one critical region each. The critical regions are protected by means of one of the mutual exclusion methods discussed so far. As before, the critical region entry code is represented by the function enter(). The following sequence of events may take place:

1. Process P_0 becomes ready for execution at t_0, while P_1 is blocked for some other reason. Supposing that P_0 is the only ready process in the system at the moment, the scheduler makes it run.

2. At t_1, P_0 wants to access the shared data; hence, it invokes the critical region entry function `enter()`. This function is nonblocking because P_1 is currently outside its critical region and allows P_0 to proceed immediately.

3. According to the fixed-priority relationship between P_0 and P_1, as soon as P_1 becomes ready, the scheduler grabs the processor from P_0 and immediately brings P_1 into the running state with an action often called *preemption*. In Figure 5.8, this happens at t_2.

4. If, at t_3, P_1 tries to enter its critical section, it will get stuck in `enter()` because P_0 was preempted before it abandoned its own critical section.

It should be noted that the last event just discussed actually hides two distinct facts. One of them is normal, the other is not:

- Process P_1, the higher-priority process, cannot enter its critical section because it is currently blocked by P_0, a lower-priority process. This condition is called *priority inversion* because a synchronization constraint due to an interaction between processes—through a set of shared data in this case—is going against the priority assignment scheme obeyed by the scheduler. In Figure 5.8, the priority inversion region is shown in dark gray.

  Even if this situation may seem disturbing, it is not a problem by itself. Our lock-based synchronization scheme is indeed working as designed and, in general, a certain amount of priority inversion is unavoidable when one of those schemes is being used to synchronize processes with different priorities. Even in this very simple example, allowing P_1 to proceed would necessarily lead to a race condition.

- Unfortunately, as long as P_1 is trapped in `enter()`, it will actively use the processor—because it is performing a busy wait—and will stay in the running state so that our fixed-priority scheduler will never switch the processor back to P_0. As a consequence, P_0 will not further proceed with execution, it will never abandon its critical region, and P_1 will stay trapped forever. More generally, we can say that the priority inversion region is *unbounded* in this case because we are unable to calculate a finite upper bound for the maximum time P_1 will be blocked by P_0.

It is also useful to remark that some priority relationships between concurrent system activities can also be "hidden" and difficult to ascertain because, for instance, they are enforced by hardware rather than software. In several operating system architectures, interrupt handling implicitly has a priority greater than any other kind of activity as long as interrupts are enabled. In those architectures, the unbounded priority situation shown in Figure 5.8 can also occur when P_0 is a regular process, whereas P_1 is an interrupt handler that, for some reason, must exchange data with P_0. Moreover, it happens regardless of the software priority assigned to P_0.

From Section 3.4 in Chapter 3, we know that the process state diagram already comprises a state, the *Blocked* state, reserved for processes that cannot proceed for some reason, for instance, because they are waiting for an I/O operation to complete. Any process belonging to this state does not proceed with execution but does so

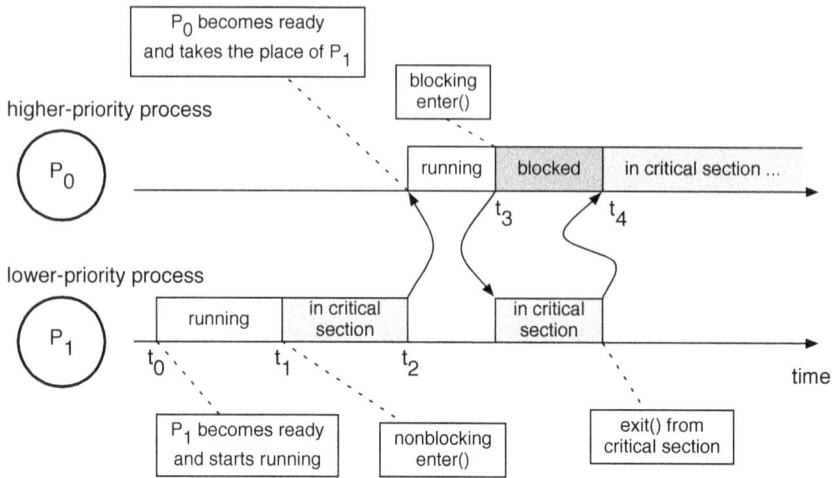

Figure 5.9   Using passive wait instead of busy wait solves the unbounded priority inversion problem in simple cases.

without wasting processor cycles and without preventing other processes from executing in its place. This kind of wait is often called *passive* wait just for this reason.

It is therefore natural to foresee a clever interaction between interprocess synchronization and the operating system scheduler so that the synchronization mechanism makes use of the *Blocked* state to prevent processes from running when appropriate. This certainly saves processor power and, at least in our simple example, also addresses the unbounded priority inversion issue. As shown in Figure 5.9, if enter() and exit() are somewhat modified to use passive wait,

1. At t_3, P_1 goes into the *Blocked* state, thus performing a passive instead of a busy wait. The scheduler gives the processor to P_0 because P_1 is no longer ready for execution.
2. Process P_0 resumes execution in its critical section and exits from it at t_4 by calling exit(). As soon as P_0 exits from the critical section, P_1 is brought back into the *Ready* state because it is ready for execution again.
3. As soon as P_1 is ready again, the scheduler reevaluates the situation and gives the processor back to P_1 itself so that P_1 is now running in its critical section.

It is easy to see that there still is a priority inversion region, shown in dark gray in Figure 5.9, but it is no longer unbounded. The maximum amount of time P_1 can be blocked by P_0 is, in fact, bounded by the maximum amount of time P_0 can possibly spend executing in its critical section. For well-behaved processes, this time will certainly be finite.

Even if settling on passive wait is still not enough to completely solve the unbounded priority inversion problem in more complex cases, as will be shown in

Chapter 14, the underlying idea is certainly a step in the right direction and led to a number of popular synchronization methods that will be discussed in the following sections.

The price to be paid is a loss of efficiency because any passive wait requires a few transition in the process state diagram, the execution of the scheduling algorithm, and a context switch. All these operations are certainly slower with respect to a tight, busy loop, which does not necessarily require any context switch when multiple physical processors are available. For this reason, busy wait is still preferred when the waiting time is expected to be small, as it happens for short critical regions, and efficiency is of paramount importance to the point that wasting several processor cycles in busy waiting is no longer significant.

This is the case, for example, when busy wait is used as a "building block" for other, more sophisticated methods based on passive wait, which must work with multiple, physical processors. Another field of application of busy wait is to ensure mutual exclusion between physical processors for performance-critical operating system objects, such as the data structures used by the scheduler itself.

Moreover, in the latter case, using passive wait would clearly be impossible anyway.

5.5 SEMAPHORES

Semaphores were first proposed as a general interprocess synchronization framework by Dijkstra [25]. Even if the original formulation was based on busy wait, most contemporary implementations use passive wait instead. Even if semaphores are not powerful enough to solve, strictly speaking, *any* arbitrary concurrent programming problem, as pointed out for example in [57], they have successfully been used to address many problems of practical significance. They still are the most popular and widespread interprocess synchronization method, also because they are easy and efficient to implement.

A semaphore is an object that comprises two abstract items of information:

1. a nonnegative, integer *value*;
2. a *queue* of processes passively waiting on the semaphore.

Upon initialization, a semaphore acquires an initial value specified by the programmer, and its queue is initially empty. Neither the value nor the queue associated with a semaphore can be read or written directly after initialization. On the contrary, the only way to interact with a semaphore is through the following two primitives that are assumed to be executed *atomically*:

1. P(s), when invoked on semaphore s, checks whether the value of the semaphore is (strictly) greater than zero.
   - If this is the case, it decrements the value by one and returns to the caller without blocking.

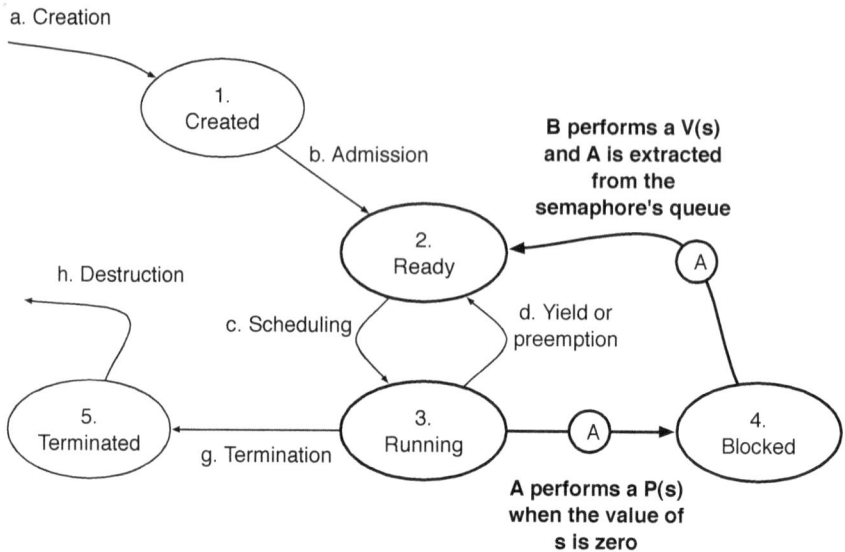

Figure 5.10 Process State Diagram transitions induced by the semaphore primitives P() and V().

- Otherwise, it puts the calling process into the queue associated with the semaphore and blocks it by moving it into the *Blocked* state of the process state diagram.
2. V(s), when invoked on semaphore s, checks whether the queue associated with that semaphore is empty or not.
   - If the queue is empty, it increments the value of the semaphore by one.
   - Otherwise, it picks one of the blocked processes found in the queue and makes it ready for execution again by moving it into the *Ready* state of the process state diagram.
   In both cases, V(s) never blocks the caller. It should also be remarked that, when V(s) unblocks a process, it does not necessarily make it running immediately. In fact, determining which processes must be run, among the *Ready* ones, is a duty of the scheduling algorithm, not of the interprocess communication mechanism. Moreover, this decision is often based on information—for instance, the relative priority of the *Ready* processes— that does not pertain to the semaphore and that the semaphore implementation may not even have at its disposal.

The process state diagram transitions triggered by the semaphore primitives are highlighted in Figure 5.10. As in the general process state diagram shown in Figure 3.4 in Chapter 3, the transition of a certain process A from the *Running* to the *Blocked* state caused by P() is voluntary because it depends on, and is caused by, a specific action performed by the process that is subjected to the transition, in this case A itself.

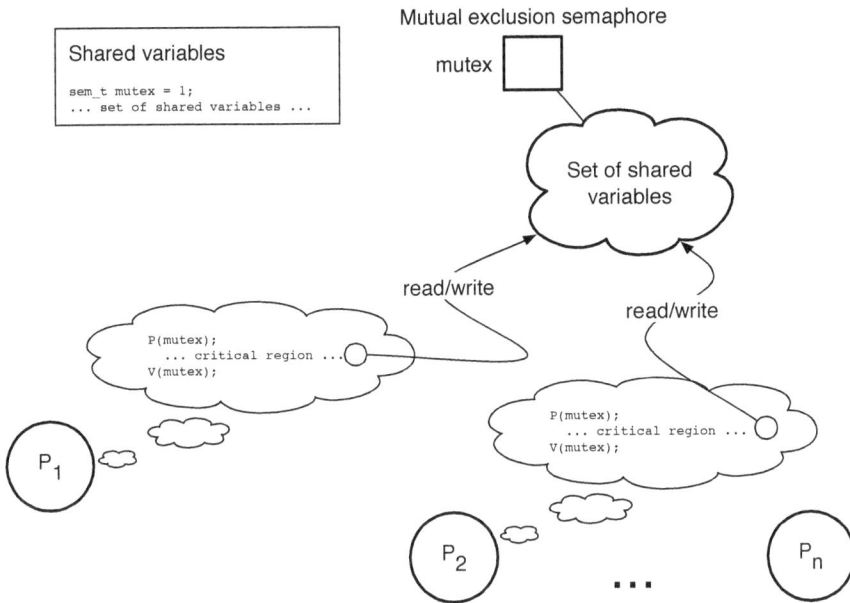

Figure 5.11   A semaphore can be used to enforce mutual exclusion.

On the other hand, the transition of a process A from the *Blocked* back into the *Ready* state is involuntary because it depends on an action performed by another process. In this case, the transition is triggered by another process B that executes a V() involving the semaphore on which A is blocked. After all, by definition, as long as A is blocked, it does not proceed with execution and cannot perform any action by itself.

Semaphores provide a simple and convenient way of enforcing mutual exclusion among an arbitrary number of processes that want to have access to a certain set of shared variables. As shown in Figure 5.11, it is enough to associate a mutual exclusion semaphore mutex to each set of global variables to be protected. The initial value of this kind of semaphore is always 1.

Then, all critical regions pertaining to that set of global variables must be surrounded by the statements P(mutex) and V(mutex), using them as a pair of "brackets" around the code, so that they constitute the critical region entry and exit code, respectively.

In this chapter, we focus on the high-level aspects of using semaphores. For this reason, in Figure 5.11, we assume that a semaphore has the abstract data type sem_t and it can be defined as any other variable. In practice, this is not the case, and some additional initialization steps are usually required. See Chapters 7 and 8 for more information on how semaphores are defined in an actual operating system.

This method certainly fulfills the first correctness conditions of Section 5.1. When one process P_1 wants to enter into its critical region, it first executes the critical region entry code, that is, P(mutex). This statement can have two different effects, depending on the value of mutex:

1. If the value of mutex is 1—implying that no other processes are currently within a critical region controlled by the same semaphore—the effect of P(mutex) is to decrement the semaphore value and allow the invoking process to proceed into the critical region.
2. If the value of mutex is 0—meaning that another process is already within a critical region controlled by the same semaphore—the invoking process is blocked at the critical region boundary.

Therefore, if the initial value of mutex is 1, and many concurrent processes $P_1, \ldots, P_n$ want to enter into a critical region controlled by that semaphore, only one of them—for example P_1—will be allowed to proceed immediately because it will find mutex at 1. All the other processes will find mutex at 0 and will be blocked on it until P_1 reaches the critical region exit code and invokes V(mutex).

When this happens, one of the processes formerly blocked on mutex will be resumed—for example P_2—and will be allowed to execute the critical region code. Upon exit from the critical region, P_2 will also execute V(mutex) to wake up another process, and so on, until the last process P_n exits from the critical region while no other processes are currently blocked on P(mutex).

In this case, the effect of V(mutex) is to increment the value of mutex and bring it back to 1 so that exactly one process will be allowed to enter into the critical region immediately, without being blocked, in the future.

It should also be remarked that no race conditions during semaphore manipulation are possible because the semaphore implementation must guarantee that P() and V() are executed atomically.

For what concerns the second correctness condition, it can easily be observed that the only case in which the mutual exclusion semaphore prevents a process from entering a critical region takes place when another process is already within a critical region controlled by the same semaphore. Hence, processes doing internal operations cannot prevent other processes from entering their critical regions in any way.

The behavior of semaphores with respect to the third and fourth correctness conditions depend on their implementation. In particular, both conditions can easily be fulfilled if the semaphore queuing policy is first-in first-out, so that V() always wakes up the process that has been waiting on the semaphore for the longest time. When a different queuing policy must be adopted for other reasons, for instance, to solve the unbounded priority inversion, as discussed in Chapter 14, then some processes may be subject to an indefinite wait, and this possibility must usually be excluded by other means.

Besides mutual exclusion, semaphores are also useful for *condition synchronization*, that is, when we want to block a process until a certain event occurs or a certain condition is fulfilled. For example, considering again the producers–consumers problem, it may be desirable to block any consumer that wants to get a data item from the shared buffer when the buffer is completely empty, instead of raising an error

```
void prod(int d) {
  P(empty);
  P(mutex);
    buf[in] = d;
    in = (in+1) % N;
  V(mutex);
  V(full);
}
```

```
Macros/Shared
variables

#define N 8
int buf[N];
int in=0, out=0;
sem_t mutex=1;
sem_t empty=N;
sem_t full=0;
```

```
int cons(void) {
  int c;
  P(full);
  P(mutex);
    c = buf[out];
    out = (out+1) % N;
  V(mutex);
  V(empty);
  return c;
}
```

Figure 5.12   Producers–consumers problem solved with mutual exclusion and condition synchronization semaphores.

indication. Of course, a blocked consumer must be unblocked as soon as a producer puts a new data item into the buffer. Symmetrically, we might also want to block a consumer when it tries to put more data into a buffer that is already completely full.

In order to do this, we need *one* semaphore for each synchronization *condition* that the concurrent program must respect. In this case, we have two conditions, and hence we need two semaphores:

1. The semaphore empty counts how many empty elements there are in the buffer. Its initial value is N because the buffer is completely empty at the beginning. Producers perform a P(empty) before putting more data into the buffer to update the count and possibly block themselves, if there is no empty space in the buffer. After removing one data item from the buffer, consumers perform a V(empty) to either unblock one waiting producer or increment the count of empty elements.

2. Symmetrically, the semaphore full counts how many full elements there are in the buffer. Its initial value is 0 because there is no data in the buffer at the beginning. Consumers perform P(full) before removing a data item from the buffer, and producers perform a V(full) after storing an additional data item into the buffer.

The full solution to the producers–consumers problem is shown in Figure 5.12. In summary, even if semaphores are all the same, they can be used in two very different ways, which should not be confused:

1. A *mutual exclusion* semaphore, like mutex in the example, is used to prevent more than one process from executing within a set of critical regions pertaining to the same set of shared data. The use of a mutual exclusion semaphore is quite stereotyped: its initial value is always 1, and P() and V() are placed, like brackets, around the critical regions code.

2. A *condition synchronization* semaphore, such as empty and full in the example, is used to ensure that certain sequences of events do or do not occur. In this particular case, we are using them to prevent a producer from storing data into a full buffer, or a consumer from getting data from an empty buffer. They are usually more difficult to use correctly because there is no stereotype to follow.

5.6 MONITORS

As discussed in the previous section, semaphores can be defined quite easily; as we have seen, their behavior can be fully described in about one page. Their practical implementation is also very simple and efficient so that virtually all modern operating systems support them. However, semaphores are also a very *low-level* interprocess communication mechanism. For this reason, they are difficult to use, and even the slightest mistake in the placement of a semaphore primitive, especially P(), may completely disrupt a concurrent program.

For example, the program shown in Figure 5.13 may seem another legitimate way to solve the producers–consumers problem. Actually, it has been derived from the solution shown in Figure 5.12 by swapping the two semaphore primitives shown in boldface. After all, the program code still "makes sense" after the swap because, as programmers, we could reason in the following way:

- In order to store a new data item into the shared buffer, a producer must, first of all, make sure that it is the only process allowed to access the shared buffer itself. Hence, a P(mutex) is needed.
- In addition, there must be room in the buffer, that is, at least one element must be free. As discussed previously, P(empty) updates the count of free buffer elements held in the semaphore empty and blocks the caller until there is at least one free element.

Unfortunately, this kind of reasoning is incorrect because the concurrent execution of the code shown in Figure 5.13 can lead to a *deadlock*. When a producer tries to store an element into the buffer, the following sequence of events may occur:

- The producer succeeds in gaining exclusive access to the shared buffer by executing a P(mutex). From this point on, the value of the semaphore mutex is zero.
- If the buffer is full, the value of the semaphore empty will be zero because its value represents the number of empty elements in the buffer. As a consequence, the execution of P(empty) blocks the producer. It should also be noted that the producer is blocked *within* its critical region, that is, *without releasing* the mutual exclusion semaphore mutex.

```
void prod(int d) {
    P(mutex);
    P(empty);
      buf[in] = d;
      in = (in+1) % N;
    V(mutex);
    V(full);
}
```

```
Macros/Shared
variables

#define N 8
int buf[N];
int in=0, out=0;
sem_t mutex=1;
sem_t empty=N;
sem_t full=0;
```

```
int cons(void) {
    int c;
    P(full);
    P(mutex);
      c = buf[out];
      out = (out+1) % N;
    V(mutex);
    V(empty);
    return c;
}
```

Figure 5.13 Semaphores may be difficult to use: even the incorrect placement of one single semaphore primitive may lead to a deadlock.

After this sequence of events takes place, the only way to wake up the blocked producer is by means of a V(empty). However, by inspecting the code, it can easily be seen that the only V(empty) is at the very end of the consumer's code. No consumer will ever be able to reach that statement because it is preceded by a critical region controlled by mutex, and the current value of mutex is zero.

In other words, the consumer will be blocked by the P(mutex) located at the beginning of the critical region itself as soon as it tries to get data item from the buffer. All the other producers will be blocked, too, as soon as they try to store data into the buffer, for the same reason.

As a side effect, the first N consumers trying to get data from the buffer will also bring the value of the semaphore full back to zero so that the following consumers will not even reach the P(mutex) because they will be blocked on P(full).

The presence of a deadlock can also be deducted in a more abstract way, for instance, by referring to the Havender/Coffman conditions presented in Chapter 4. In particular,

- As soon as a producer is blocked on P(empty) and a consumer is blocked on P(mutex), there is a *circular wait* in the system. In fact, the consumer waits for the producer to release the resource "empty space in the buffer," an event represented by V(empty) in the code. Symmetrically, the consumer waits for the producer to release the resource "mutual exclusion semaphore," an event represented by V(mutex).
- Both resources are subject to a *mutual exclusion*. The mutual exclusion semaphore, by definition, can be held by only one process at a time. Similarly, the buffer space is either empty or full, and hence may belong to either the producers or the consumers, but not to both of them at the same time.
- The *hold and wait* condition is satisfied because both processes hold a resource—either the mutual exclusion semaphore or the ability to provide more empty buffer space—and wait for the other.
- Neither resource can be *preempted*. Due to the way the code has been designed, the producer cannot be forced to relinquish the mutual exclusion semaphore before it gets some empty buffer space. On the other hand, the consumer cannot be forced to free some buffer space without first passing through its critical region controlled by the mutual exclusion semaphore.

To address these issues, a higher-level and more structured interprocess communication mechanism, called *monitor*, was proposed by Brinch Hansen [16] and Hoare [41]. It is interesting to note that, even if these proposals date back to the early '70s, they were already based on concepts that are common nowadays and known as object-oriented programming. In its most basic form, a monitor is a composite object and contains

- a set of *shared data*;
- a set of *methods* that operate on them.

With respect to its components, a monitor guarantees the following two main properties:

- *Information hiding*, because the set of shared data defined in the monitor is accessible only through the monitor methods and cannot be manipulated directly from the outside. At the same time, monitor methods are not allowed to access any other shared data. Monitor methods are not hidden and can be freely invoked from outside the monitor.
- *Mutual exclusion* among monitor methods, that is, the monitor implementation, must guarantee that only one process will be actively executing within any monitor method at any given instant.

Both properties are relatively easy to implement in practice because monitors—unlike semaphores—are a *programming language* construct that must be known to, and supported by, the language compiler. For instance, mutual exclusion can be implemented by forcing a process to wait when it tries to execute a monitor method while another process is already executing within the same monitor.

This is possible because the language compiler knows that a call to a monitor method is not the same as a regular function call and can therefore handle it appropriately. In a similar way, most language compilers already have got all the information they need to enforce the information-hiding rules just discussed while they are processing the source code.

The two properties just discussed are clearly powerful enough to avoid any race condition in accessing the shared data enclosed in a monitor and are a valid substitute for mutual exclusion semaphores. However, an attentive reader would certainly remember that *synchronization semaphores* have an equally important role in concurrent programming and, so far, no counterpart for them has been discussed within the monitor framework.

Unsurprisingly, that counterpart does exist and takes the form of a third kind of component belonging to a monitor: the *condition variables*. Condition variables can be used only by the methods of the monitor they belong to, and cannot be referenced in any way from outside the monitor boundary. They are therefore hidden exactly like the monitor's shared data. The following two primitives are defined on a condition variable c:

- `wait(c)` blocks the invoking process and releases the monitor in a single, *atomic* action.
- `signal(c)` wakes up one of the processes blocked on c; it has no effect if no processes are blocked on c.

The informal reasoning behind the primitives is that, if a process starts executing a monitor method and then discovers that it cannot finish its work immediately, it invokes `wait` on a certain condition variable. In this way, it blocks and allows other processes to enter the monitor and perform their job.

When one of those processes, usually by inspecting the monitor's shared data, detects that the first process can eventually continue, it calls `signal` on the same

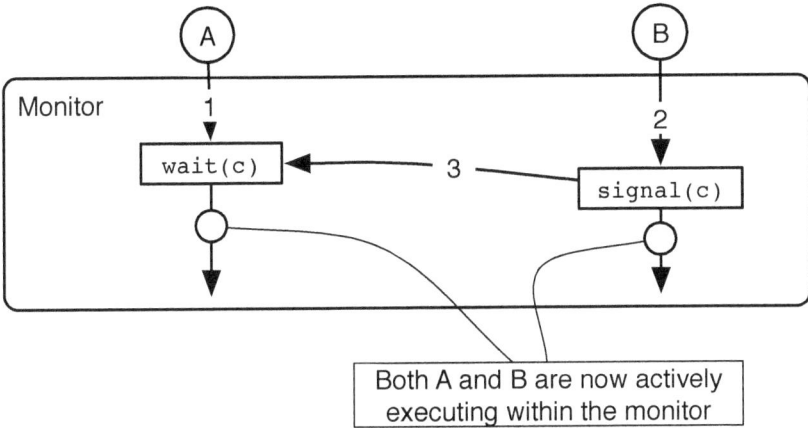

Figure 5.14 After a `wait`/`signal` sequence on a condition variable, there is a race condition that must be adequately addressed.

condition variable. The provision for multiple condition variables stems from the fact that, in a single monitor, there may be many, distinct reasons for blocking. Processes can easily be divided into groups and then awakened selectively by making them block on distinct condition variables, one for each group.

However, even if the definition of `wait` and `signal` just given may seem quite good by intuition, it is nonetheless crucial to make sure that the synchronization mechanism does not "run against" the mutual exclusion property that monitors must guarantee in any case, leading to a race condition. It turns out that, as shown in Figure 5.14, the following sequence of events involving two processes *A* and *B* may happen:

1. Taking for granted that the monitor is initially free—that is, no processes are executing any of its methods—process *A* enters the monitor by calling one of its methods, and then blocks by means of a `wait(c)`.

2. At this point, the monitor is free again, and hence, another process *B* is allowed to execute within the monitor by invoking one of its methods. There is no race condition because process *A* is still blocked.

3. During its execution within the monitor, *B* may invoke `signal(c)` to wake up process *A*.

After this sequence of events, *both A and B* are actively executing within the monitor. Hence, they are allowed to manipulate its shared data concurrently in an uncontrolled way. In other words, the mutual exclusion property of monitors has just been violated. The issue can be addressed in three different ways:

1. As proposed by Brinch Hansen [16], it is possible to work around the issue by constraining the placement of `signal` within the monitor methods: in particular, if a `signal` is ever invoked by a monitor method, it must be

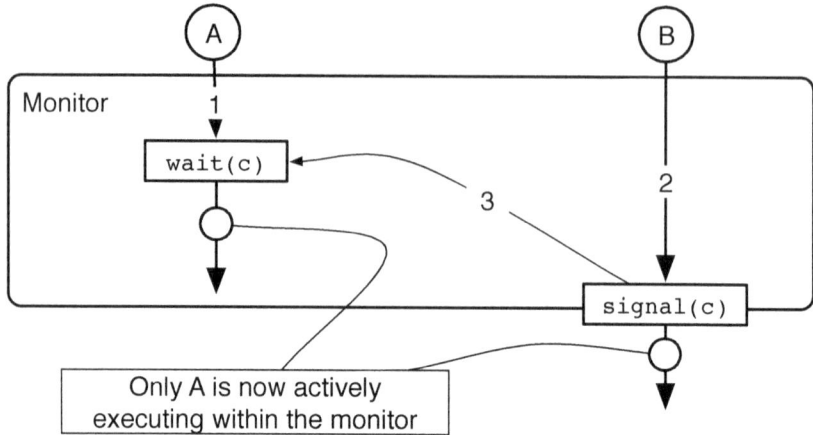

Figure 5.15 An appropriate constraint on the placement of `signal` in the monitor methods solves the race condition issue after a `wait`/`signal` sequence.

its *last* action, and implicitly makes the executing process exit from the monitor. As already discussed before, a simple scan of the source code is enough to detect any violation of the constraint.

In this way, only process *A* will be executing within the monitor after a `wait`/`signal` sequence, as shown in Figure 5.15, because the `signal` must necessarily be placed right at the monitor boundary. As a consequence, process *B* will indeed keep running concurrently with *A*, but *outside* the monitor, so that no race condition occurs. Of course, it can also be shown that the constraint just described solves the problem in general, and not only in this specific case.

The main advantage of this solution is that it is quite simple and efficient to implement. On the other hand, it leaves to the programmer the duty of designing the monitor methods so that `signal` only appears in the right places.

2. Hoare's approach [41] imposes no constraints at all on the placement of `signal`, which can therefore appear and be invoked anywhere in monitor methods. However, the price to be paid for this added flexibility is that the semantics of `signal` become somewhat less intuitive because it may now *block* the caller.

In particular, as shown in Figure 5.16, the signaling process *B* is blocked when it successfully wakes up the waiting process *A* in step 3. In this way, *A* is the only process actively executing in the monitor after a `wait`/`signal` sequence and there is no race condition. Process *B* will be resumed when process *A* either exits from the monitor or waits again, as happens in step 4 of the figure.

Any other process cannot enter
the monitor in the meanwhile

Monitor

wait(c)

signal(c)

Only A is actively executing
within the monitor after `signal`

B can proceed only when A
exits from the monitor (or
performs another `wait`)

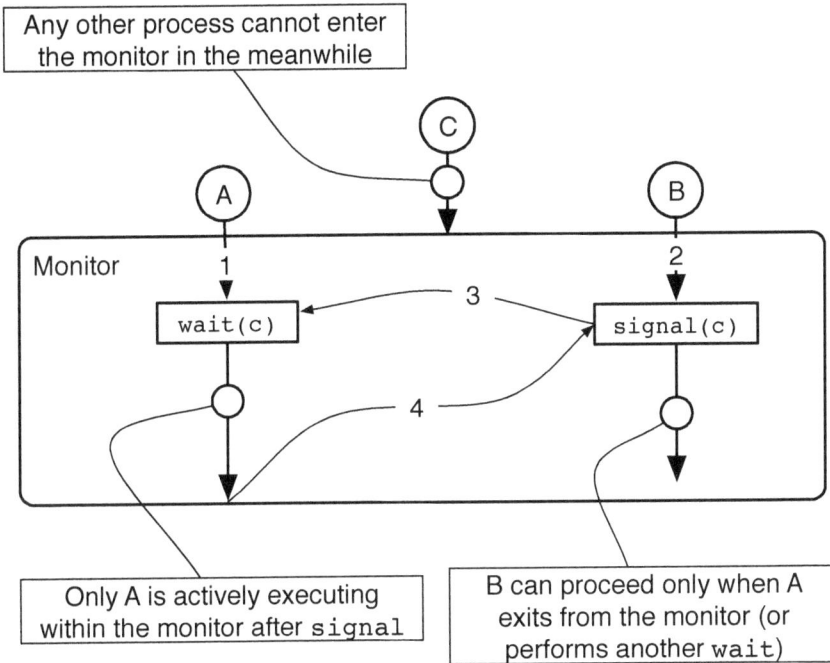

Figure 5.16 Another way to eliminate the race condition after a `wait/signal` sequence is to block the signaling process until the signaled one ceases execution within the monitor.

In any case, processes like *B*—that entered the monitor and then blocked as a consequence of a `signal`—take precedence on processes that want to enter the monitor from the outside, like process *C* in the figure. These processes will therefore be admitted into the monitor, one at a time, only when the process actively executing in the monitor leaves or waits, and no processes are blocked due to a `signal`.

3. The approach adopted by the POSIX standard [52] differs from the previous two in a rather significant way. Application developers must keep these differences in mind because writing code for a certain "flavor" of monitors and then executing it on another will certainly lead to unexpected results.

The reasoning behind the POSIX approach is that the process just awakened after a `wait`, like process *A* in Figure 5.17, must acquire the monitor's mutual exclusion lock *before* proceeding, whereas the signaling process (*B* in the figure) continues immediately. In a sense, the POSIX approach is like Hoare's, but it solves the race condition by postponing the *signaled* process, instead of the *signaling* one.

When the signaling process *B* eventually leaves the monitor or blocks in a `wait`, one of the processes waiting to start or resume executing in the monitor is chosen for execution. This may be

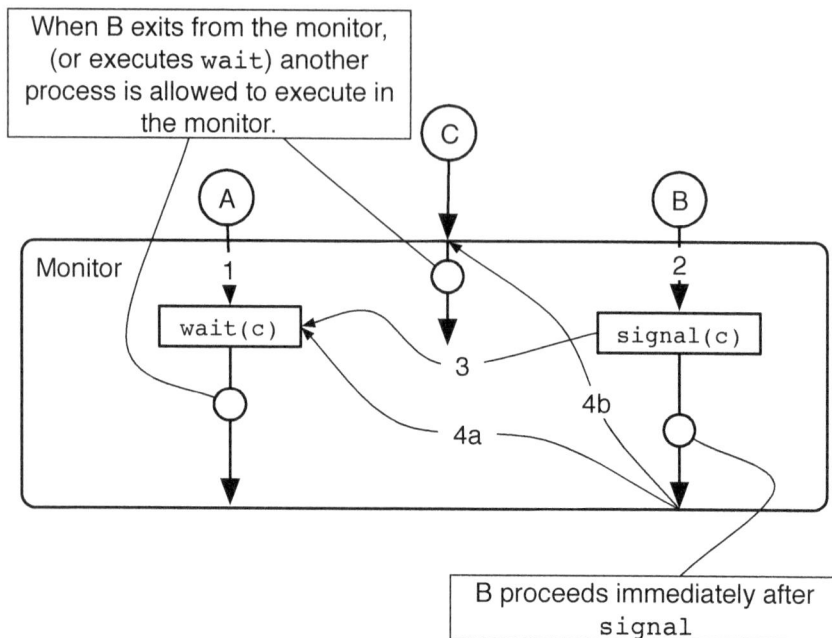

Figure 5.17   The POSIX approach to eliminate the race condition after a `wait`/`signal` sequence is to force the process just awakened from a `wait` to reacquire the monitor mutual exclusion lock before proceeding.

- one process waiting to reacquire the mutual exclusion lock after being awakened from a `wait` (process *A* and step 4a in the figure); or,
- one process waiting to enter the monitor from the outside (process *C* and step 4b in the figure).

The most important side effect of this approach from the practical standpoint is that, when process *A* waits for a condition and then process *B* signals that the condition has been fulfilled, process *A* cannot be 100% sure that the condition it has been waiting for will still be true when it will eventually resume executing in the monitor. It is quite possible, in fact, that another process *C* was able to enter the monitor in the meantime and, by altering the monitor's shared data, make the condition false again.

To conclude the description, Figure 5.18 shows how the producers–consumers problem can be solved by means of a Brinch Hansen monitor, that is, the simplest kind of monitor presented so far. Unlike the previous examples, this one is written in "pseudo C" because the C programming language, by itself, does not support monitors. The fake keyword `monitor` introduces a monitor. The monitor's shared data and methods are syntactically grouped together by means of a pair of braces. Within the monitor, the keyword `condition` defines a condition variable.

```
#define N 8
monitor ProducersConsumers
{
    int buf[N];
    int in = 0, out = 0;
    condition full, empty;
    int count = 0;

    void produce(int v)
    {
        if(count == N)  wait(empty);
        buf[in] = v;
        in = (in + 1) % N;
        count = count + 1;
        if(count == 1)  signal(full);
    }

    int consume(void)
    {
        int v;
        if(count == 0)  wait(full);
        v = buf[out];
        out = (out + 1) % N;
        count = count - 1;
        if(count == N-1)  signal(empty);
        return v;
    }
}
```

Figure 5.18 Producers–consumers problem solved by means of a Brinch Hansen monitor.

The main differences with respect to the semaphore-based solution of Figure 5.12 are

- The mutual exclusion semaphore `mutex` is no longer needed because the monitor construct already guarantees mutual exclusion among monitor methods.
- The two synchronization semaphores `empty` and `full` have been replaced by two condition variables with the same name. Indeed, their role is still the same: to make producers wait when the buffer is completely full, and make consumers wait when the buffer is completely empty.
- In the semaphore-based solution, the value of the synchronization semaphores represented the number of empty and full elements in the buffer. Since condition variables have no memory, and thus have no value at all, the monitor-based solution keeps that count in the shared variable `count`.
- Unlike in the previous solution, all `wait` primitives are executed conditionally, that is, only when the invoking process must certainly wait. For example, the producer's `wait` is preceded by an `if` statement checking whether `count` is equal to N or not, so that `wait` is executed only when the buffer is completely full. The same is also true for `signal`, and is due to the semantics differences between the semaphore and the condition variable primitives.

5.7   SUMMARY

In order to work together toward a common goal, processes must be able to *communicate*, that is, exchange information in a meaningful way. A set of shared variables is undoubtedly a very effective way to pass data from one process to another. However, if multiple processes make use of shared variables in a careless way, they will likely incur a *race condition*, that is, a harmful situation in which the shared variables are brought into an inconsistent state, with unpredictable results.

Given a set of shared variables, one way of solving this problem is to locate all the regions of code that make use of them and force processes to execute these *critical regions* one at a time, that is, in *mutual exclusion*. This is done by associating a sort of *lock* to each set of shared variables. Before entering a critical region, each process tries to acquire the lock. If the lock is unavailable, because another process holds it at the moment, the process waits until it is released. The lock is released at the end of each critical region.

The lock itself can be implemented in several different ways and at different levels of the system architecture. That is, a lock can be either *hardware-* or *software-based*. Moreover, it can be based on *active* or *passive wait*. Hardware-based locks, as the name says, rely on special CPU instructions to realize lock acquisition and release, whereas software-based locks are completely implemented with ordinary instructions.

When processes perform an active wait, they repeatedly evaluate a predicate to check whether the lock has been released or not, and consume CPU cycles doing so. On the contrary, a passive wait is implemented with the help of the operating system scheduler by moving the waiting processes into the *Blocked* state. This is a dedicated state of the Process State Diagram, in which processes do not compete for CPU usage and therefore do not proceed with execution.

From a practical perspective, the two most widespread ways of supporting mutual exclusion for shared data access in a real-time operating system are *semaphores* and, at a higher level of abstraction, *monitors*. Both of them are based on passive wait and are available in most modern operating systems.

Moreover, besides mutual exclusion, they can both be used for *condition synchronization*, that is, to establish timing constraints among processes, which are not necessarily related to mutual exclusion. For instance, a synchronization semaphore can be used to block a process until an external event of interest occurs, or until another process has concluded an activity.

Last, it should be noted that adopting a lock to govern shared data access is not the only way to proceed. Indeed, it is possible to realize shared objects that can be concurrently manipulated by multiple processes without using any locks [2, 3, 4, 38, 39, 56].

5.8   EXERCISES

EXERCISE 1

Consider these fragments of pseudocode, invoked repeatedly by two concurrent processes P_1 and P_2. Variables a and b are shared and c is local to P_2.

```
sem_t mutex = 1;
int a = 0;
int b = 0;
```

P_1:

```
P(mutex);
    b = b + 1;
V(mutex);
P(mutex);
    a = a + 1;
V(mutex);
```

P_2:

```
int c;
P(mutex);
    c = a + b;
V(mutex);
... use c ...
```

- Is it possible that P_2 gets an invalid value for a?
- Are the values of c used by P_2 always even?

EXERCISE 2

Consider the following classic implementation of a monitor with one single condition variable, using two semaphores and a shared integer variable. Processes call the enter() and exit() functions to enter and exit the monitor, respectively. The function wait() makes the calling process wait on the condition variable and signal() signals it. Does the code implement a Hoare's, Brinch Hansen's, or POSIX monitor?

```
sem_t m = 1;
sem_t c = 0;
int    nw = 0;
```

```
void enter(void)
{
    P(m);
}

void exit(void)
{
    V(m);
}
```

```
void wait(void)
{
    nw = nw + 1;
    V(m);
    P(c)
    nw = nw - 1;
}

void signal(void)
{
    if(nw > 0)
        V(c);
    else
        V(m);
}
```

6 Interprocess Communication Based on Message Passing

All interprocess communication methods presented in Chapter 5 are essentially able to pass *synchronization* signals from one process to another. They rely on *shared memory* to transfer *data*. Informally speaking, we know that it is possible to meaningfully transfer data among a group of producers and consumers by making them read from, and write into, a shared memory buffer "at the right time." We use one or more semaphores to make sure that the time is indeed right, but they are not directly involved in the data transfer.

It may therefore be of interest to look for a different interprocess communication approach in which one *single* supporting mechanism accomplishes both data transfer *and* synchronization, instead of having two distinct mechanisms for that. In this way, we would not only have a higher-level interprocess communication mechanism at our disposal but we will be able to use it even if there is no shared memory available. This happens, for example, when the communicating processes are executed by distinct computers.

Besides being interesting from a theoretical perspective, this approach, known as *message passing*, is very important from the practical standpoint, too. In Chapters 7 and 8, it will be shown that most operating systems, even very simple ones, provide a message-passing facility that can easily be used by threads and processes residing on the same machine. Then, in Chapter 9, we will see that a message-passing interface is also available among processes hosted on different computers linked by a communication network.

6.1 BASICS OF MESSAGE PASSING

In its simplest, and most abstract, form a message-passing mechanism involves two basic primitives:

- a *send* primitive, which sends a certain amount of information, called a *message*, to another process;
- a *receive* primitive, which allows a process to block waiting for a message to be sent by another process, and then retrieve its contents.

Even if this definition still lacks many important details that will be discussed later, it is already clear that the most apparent effect of message passing primitives is to transfer a certain amount of information from the sending process to the receiving

DOI: 10.1201/9781003593416-6

one. At the same time, the arrival of a message to a process also represents a synchronization signal because it allows the process to proceed after a blocking receive.

The last important requirement of a satisfactory interprocess communication mechanism, mutual exclusion, is not a concern here because messages are never shared among processes, and their ownership is passed from the sender to the receiver when the message is transferred. In other words, the mechanism works as if the message were instantaneously copied from the sender to the receiver even if real-world message passing systems do their best to avoid actually copying a message for performance reasons.

In this way, even if the sender alters a message after sending it, it will merely modify its local copy, and this will therefore not influence the message sent before. Symmetrically, the receiver is allowed to modify a message it received, and this action will not affect the sender in any way.

Existing message-passing schemes comprise a number of variations around this basic theme, which will be the subject of the following sections. The main design choices left open by our summary description are

1. For a sender, how to identify the intended recipient of a message. Symmetrically, for a receiver, how to specify from which other processes it is interested in receiving messages. In more abstract terms, a process *naming scheme* must be defined.
2. The *synchronization model*, that is, under what circumstances communicating processes shall be blocked, and for how long, when they are engaged in message passing.
3. How many *message buffers*, that is, how much space to hold messages already sent but not received yet, is provided by the system.

6.2   NAMING SCHEME

The most widespread naming schemes differ for two important aspects:

1. how the `send` and `receive` primitives are associated to each other;
2. their symmetry (or asymmetry).

About the first aspect, the most straightforward approach is for the sending process to name the receiver *directly*, for instance, by passing its process identifier to `send` as an argument. On the other hand, when the software gets more complex, it may be more convenient to adopt an *indirect* naming scheme in which the `send` and `receive` primitives are associated because they both name the same intermediate entity. In the following, we will use the word *mailbox* for this entity, but in the operating system jargon, it is also known under several other names, such as *channel* or *message queue*.

As shown in Figure 6.1, an indirect naming scheme is advantageous to software modularity and integration. If, for example, a software module *A* wants to send a message to another module *B*, the process *P* (of module *A*) responsible for the communication must know the identity of the intended recipient process *Q* within module

Direct naming scheme

Indirect naming scheme

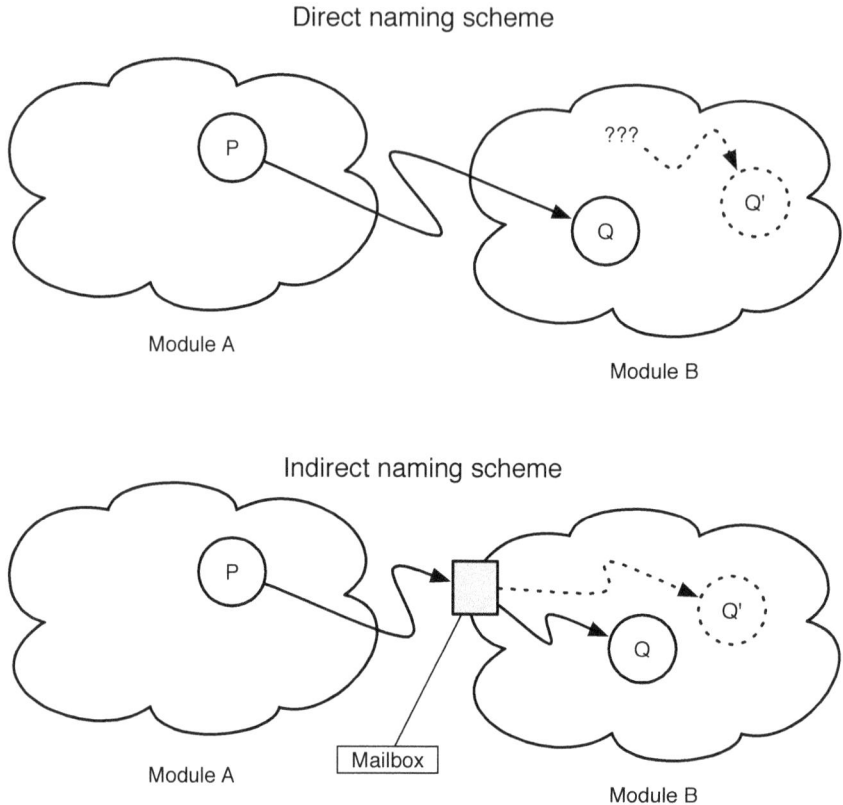

Figure 6.1 Direct versus indirect naming scheme; the direct scheme is simpler, the other one makes software integration easier.

B. If the internal architecture of module B is later changed, so that the intended recipient becomes Q′ instead of Q, module A must be updated accordingly, or otherwise communication will no longer be possible.

In other words, module A becomes dependent not only upon the *interface* of module B—that would be perfectly acceptable—but also upon its internal design and implementation. In addition, if process identifiers are used to name processes, as it often happens, even more care is needed because there is usually no guarantee that the identifier of a certain process will still be the same across reboots even if the process itself was not changed at all.

On the contrary, if the communication is carried out with an indirect naming scheme, depicted in the lower part of the figure, module A and process P must only know the name of the *mailbox* that module B is using for incoming messages. The name of the mailbox is part of the external interface of module B and will likely stay the same even if B's implementation and internal design change with time, unless the external interface of the module is radically redesigned, too.

Another side effect of indirect naming is that the relationship among communicating processes becomes more complex. For both kinds of naming, we can already have

- a *one-to-one* structure, in which one process sends messages to another;
- a *many-to-one* structure, in which many processes send messages to a single recipient.

With indirect naming, since multiple processes can receive messages from the same mailbox, there may also be a *one-to-many* or a *many-to-many* structure, or in which one or more processes send messages to a group of recipients, without caring about which of them will actually get the message.

This may be useful to conveniently handle concurrent processing in a server. For example, a web server may comprise a number of "worker" processes (or threads), all equal and able to handle a single HTTP request at a time. All of them will be waiting for requests through the same intermediate entity (which will most likely be a network communication endpoint in this case).

When a request eventually arrives, one of the workers will get it, process it, and provide an appropriate reply to the client. Meanwhile, the other workers will still be waiting for additional requests and may start working on them concurrently.

This example also brings us to discussing the second aspect of naming schemes, that is, their symmetry or asymmetry. If the naming scheme is *symmetric*, the sender process names either the receiving process or the destination mailbox, depending on whether the naming scheme is direct or indirect. Symmetrically, the receiver names either the sending process or the source mailbox.

If the naming scheme is asymmetric, the receiver does not name the source of the message in any way; it will accept messages from any source, and it will usually be informed about which process or mailbox the received message comes from. This scheme fits the client–server paradigm better because, in this case, the server is usually willing to accept requests from any of its clients and may not ever know their name in advance.

Regardless of the naming scheme being adopted, another very important issue is to guarantee that the named processes actually are what they say they are. In other words, when a process sends a message to another, it must be reasonably sure that the data will actually reach the intended destination instead of a malicious process. Similarly, no malicious processes should be able to look at or, even worse, alter the data while they are in transit.

In the past, this design aspect was generally neglected in most real-time, embedded systems because the real-time communication network was completely closed to the outside world and it was very difficult for a mischievous agent to physically connect to that network and do some damage. Nowadays this is no longer the case because many embedded systems are connected to the public Internet on purpose, for example, for remote management, maintenance, and software updates.

Besides its obvious advantages, this approach has the side effect of opening the real-time network and its nodes to a whole new lot of security threats, which are already well known to most Internet users. Therefore, even if network security as a topic is well beyond the scope of this book and will not be further discussed, it is nonetheless important for embedded system designers to be warned about the issue.

6.3   SYNCHRONIZATION MODEL

As said in the introduction to this chapter, message passing incorporates both data transfer *and* synchronization within the same communication primitives. In all cases, data transfer is accomplished by moving a message from the source to the destination process. However, the synchronization aspects are more complex and subject to variations from one implementation to another.

The most basic synchronization constraint that is always supported is that the `receive` primitive must be able to wait for a message if it is not already available. In most cases, there is also a nonblocking variant of `receive`, which basically checks whether a message is available and, in that case, retrieves it, but never waits if it is not. On the sending side, the establishment of additional synchronization constraints proceeds, in most cases, along three basic schemes:

1. As shown in Figure 6.2, a message transfer is *asynchronous* if the sending process is never blocked by `send` even if the receiving process has not yet executed a matching `receive`. This kind of message transfer gives rise to two possible scenarios:
   - If, as shown in the upper part of the figure, the receiving process *B* executes `receive` before the sending process *A* has sent the message, it will be blocked and it will wait for the message to arrive. The message transfer will take place when *A* eventually sends the message.
   - If the sending process *A* sends the message before the receiving process *B* performs a matching `receive`, the system will buffer the message (typically up to a certain maximum capacity as detailed in Section 6.4), and *A* will continue right away. As shown in the lower part of the figure, the `receive` later performed by *B* will be satisfied immediately in this case.

   The most important characteristic to keep in mind about an asynchronous message transfer is that, when *B* eventually gets a messages from *A*, it does not get any information about what *A* is currently doing because *A* may be executing well beyond its `send` primitive. In other words, an asynchronous message transfer always conveys "out of date" information to the receiver.

2. In a *synchronous* message transfer, also called *rendezvous* and shown in Figure 6.3, there is an additional synchronization constraint, highlighted by a grey oval in the lower part of the figure: if the sending process *A* invokes the `send` primitive when the receiving process *B* has not called `receive` yet, *A* is blocked until *B* does so.

   When *B* is eventually ready to receive the message, the message transfer takes place, and *A* is allowed to continue. As shown in the upper part of the figure, nothing changes with respect to the asynchronous model if the

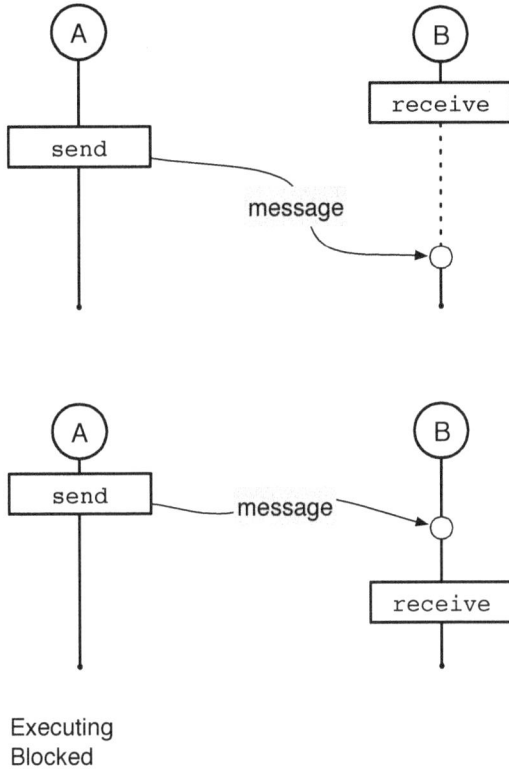

Figure 6.2 Asynchronous message transfer. The sender is never blocked by send even if the receiver is not ready for reception.

receiver is ready for reception when the sender invokes send. In any case, with this kind of message transfer, the receiver *B* can rest assured that the sending process *A* will not proceed beyond its send before *B* has actually received the message.

This difference about the synchronization model has an important impact for what concerns message buffering, too: since in a rendezvous the message sender is forced to wait until the receiver is ready, the system must not necessarily provide any form of intermediate buffering to handle this case. The message can simply be kept by the sender until the receiver is ready and then transferred directly from the sender to the receiver address space.

3. A *remote invocation* message transfer, also known as *extended rendezvous*, is even stricter for what concerns synchronization. As depicted in Figure 6.4, when process *A* sends a request message to process *B*, it is blocked until a reply message is sent back from *B* to *A*.

As the name suggests, this synchronization model is often used to imitate a function call, or invocation, using message passing. As in a regular

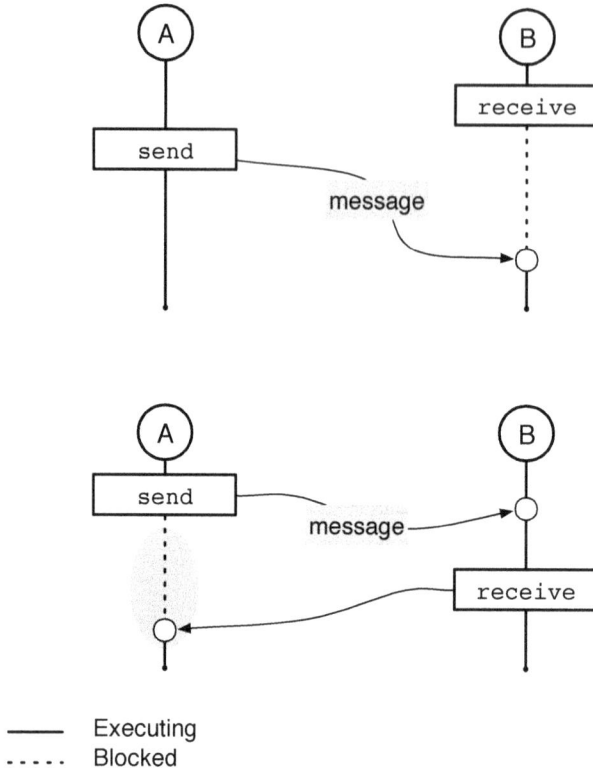

Figure 6.3 Synchronous message transfer, or rendezvous. The sender is blocked by `send` when the receiver is not ready for reception.

function call, the requesting process *A* prepares the arguments of the function it wants process *B* to execute. Then, it puts them into a request message and sends the message to process *B*, often called the *server*, which will be responsible to execute it.

At the same time, and often with the same message passing primitive entailing a combination of both `send` and `receive`, *A* also blocks, waiting for a reply from *B*. The reply will contain any return values resulting from the function execution.

Meanwhile, *B* has received the request and performs a local computation in order to execute the request, compute its results, and eventually generate the reply message. When the reply is ready, *B* sends it to *A* and unblocks it. It should also be noted that the last message is not sent asynchronously, but *B* blocks until the message has been received by *A*. In this way, *B* can make sure that the reply has reached its intended destination, or at least be notified if there was an error.

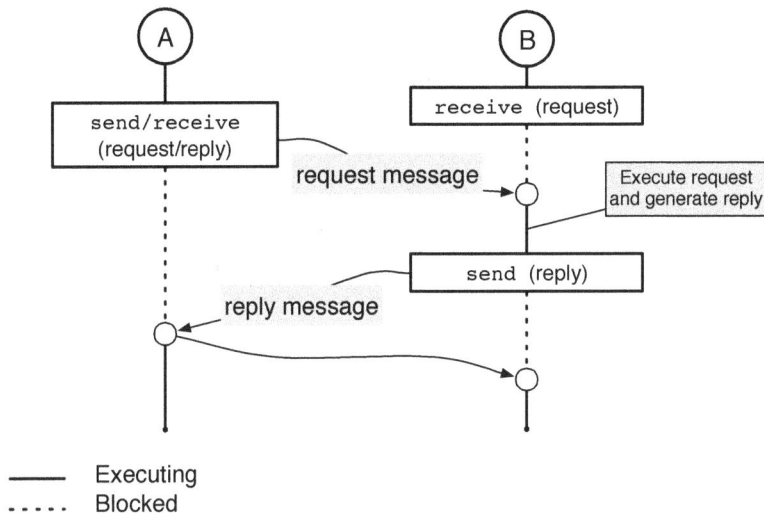

Figure 6.4 Remote invocation message transfer, or extended rendezvous. The sender is blocked until it gets a reply from the receiver. Symmetrically, the receiver is blocked until the reply has successfully reached the original sender.

The synchronization models discussed so far are clearly related to each other. In particular, it is easy to see that all synchronization models can be implemented starting from the first one, that is

- A synchronous message transfer from a process A to another process B can be realized by means of two asynchronous message transfers going in opposite directions. The first transfer (from A to B) carries the actual message to be transferred, and the second one (from B to A) holds an acknowledgment. It should be noted that the second message transfer is not used to actually transfer data between processes but only for synchronization. Its purpose is to block A until B has successfully received the data message.
- A remote invocation from A to B can be based on two synchronous message transfer going in opposite directions as before. The first transfer (from A to B) carries the request, and the second one (from B to A) the corresponding reply. Both being synchronous, the message transfers ensure that neither A nor B is allowed to continue before both the request and the reply have successfully reached their intended destination.

At first sight it may seem that, since an asynchronous message transfer can be used as the "basic building block" to construct all the others, it is the most useful one. For this reason, as will be discussed in Chapters 7 and 8, most operating systems provide just this synchronization model. However, it has been remarked [18] that it has a few drawbacks, too:

- The most important concern is perhaps that asynchronous message transfers give "too much freedom" to the programmer, somewhat like the "goto" statement of unstructured sequential programming. The resulting programs are therefore more complex to understand and check for correctness, also due to the proliferation of explicit message passing primitives in the code.
- Moreover, the system is also compelled to offer a certain amount of buffer for messages that have already been sent but have not been received yet; the amount of buffering is potentially infinite because, in principle, messages can be sent and never received. Most systems only offer a limited amount of buffer, as described in Section 6.4, and hence the kind of message transfer they implement is not truly asynchronous.

6.4 MESSAGE BUFFERS

In most cases, even if message passing occurs among processes being executed on the same computer, the operating system must provide a certain amount of buffer space to hold messages that have already been sent but have not been received yet. As seen in Section 6.3, the only exception occurs when the message transfer is completely synchronous so that the message can be moved directly from the sender to the recipient address space.

The role of buffers becomes even more important when message passing occurs on a communication network. Most network equipment, for example, switches and routers, works according to the store and forward principle in which a message is first received completely from a certain link, stored into a buffer, and then forwarded to its destination through another link. In this case, dealing with one or more buffers is simply unavoidable.

It also turns out that it is not always possible to decide whether a buffer will be useful or not, and how large it should be, because it depends on the application at hand. The following is just a list of the main aspects to be considered for a real-time application.

- Having a large buffer between the sender and the receiver decouples the two processes and, *on average*, makes them less sensitive to any variation in execution and message passing speed. Thus, it increases the likelihood of executing them concurrently without unnecessarily waiting for one another.
- The interposition of a buffer increases the message transfer *delay* and makes it less predictable. As an example, consider the simple case in which we assume that the message transfer time is negligible, the receiver consumes messages at a fixed rate of k messages per second, and there are already m messages in the buffer when the $m + 1$ message is sent. In this case, the receiver will start processing the $m + 1$ message after m/k seconds. Clearly, if m becomes too large for any reason, the receiver will work on "stale" data.
- For some synchronization models, the amount of buffer space required at any given time to fulfill the model may depend on the processes' behavior

and be very difficult to predict. For the purely asynchronous model, the maximum amount of buffer space to be provided by the system may even be *unbounded* in some extreme cases. This happens, for instance, when the sender is faster than the receiver so that it systematically produces more messages than the receiver is able to consume.

For these and other reasons, the approach to buffering differs widely from one message passing implementation to another. Two extreme examples are provided by

1. The local message-passing primitives, discussed in Chapters 7 and 8. Those are intended for use by real-time processes all executing on the same computer.
2. The network communication primitives, discussed in Chapter 9 and intended for processes with weaker real-time requirements, but possibly residing on distinct computers.

In the first case, the focus is on the *predictability* of the mechanism from the point of view of its worst-case communication delay and amount of buffer space it needs. Accordingly, those systems require the user to declare in advance the maximum number of messages a certain mailbox can hold and their maximum size right when the mailbox itself is created.

Then, they implement a variant of the asynchronous communication model, in which the send primitive blocks the caller when invoked on a mailbox that is completely full at the moment, waiting for some buffer space to be available in the future. Since this additional synchronization constraint is not always desirable, they also provide a nonblocking variant of send that immediately returns an error indication instead of waiting.

In the second case, the goal is instead to hide any anomaly in network communication and provide a smooth *average* behavior of the message-passing mechanism. Therefore, each network equipment makes its "best effort" to provide an appropriate buffering, but without giving any absolute guarantee. The most important consequence is that, at least for long-distance connections, it may be very difficult to know for sure how much buffer is being provided, and the amount of buffer may change with time.

6.5 MESSAGE STRUCTURE AND CONTENTS

Regardless of the naming scheme, synchronization model, and kind of buffering being used, understanding *what* kind of data can actually be transmitted within a message with meaningful results is of paramount importance. In an ideal world it would be possible to directly send and receive any kind of data, even of a user-defined type, but this is rarely the case in practice.

The first issue is related to *data representation*: the same data type, for instance the int type of the C language, may be represented in very different ways by the sender and the receiver, especially if they reside on different hosts. For instance, the number of bits may be different, as well as the endianness, depending on the processor architecture. When this happens, simply moving the bits that made up an int data item from one host to another is clearly not enough to ensure a meaningful communication.

A similar issue also occurs if the data item to be exchanged contains *pointers*. Even if we take for granted that pointers have the same representation in both the sending and receiving hosts, a pointer has a well-defined meaning only within its own address space, as discussed in Chapter 2. Hence, a pointer may or may not make sense after message passing, depending on how the sending and receiving agents are related to each other:

1. If they are two threads belonging to the same process (and, therefore, they necessarily reside on the same host), they also live within the same address space, and the pointer will still reference the same underlying memory object.

2. If they are two processes residing on the same host, the pointer will still be meaningful after message passing only under certain very specific conditions, that is, only if their programmers were careful enough to share a memory segment between the two processes, make sure that it is mapped at the same virtual address in both processes, and allocate the referenced object there.

3. If the processes reside on different hosts, there is usually no way to share a portion of address spaces between them, and the pointer will definitely lose its meaning after the transfer.

   Even worse, it may happen that the pointer will still be formally valid in the receiver's context—that is, it will not be flagged as invalid by the memory management subsystem because it falls within the legal boundaries of the address space—but will actually point to a different, and unrelated, object.

In any case, it should also be noted that, even if passing a pointer makes sense (as in cases 1 and 2 above), it implies further memory management issues, especially if memory is dynamically allocated. For instance, programmers must make sure that, when a pointer to a certain object is passed from the sender to the receiver, the object is not freed (and its memory reused) before the receiver is finished with it.

This fact may not be trivial to detect for the sender, which in a sense can be seen as the "owner" of the object when asynchronous or synchronous transfers are in use. This is because, as discussed in Section 6.3, the sender is allowed to continue after the execution of a send primitive even if the receiver either did not get the message (asynchronous transfer) or did not actually work on the message (synchronous transfer) yet.

Since the problem is very difficult to solve in general terms, most operating systems and programming languages leave this burden to the programmer. In other words, in many cases, the message-passing primitives exported by the operating system and available to the programmer are merely able to move a *sequence of bytes* from one place to another.

The programmer is then entirely responsible for making sure that the sequence of bytes can be interpreted by the receiver. This is the case for both POSIX/Linux and FreeRTOS operating systems (discussed in Chapters 7 and 8), as well as the socket programming interface for network communication (outlined in Chapter 9).

6.6 PRODUCER–CONSUMER PROBLEM WITH MESSAGE PASSING

The most straightforward solution to the producer–consumer problem using message passing is shown in Figure 6.5. For simplicity, the example only deals with one producer P and one consumer C, exchanging integer data items no larger than 32 bits. For the same reason, the operations performed to set up the communication path and error checks have been omitted, too.

Despite of the simplifications, the example still contains all the typical elements of message passing. In particular, when the producer P wants to send a certain data item d, it calls the function prod with d as argument to perform the following operations:

- Convert the data item to be sent, d, from the host representation to a neutral representation that both the sender and the receiver understand. This operation is represented in the code as a call to the abstract function host_to_neutral(). For a single, 32-bit integer variable, one sensible choice for a C-language program conforming to the POSIX standard would be, for instance, the function htonl().
- Send the message to the consumer C. A direct, symmetric naming scheme has been adopted in the example, and hence the send primitive names the intended receiver directly with its first argument. The next two arguments are the memory address of the message to be sent and its size.

On the other side, the consumer C invokes the function cons() whenever it is ready to retrieve a message:

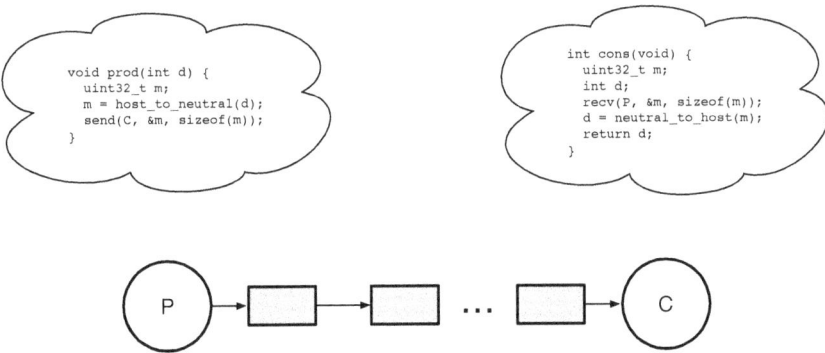

Figure 6.5 A straightforward solution to the producer–consumer problem with synchronous message passing. The same approach also works with asynchronous message passing with a known, fixed amount of buffering.

- The function waits until a message arrives, by invoking the `recv` message-passing primitive. Since the naming scheme is direct and symmetric, the first argument of `recv` identifies the intended sender of the message, that is, *P*. The next two arguments locate a memory buffer in which `recv` is expected to store the received message and its size.
- Then, the data item found in the message just received is converted to the host representation by means of the function `neutral_to_host()`. For a single, 32-bit integer variable, a suitable POSIX function would be `ntohl()`. The result d is returned to the caller.

Upon closer examination of Figure 6.5, it can be seen that the code just described gives rise to a unidirectional flow of messages, depicted as light grey boxes, from *P* to *C*, each carrying one data item. The absence of messages represents a synchronization condition because the consumer *C* is forced to wait within `cons()` until a message from *P* is available.

However, if we compare this solution with, for instance, the semaphore-based solution shown in Figure 5.12 in Chapter 5, it can easily be noticed that another synchronization condition is amiss. In fact, in the original formulation of the producer–consumer problem, the producer *P* must wait if there are "too many" messages already enqueued for the consumer. In Figure 5.12, the exact definition of "too many" is given by N, the size of the buffer interposed between producers and consumers.

Therefore, the solution just proposed is completely satisfactory—and matches the previous solutions, based on other interprocess synchronization mechanisms—only if the second synchronization condition is somewhat provided implicitly by the message-passing mechanism itself. This happens when the message transfer is synchronous, implying that there is no buffer at all between *P* and *C*.

An asynchronous message transfer can also be adequate if the maximum amount of buffer provided by the message-passing mechanism is known and fixed, and the send primitive blocks the sender when there is no buffer space available.

If only asynchronous message passing is available, the second synchronization condition must be implemented explicitly. Assuming that the message-passing mechanism can successfully buffer at least N messages, a second flow of empty messages that goes from *C* to *P* and only carries synchronization information is adequate for this, as shown in Figure 6.6. In the figure, the additional code with respect to Figure 6.5 is highlighted in bold. The data type `empty_t` represents an empty message. With respect to the previous example,

- The consumer *C* sends an empty message to *P* after retrieving a message from *P* itself.
- The producer *P* waits for an empty message from the consumer *C* before sending its own message to it.
- By means of the initialization function `cons_init()`, the consumer injects N empty messages into the system at startup.

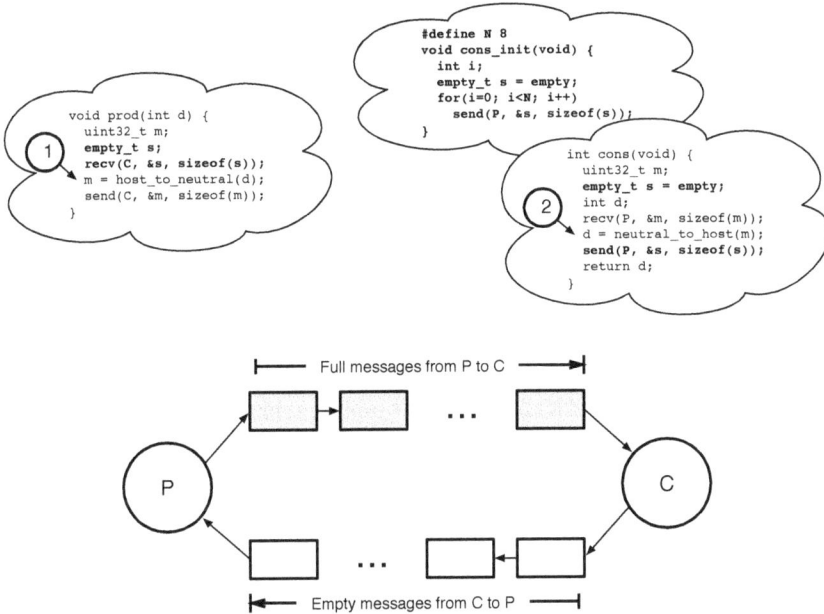

Figure 6.6 A more involved solution to the producer–consumer problem based on asynchronous message passing. In this case, the synchronization condition for the producer P is provided explicitly rather than implicitly.

At startup, there are therefore N empty messages. As the system evolves, the total number of empty plus full messages is constant and equal to N because one empty (full) message is sent whenever a full (empty) message is retrieved. The only transient exception happens when the producer or the consumer are executing at locations 1 and 2 of Figure 6.6, respectively. In that case, the total number of messages can be $N-1$ or $N-2$ because one or two messages may have been received by P and/or C and have not been sent back yet.

In this way, C still waits if there is no full message from P at the moment, as before. In addition, P also waits if there is no empty message from C. The total number of messages being constant, this also means that P already sent N full messages that have not yet been handled by C.

6.7 SUMMARY

In this chapter we learnt that *message passing* is a valid alternative to interprocess communication based on shared variables and synchronization devices because it encompasses both data transfer and synchronization in the same set of primitives.

Although the basics of message passing rely on two intuitive and simple primitives, *send* and *receive*, there are several design and implementation variations worthy of attention. They fall into three main areas:

1. How to identify, or *name*, message senders and recipients;
2. What kind of *synchronization* constraints the send and receive primitives enforce;
3. How much *buffer* space, if any, is provided by the message-passing mechanism.

Moreover, to use message passing in a correct way, it is of paramount importance to ensure that messages retain their meaning after they are transferred from one process or thread to another. Especially when working with a distributed system in which the application code is executed by many agents spread across multiple hosts, issues such as *data representation* discrepancies among computer architectures as well as loss of *pointer validity* across distinct address spaces cannot be neglected.

Then, message passing has been applied to the well-known producer–consumer problem to show that its use leads to a quite simple and intuitive solution. The example also highlighted that, in some cases, it may be appropriate to introduce a message stream between processes, even if no data transfer is required, as a way to guarantee that they synchronize in the right way.

For the sake of completeness, it should also be remarked that the message-addressing scheme presented in this chapter, based on explicitly naming the source and recipient of a message, is not the only possible one. A popular alternative— quite common in real-time networks based on an underlying communication medium that supports broadcast transmission—is to adopt the so-called *Publish/Subscribe* scheme.

With this approach, the sending processes do not explicitly name any intended receiver. Rather, they attach a tag to each message that specifies the message *class* or *contents* rather than recipients. The message is then *published*, often by broadcasting it on the network, so that any interested party can get it. In turn, each receiving process *subscribes* to the message classes it is interested in so that it only receives and acts upon messages belonging to those specific classes.

A full discussion of the Publish/Subscribe scheme is beyond the scope of this book. Interested readers can refer, for example, to Reference [32] for a thorough discussion of this addressing scheme in the context of the Controller Area Network [53, 54].

6.8 EXERCISES

EXERCISE 1

Show how message passing can be used to let a process P wait for events generated by n other processes $Q_1, \ldots, Q_n$. The function that P invokes to wait for events must also return the identity k of the process Q_k that generated the event.

EXERCISE 2

Describe how a message queue with a 1-element buffer can be used to implement mutual exclusion among n processes.

EXERCISE 3

Suppose that process G inserts a message into a message queue periodically, every t_G. At time t_S another process S extracts a message M from the queue and learns that, immediately before the extraction, there were $k \geq 1$ messages in the queue. When did G generate M, assuming there were no queue overflows?

7 Interprocess Communication Primitives in POSIX/Linux

Chapters 5 and 6 have introduced the basic concepts of interprocess communication, called **IPC** in the following text. These mechanisms are implemented by operating systems to allow the correct execution of processes. Different operating systems define different IPC interfaces, often making the porting of applications to different platforms a hard task. In this chapter, the interprocess communication primitives are presented for the Linux operating system, for which the Application Programming Interface (API) has been standardized in POSIX. POSIX, which stands for "Portable Operating System Interface [for Unix]," is a family of standards specified by the IEEE to define common APIs along variants of the Unix operating system, including Linux. Until recently, the POSIX API was regulated by IEEE Standard 1003.1 [46]. It was later replaced by the joint ISO/IEC/IEEE Standard 9945 [52].

In the following sections we shall see how semaphores, message queues, and other interprocess communication mechanisms are presented in Linux under two different contexts: *process* and *thread*. This is a fundamental distinction that has many implications in the way IPC is programmed and that may heavily affect performance. The first section of this chapter will describe in detail what the differences are between the two configurations, as well as the pros and cons of each solution. The following two sections will present the interprocess mechanisms for Linux threads and processes and the last section will then introduce some Linux primitives for the management of clocks and timers, an important aspect when developing programs that interact with the outside world in embedded applications.

7.1 THREADS AND PROCESSES

Chapter 3 introduced the concept of process, which can be considered an independent flow of execution for a program. The operating system is able to manage multiple processes, that is, the concurrent execution of multiple programs, even if the underlying computer has a single processor. The management of multiple processes on a single processor computer relies on two main facts:

1. A program does not always require the processor: we have seen in Chapter 2 that, when performing an I/O operation, the processor must await the termination of the data transfer between the device and memory. In the meantime, the operating system can assign the processor to another process that is ready for computation.

DOI: 10.1201/9781003593416-7

2. Even in the case where a program does not make I/O operations, not releasing the processor, the operating system can decide to reclaim the processor and assign it to another ready program in order to guarantee the fair execution of the active processes.

The **Scheduler** is the component of the operating system that supervises the assignment of the processor to processes. Chapter 11 will describe in detail the various scheduling algorithms that represent a very important aspect of the system behavior since it determines how the computer reacts to external events. The transfer of processor ownership is called *Context Switch*, and we have already seen in Chapter 3 that there exists a set of information that needs to be saved/restored every time the processor is moved from one process to another, among which,

1. The saved value of the processor registers, including
   - the *Program Counter*, that is, the address of the next machine instruction to be executed by the program;
   - the *Stack Pointer*, that is, the address of the stack in memory containing the local program variables and the arguments of all the active procedures of the program at the time the scheduler reclaims the processor.
2. The descriptors of the files and devices currently opened by the program.
3. The page table content for that program. We have seen in Chapter 2 that, when virtual memory is supported by the system, the memory usage of the program is described by a set of page table entries that specify how virtual addresses are translated into physical addresses. In this case, the context switch changes the memory mapping and avoids the new process overwriting sections of memory used by the previous one.
4. Process-specific data structures maintained by the operating system to manage the process.

The amount of information to be saved for the process losing the processor and to be restored for the new process can be large, and therefore many processor cycles may be spent at every context switch. Very often, most of the time spent at the context switch is due to saving and restoring the page table since the page table entries describe the possibly large number of memory pages used by the process. For the same reason, creating new processes involves the creation of a large set of data structures.

The above facts are the main reason for a new model of computation represented by *threads*. Conceptually, threads are not different from processes because both entities provide an independent flow of execution for programs. This means that all the problems, strategies, and solutions for managing concurrent programming apply to processes as well as to threads. There are, however, several important differences due to the amount of information that is saved by the operating system in context switches. Threads, in fact, live in the context of a process and share most process-specific information, in particular memory mapping. This means that the threads that are activated within a given process share the same memory space and the same files and devices. For this reason, threads are sometimes called "lightweight processes." Figure 7.1 shows on the left the information forming the process context.

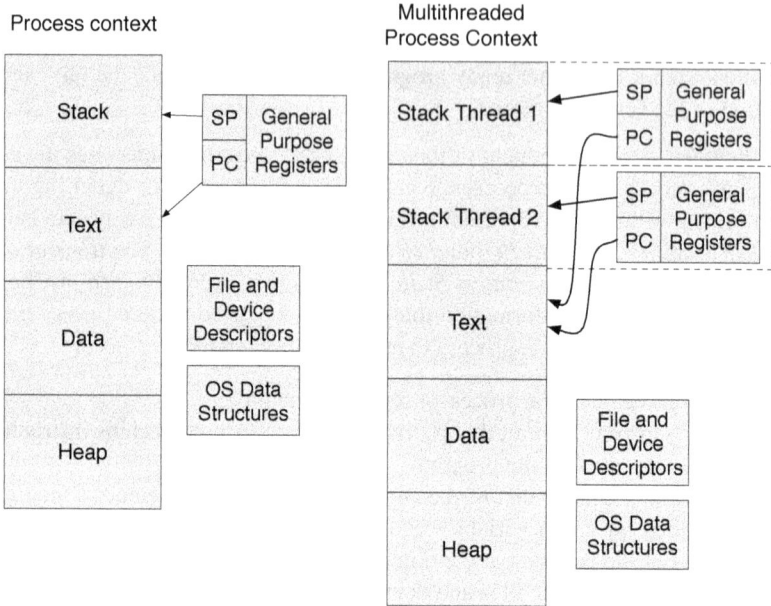

Figure 7.1   Process and Thread contexts.

The memory assigned to the process is divided into

- **Stack**, containing the private (sometimes called also automatic) variables and the arguments of the currently active routines. Normally, a processor register is designated to hold the address of the top of the stack;
- **Text**, containing the machine code of the program being executed. This area is normally only read;
- **Data**, containing the data section of the program. Static C variables and variables declared outside the routine body are maintained in the data section;
- **Heap**, containing the dynamically allocated data structures. Memory allocated by C *malloc()* routine or by the *new* operator in C++ belong to the heap section.

In addition to the memory used by the program, the process context is formed by the content of the registers, the descriptors for the open files and devices, and the other operating system structures maintained for that process. On the right of Figure 7.1 the set of information for a process hosting two threads is shown. Note that the Text, Data, and Heap sections are the same for both threads. Only the Stack memory is replicated for each thread, and the thread-specific context is only formed by the register contents. The current content of the processor registers, in fact, represents a snapshot of the program activity at the time the processor is removed by the scheduler from one thread to be assigned to another one. In particular, the stack pointer register contains the address of the thread-specific stack, and the program counter contains the address of the next instruction to be executed by the program. As the memory-mapping information is shared among the threads belonging to the same process as well as the open files and devices, the set of registers basically represents the only information to be saved in a context switch. Therefore, unless a thread from a different process is activated, the time required for a context switch between threads is much shorter compared to the time required for a context switch among processes.

7.1.1 CREATING THREADS

Historically, hardware vendors have implemented proprietary versions of threads, making it difficult for programmers to develop threaded applications that could be portable across different systems. For this reason, a standardized interface has been specified by the IEEE POSIX 1003.c standard in 1995, and an API for POSIX threads, called *Pthreads*, is now available on every UNIX system including Linux. The C types and routine prototypes for threads are defined in the pthread.h header file.
The most important routine is:

```
int pthread_create(thread_t *thread, pthread_attr_t *attr,
        void *(*start_routine)(void*), void *arg)
```

which creates and starts a new thread. Its arguments are the following:

- thread: the returned identifier of the created thread to be used for subsequent operations. This is of type thread_t which is opaque, that is, the programmer has no knowledge of its internal structure, this being only meaningful to the pthread routines that receive it as argument.
- attr: the attributes of the thread. Attributes are represented by the opaque type pthread_attr_t.
- start_routine: the routine to be executed by the thread.
- arg: the pointer argument passed to the routine.

All the pthread routines return a status that indicates whether the required action was successful: all functions return 0 on success and a nonzero error code on error. Since the data type for the attribute argument is opaque, it is not possible to define directly its attribute fields, and it is necessary to use specific routines for this purpose. For example, one important attribute of the thread is the size of its stack: if the stack is not large enough, there is the risk that a stack overflow occurs especially when the program is using recursion. To prepare the attribute argument specifying a given stack size, it is necessary first to initialize a pthread_attr_t parameter with default setting and then use specific routines to set the specific attributes. After having been used, the argument should be disposed. For example, the following code snippet initializes a pthread_attr parameter and then sets the stack size to 4 MByte (the default stack size on Linux is normally 1 MByte for 32 bit architectures, and 2 MByte for 64 bit architectures).

```
pthread_attr_t atrr;
//Attribute initialization
pthread_attr_init (&attr);
//Set stack size to 4 MBytes
pthread_attr_setstacksize(&attr,  0x00400000);
...
//Use attr in thread creation
...
//Dispose attribute parameter
pthread_attr_destroy(&attr);
```

When NULL is passed as the second argument of `pthread_create()`, the default setting for thread arguments is used: this is the most common case in practice unless specific settings are required.

Only one pointer parameter can be passed to the thread routine. When more than one argument have to be passed, the common practice is to allocate in memory a structure containing all the information to be passed to the routine thread and then to pass the pointer to such structure.

As soon as a thread has been created, it starts execution in parallel with the process, or thread, that called `pthread_create()`. It is often necessary to synchronize the program with the other threads, making sure that all the created threads have finished their execution before a given point in the code is reached. For example, it is necessary to know when the threads have terminated before starting using the results computed by them. The following routine allows one to wait for the termination of a given thread specified by its thread identifier:

```
int pthread_join(pthread_t thread, void **value_ptr);
```

The second argument, when non-NULL, is the pointer to the returned value of the thread. A thread may return a value either when the code terminates with a `return` statement or when `pthread_exit(void *value)` is called. The latter is preferable especially when many threads are created and terminated because `pthread_exit()` frees the internal resources allocated for the thread.

Threads can either terminate spontaneously or be canceled. Extreme care is required when canceling threads because an abrupt termination may lead to inconsistent data, especially when the thread is sharing data structures. Even worse, a thread may be canceled in a critical section: If this happens, no other thread will ever be allowed to enter that section. For this reason, POSIX defines the following routines to handle thread cancelation:

```
int pthread_setcancelstate(int state, int *oldstate)
```

```
void pthread_cleanup_push(void (*routine)(void*), void *arg)
```

`pthread_setcancelstate()` enables or disables run time the possibility of canceling the calling thread, depending on the value of the passed state argument which can be either PTHREAD_CANCEL_ENABLE or PTHREAD_CANCEL_DISABLE. The previous cancelability state is returned in oldstate. For example, when entering a critical section, a thread may disable cancelation in order to avoid preventing that critical section to other threads.

pthread_cleanup_push() allows registering a routine that is then automatically invoked upon thread cancelation. This represents another way to handle the proper release of the allocated resources in case a thread is canceled.

Finally, a thread is canceled by routine:

```
int pthread_cancel(pthread_t thread)
```

By means of a proper organization of the code, it is possible to avoid using the above routines for terminating threads. For example, it is possible to let the thread routine periodically check the value of some shared flag indicating the request to kill the thread: whenever the flag becomes true, the thread routine exits, after the proper cleanup actions.

The following code example creates a number of threads to carry out the computation of the sum of all the elements of a very large square matrix. This is achieved by assigning each thread a different portion of the input matrix. After creating all the threads, the main program waits the termination of all of them and makes the final summation of all the partial results reported by the different threads. The code is listed below:

```
#include <pthread.h>
#include <stdio.h>
#include <stdlib.h>
#include <sys/time.h>

#define MAX_THREADS 256
#define ROWS 10000
#define COLS 10000

/* Arguments exchanged with threads */
struct argument{
  int startRow;
  int nRows;
  long partialSum;
} threadArgs[MAX_THREADS];

/* Matrix pointer: it will be dynamically allocated */
long *bigMatrix;

/* Thread routine: make the summation of all the elements of the
   assigned matrix rows */
static void *threadRoutine(void *arg)
{
  int i, j;
/* Type-cast passed pointer to expected structure
   containing the start row, the number of rows to be summed
   and the return sum argument */
  struct argument *currArg = (struct argument *)arg;
  long sum = 0;
  for(i = 0; i < currArg->nRows; i++)
    for(j = 0; j < COLS; j++)
      sum += bigMatrix[(currArg->startRow + i) * COLS + j];
  currArg->partialSum = sum;
  return NULL;
```

```
}
int main(int argc, char *args[])
{
/* Array of thread identifiers */
  pthread_t threads[MAX_THREADS];
  long totalSum;
  int i, j, nThreads, rowsPerThread, lastThreadRows;
/* Get the number of threads from command parameter */
  if(argc != 2)
  {
    printf("Usage: threads <numThreads>\n");
    exit(0);
  }
  sscanf(args[1], "%d", &nThreads);
/* Allocate  the matrix M */
  bigMatrix = malloc(ROWS*COLS*sizeof(long));
/* Fill the matrix with some values */
  ...

/* If the number of rows cannot be divided exactly by the number of
   threads, let the last thread handle also the remaining rows */
  rowsPerThread = ROWS / nThreads;
  if(ROWS % nThreads == 0)
    lastThreadRows = rowsPerThread;
  else
    lastThreadRows = rowsPerThread + ROWS % nThreads;

/* Prepare arguments for threads */
  for(i = 0; i < nThreads; i++)
  {
/* Prepare Thread arguments */
    threadArgs[i].startRow = i*rowsPerThread;
    if(i == nThreads - 1)
      threadArgs[i].nRows = lastThreadRows;
    else
      threadArgs[i].nRows = rowsPerThread;
  }
/* Start the threads using default thread attributes */
  for(i = 0; i < nThreads; i++)
    pthread_create(&threads[i], NULL, threadRoutine, &threadArgs[i]);

/* Wait thread termination and use the corresponding
   sum value for the final summation */
  totalSum = 0;
  for(i = 0; i < nThreads; i++)
  {
    pthread_join(threads[i], NULL);
    totalSum += threadArgs[i].partialSum;
  }
}
```

In the foregoing code there are several points worth examining in detail. First of all, the matrix is declared outside the body of any routine in the code. This means that the memory for it is not allocated in the Stack segment but in the Heap segment, being dynamically allocated in the main program. This segment is shared by every thread (only the stack segment is private for each thread). Since the matrix is accessed only in read mode, there is no need to consider synchronization. The examples in the next section will present applications where the shared memory is accessed for both reading and writing, and, in this case, additional mechanisms for ensuring data coherence will be required. Every thread needs two parameters: the row number of the first element of the set of rows assigned to the thread, and the number of rows to be considered. Since only one pointer argument can be passed to threads, the program creates an array of data structures in shared memory, each containing the two arguments for each thread, plus a third return argument that will contain the partial sum, and then passes the pointer of the corresponding structure to each thread . Finally, the program awaits the termination of the threads by calling in `pthread_join()` in a loop with as many iterations as the number of activated threads. Note that this works also when the threads terminate in an order that is different from the order `pthread_join()` is called. In fact, if `pthread_join()` is called for a thread that has already terminated, the routine will return soon with the result value passed by the thread to `pthread_exit()` and maintained temporarily by the system. In the program, the partial sum computed by each thread is stored in the data structure used to exchange the thread routine argument, and therefore the second parameter of `pthread_join()` is null, and `pthread_exit()` is not used in the thread routine.

In the above example, the actions carried out by each thread are purely computational. So, with a single processor, there is no performance gain in carrying out computation either serially or in parallel because every thread requires the processor 100% of its time and therefore cannot proceed when the processor is assigned to another thread. Modern processors, however, adopt a multicore architecture, that is, host more than one computing unit in the processor, and therefore, there is a true performance gain in carrying out computation concurrently. Figure 7.2 shows the execution time for the above example at an increasing number of threads on an 8-core processor. The execution time halves passing from one to two threads, and the performance improves introducing additional threads. When more than 8 threads are used, the performance does not improve any further; rather it worsens slightly. In fact, when more threads than available cores are used in the program, there cannot be any gain in performance because the thread routine does not make any I/O operation and requires the processor (core) during all its execution. The slight degradation in

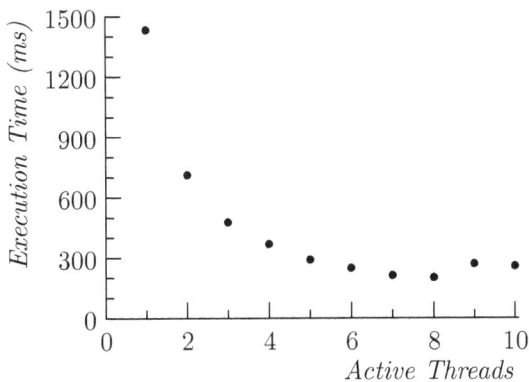

Figure 7.2 Execution time of the large matrix summation for an increasing number of executor threads on an 8-core processor.

performance is caused by the added overhead in the context switch due to the larger number of active threads.

The improvement in execution speed due to multithreading becomes more evident when the program being executed by threads makes I/O operations. In this case, the operating system is free to assign the processor to another thread when the current thread starts an I/O operation and needs to await its termination. For this reason, if the routines executed by threads are I/O intensive, adding new threads still improves performance because this reduces the chance that the processor idles awaiting the termination of some I/O operation. Observe that even if no I/O operation is executed by the thread code, there is a chance that the program blocks itself awaiting the completion of an I/O operation in systems supporting memory paging. When paging in memory, pages of the active memory for processes can be held in secondary memory (i.e., on disk), and are transferred (swapped in) to RAM memory whenever they are accessed by the program, possibly copying back (swapping out) other pages in memory to make room for them. Paging allows handling a memory that is larger than the RAM memory installed in the computer, at the expense of additional I/O operations for transferring memory pages from/to the disk.

Threads represent entities that are handled by the scheduler and, from this point of view, do not differ from processes. In fact, the difference between processes and threads lies only in the actions required for the context switch, which is only a subset of the process-specific information if the processing unit is exchanged among threads of the same process. The following chapters will describe in detail how a scheduler works, but here we anticipate a few concepts that will allow us to understand the pthread API for controlling thread scheduling.

We have already seen in Chapter 3 that, at any time, the set of active processes (and threads) can be partitioned in two main categories:

- *Ready processes*, that is, processes that could use the processor as soon as it is assigned to them;
- *Waiting processes*, that is, processes that are waiting for the completion of some I/O operation, and that could not make any useful work in the meantime.

Processes are assigned a priority: higher-priority processes are considered "more important," and are therefore eligible for the possession of the processor even if other ready processes with lower priority are present. The scheduler organizes ready processes in queues, one for every defined priority, and assigns the processor to a process taken from the nonempty queue with the highest priority. Two main scheduling policies are defined:

1. **First In/First Out (FIFO)**: The ready queue is organized as a FIFO queue, and when a process is selected to run it will execute until it terminates or enters in wait state due to a I/O operation, or a higher priority process becomes ready.

2. **Round Robin (RR)**: The ready queue is still organized as a FIFO queue, but after some amount of time (often called *time slice*), the running process is preempted by the scheduler even if no I/O operation is performed and no higher priority process is ready, and inserted at the tail of the corresponding queue. With regard to the FIFO policy, this policy ensures that all the processes with the highest priority have a chance of being assigned processor time, at the expense, however, of more overhead due to the larger number of context switches.

Scheduling policy represents one of the elements that compose the thread's attributes, passed to routine `pthread_create()`. We have already seen that the thread's attributes are represented by an opaque type and that a set of routines are defined to set individual attributes. The following routine allows for defining the scheduling policy:

```
int pthread_attr_setschedpolicy(pthread_attr_t *attr, int policy);
```

where policy is either SCHED_FIFO, SCHED_RR, or SCHED_OTHER. SCHED_OTHER can only be used at static priority 0 and represents the standard Linux time-sharing scheduler that is intended for all processes that do not require special static priority real-time mechanisms. The above scheduling policies do not represent the only possible choices and the second part of this book will introduce different techniques for scheduling processes in real-time systems.

Thread priority is finally defined for a given thread by routine:

```
int pthread_setschedprio(pthread_t thread, int prio);
```

7.1.2 CREATING PROCESSES

The API for creating Linux processes is deceptively simple, formed by one system routine with no arguments:

```
pid_t fork()
```

If we compare this with the much richer pthreads API, we might be surprised from the fact that there is no way to define a specific program to be executed and to pass any arguments to it. What `fork()` actually does is just to create an exact clone of the calling process by replicating the memory content of the process and the associated structures, including the current value of the processor registers. When `forks()` returns, two identical processes at the same point of execution are present in the system (one of the duplicated processor registers is, in fact, the Program Counter that holds the address of the next instruction in the program to be executed). There is only one difference between the two: the return value of routine `fork()` is set to 0 in the created process, and to the identifier of the new process in the original process. This allows discriminating in the code between the calling and the created process, as shown by the following code snippet:

```
#include <sys/types.h>
#include <unstd.h>
//Required include files
...
pid_t pid;
...
pid = fork();
if(pid == 0)
{
//Actions for the created process
}
else
{
//Actions for the calling process
}
```

The created process is a *child* process of the creating one and will proceed in parallel with the latter. As for threads, if processes are created to carry out a collaborative work, it is necessary that, at a certain point, the creator process synchronizes with its child processes. The following system routine will suspend the execution of the process until the child process, identified by the process identifier returned by fork(), has terminated.

```
pid_t wait(pid_t pid, int *status, int options)
```

Its argument status, when non-NULL, is a pointer to an integer variable that will hold the status of the child process (e.g., if the child process terminated normally or was interrupted). Argument options, when different from 0, specifies more specialized wait options.

If processes are created to carry out collaborative work, it is necessary that they share memory segments in order to exchange information. While with threads every memory segment different from the stack was shared among threads, and therefore it suffices to use static variables to exchange information, the memory allocated for the child process is by default separate from the memory used by the calling process. We have, in fact, seen in Chapter 2 that in operating systems supporting virtual memory (e.g., Linux), different processes access different memory pages even if using the same virtual addresses, and that this is achieved by setting the appropriate values in the Page Table at every context switch. The same mechanism can, however, be used to provide controlled access to segments of shared memory by setting appropriate values in the page table entries corresponding to the shared memory pages, as shown in Figure 2.8 in Chapter 2. The definition of a segment of shared memory is done in Linux in two steps:

1. A segment of shared memory of a given size is created via system routine shmget();

2. A region of the virtual address space of the process is "attached" to the shared memory segment via system routine shmat().

The prototype of shmget() routine is

```
int shmget(key_t key, size_t size, int shmflg)
```

where key is the unique identifier of the shared memory segment, size is the dimension of the segment, and shmflags defines the way the segment is created or accessed. When creating a shared memory segment, it is necessary to provide a unique identifier to it so that the same segment can be referenced by different processes. Moreover, the shared memory segment has to be created only the first time shmget() is called, and the following times it is called by different processes with the same identifier, the memory segment is simply referenced. It is, however, not always possible to know in advance if the specified segment of shared memory has already been created by another process. The following code snippet shows how to handle such a situation. It shows also the use of system routine ftok() to create an identifier for shmget() starting from a numeric value, and the use of shmat() to associate a range of virtual addresses with the shared memory segment.

```
#include <sys/ipc.h>
#include <sys/shm.h>
# include <sys/types.h>

/* The numeric identifier of the shared memory segment
   the same value must be used by all processes sharing the segment */
#define MY_SHARED_ID 1
...
key_t key;        //Identifier to be passed to shmget()
int memId;        //The id returned by shmget() to be passed to shmat()
void *startAddr;  //The start address of the shared memory segment
...
/* Creation of the key. Routine ftok() function uses the identity
   of the file path passed as first argument (here /tmp is used, but it
   may refer to any existing file in the system) and the least
   significant 8 bits of the second argument */
key_t key = ftok("/tmp", MY_SHARED_ID);

/* First try to create a new memory segment. Flags define exclusive
   creation, i.e. if the shared memory segment already exists, shmget()
   returns with an error */
memId = shmget(key, size, IPC_CREAT | IPC_EXCL);
if(memId == -1)
/* Exclusive creation failed, the segment was already created by
   another process */
{
/* shmget() is called again without the CREATE option */
    memId = shmget(key, size, 0);
}
/* If memId == -1 here, an error occurred in the creation of
   the shared memory segment */
if(memId != -1)
{
/* Routine shmat() maps the shared memory segment to a range
   of virtual addresses */
    startAddr = (char *)shmat(memId, NULL, 0666);
/* From now, memory region pointed by startAddr
   is the shared segment */
...
```

Table 7.1
Protection bitmask

Operation and permissions	Octal value
Read by user	00400
Write by user	00200
Read by group	00040
Write by group	00020
Read by others	00004
Write by others	00002

In the case where the memory region is shared by a process and its children processes, it is not necessary to explicitly define shared memory identifiers. In fact, when a child process is created by fork(), it inherits the memory segments defined by the parent process. So, in order to share memory with children processes, it suffices, before calling fork(), to create and map a new shared memory segment passing constant ICP_PRIVATE as the first argument of shmget(). The memory Identifier returned by shmget() will then be passed to shmat(), which will in turn return the starting address of the shared memory. When the second argument of shmat() is NULL (the common case), the operating system is free to choose the virtual address range for the shared memory. The third argument passed to shmat() specifies in a bitmask the level of protection of the shared memory segment, and is normally expressed in octal value as shown in Table 7.1. Octal value 0666 will specify read-and-write access for all processes. The following example, performing the same computation of the example based on threads in the previous section, illustrates the use of shared memory among children processes.

```
#include <stdio.h>
#include <stdlib.h>
#include <sys/time.h>
#include <sys/ipc.h>
#include <sys/shm.h>
#include <sys/wait.h>
#define MAX_PROCESSES 256
#define ROWS 10000L
#define COLS 10000L

/* Arguments exchanged with child processes */
struct argument{
  int startRow;
  int nRows;
  long partialSum;
};
/* The shared memory contains the arguments exchanged between parent
   and child processes and is pointer by processArgs */
struct argument *processArgs;

/* Matrix pointer: it will be dynamically allocated */
long *bigMatrix;

/* The current process index, incremented by the parent process before
```

```
    every fork() call. */
int currProcessIdx;

/* Child process routine: make the summation of all the elements of the
   assigned matrix rows. */
static void processRoutine()
{
  int i, j;
  long sum = 0;

/* processArgs is the pointer to the shared memory inherited by the
   parent process. processArg[currProcessIdx] is the argument
   structure specific to the child process */
  for(i = 0; i < processArgs[currProcessIdx].nRows; i++)
    for(j = 0; j < COLS; j++)
      sum += bigMatrix[(processArgs[currProcessIdx].startRow + i) * COLS
                        + j];
/* Report the computed sum into the argument structure */
  processArgs[currProcessIdx].partialSum = sum;
}


int main(int argc, char *args[])
{
  int memId;
  long totalSum;
  int i, j, nProcesses, rowsPerProcess, lastProcessRows;
/* Array of process identifiers used by parent process in the wait cycle */
  pid_t pids[MAX_PROCESSES];

/* Get the number of processes from command parameter */
  if(argc != 2)
  {
    printf("Usage: processs <numProcesses>\n");
    exit(0);
  }
  sscanf(args[1], "%d", &nProcesses);
/* Create a shared memory segment to contain the argument structures
   for all child processes. Set Read/Write permission in flags argument. */
  memId = shmget(IPC_PRIVATE, nProcesses * sizeof(struct argument), 0666);
  if(memId == -1)
  {
    perror("Error in shmget");
    exit(0);
  }
/* Attach the shared memory segment. Child processes will inherit the
   shared segment already attached */
  processArgs = shmat(memId, NULL, 0);
  if(processArgs == (void *)-1)
  {
    perror("Error in shmat");
    exit(0);
  }

/* Allocate  the matrix M */
  bigMatrix = malloc(ROWS*COLS*sizeof(long));
/* Fill the matrix with some values */
  ...

/* If the number of rows cannot be divided exactly by the number of
   processs, let the last thread handle also the remaining rows  */
  rowsPerProcess = ROWS / nProcesses;
  if(ROWS % nProcesses == 0)
    lastProcessRows = rowsPerProcess;
  else
    lastProcessRows = rowsPerProcess + ROWS % nProcesses;
```

```
/* Prepare arguments for processes */
  for(i = 0; i < nProcesses; i++)
  {
    processArgs[i].startRow = i*rowsPerProcess;
    if(i == nProcesses - 1)
      processArgs[i].nRows = lastProcessRows;
    else
      processArgs[i].nRows = rowsPerProcess;
  }

/* Spawn child processes */
  for(currProcessIdx = 0; currProcessIdx < nProcesses; currProcessIdx++)
  {
    pids[currProcessIdx] = fork();
    if(pids[currProcessIdx] == 0)
    {
/* This is the child process which inherits a private copy of all
   the parent process memory except for the region pointed by
   processArgs which is shared with the parent process */
      processRoutine();
/* After computing partial sum the child process exits */
      exit(0);
    }
  }
/* Wait termination of child processes and perform final summation */
  totalSum = 0;
  for(currProcessIdx = 0; currProcessIdx < nProcesses; currProcessIdx++)
  {
/* Wait child process termination */
    waitpid(pids[currProcessIdx], NULL, 0);
    totalSum += processArgs[currProcessIdx].partialSum;
  }
}
```

From a programming point of view, the major conceptual difference with the thread-based example is that parameters are not explicitly passed to child processes. Rather, a variable within the program (currProcessIdx) is set to the index of the child process just before calling fork() so that it can be used in the child process to select the argument structure specific to it.

The attentive reader may be concerned about the fact that, since fork() creates a clone of the calling process including the associated memory, the amount of required memory in the above example may be very high due to the fact that the main process has allocated in memory a very large matrix which is duplicated in every create subprocess. Fortunately this is not the case because the memory pages in the child process are not physically duplicated. Rather, the corresponding page table entries in the child process refer to the same physical pages of the parent process and are marked as *Copy On Write*. This means that, whenever the page is accessed in read mode, both the parent and the child process refer to the same physical page, and only upon a write operation is a new page in memory created and mapped to the child process. So, pages that are only read by the parent and child processes, such as the memory pages containing the program code, are not duplicated at all. In our example, the big matrix is written only before creating child processes, and therefore, the memory pages for it are never duplicated, even if they are conceptually replicated for every process. Nevertheless, process creation and context switches require more time in respect of threads because more information, including the page table, has to be saved and restored at every context switch.

Routines `shmget()` and `shmat()`, now incorporated into POSIX, derive from the System V interface, one of the two major "flavors" of UNIX, the other being Berkeley Unix (BSD). POSIX defines also a different interface for creating named shared memory objects, that is, the routine `sem_open()`. The arguments passed to `sem_open()` specify the systemwide name of the shared memory object and the associated access mode and protection. In this case, routine `mmap()`, which has been encountered in Chapter 2 for mapping I/O into memory, is used to map the shared memory object onto a range of process-specific virtual addresses.

7.2 INTERPROCESS COMMUNICATION AMONG THREADS

In the previous example, the threads and processes were either reading the shared memory or writing it at disjoint addresses (the shared arguments containing the partial sums computed by threads/processes). For this reason, there was no need to ensure synchronization because the shared information was correctly managed regardless of the possible interleaving in read actions by means of threads/processes. We have seen in Chapter 5 that, in the more general case in which shared data are also *written* by threads/processes, using shared memory alone does not guarantee against possible errors due to the interleaved access to the shared data structures. Therefore, it is necessary to provide some sort of mutual exclusion in order to protect critical data structures against concurrent access. The POSIX `pthread` interface provides two mechanisms to manage synchronization among threads: *Mutexes* and *Condition Variables*.

7.2.1 MUTEXES AND CONDITION VARIABLES

Mutex is an abbreviation for "mutual exclusion," and mutex variables are used for protecting shared data when multiple writes occur by letting at the most one thread at a time execute critical sections of code in which shared data structures are modified. A mutex variable acts like a "lock" protecting access to a shared data resource. Only one thread can lock (or own) a mutex variable at any given time. Thus, even if several threads try to lock a mutex concurrently, only one thread will succeed, and no other thread can own that mutex until the owning thread unlocks it. The operating system will put any thread trying to lock an already locked mutex in wait state, and such threads will be made ready as soon as the mutex is unlocked. If more than one thread is waiting for the same mutex, they will compete for it, and only one will acquire the lock this turn.

Mutex variables are declared to be of type `pthread_mutex_t` and must be initialized before being used, using the following function:

```
pthread_mutex_init(pthread_mutex_t *mutex,
  pthread_mutex_attr_t *attr)
```

where the first argument is the pointer of the mutex variable, and the second one, when different from 0, is a pointer of a variable holding the attributes for the mutex.

Such attributes will be explained later in this book, so, for the moment, we will use the default attributes. Once initialized, a thread can lock and unlock the mutex via routines

```
pthread_mutex_lock(pthread_mutex_t *mutex)
pthread_mutex_unlock(pthread_mutex_t *mutex)
```

Routine `pthread_mutex_lock()` is blocking, that is, the calling thread is possibly put in wait state. Sometimes it is more convenient just to check the status of the mutex and, if the mutex is already locked, return immediately with an error rather than returning only when the thread has acquired the lock. The following routine does exactly this:

```
int pthread_mutex_trylock(pthread_mutex_t *mutex)
```

Finally, a mutex should be destroyed, that is, the associated resources released, when it is no more used:

```
pthread_mutex_destroy(pthread_mutex_t *mutex)
```

Recalling the producer/consumer example of Chapter 5, we can see that mutexes are well fit to ensure mutual exclusion for the segments of code that update the circular buffer and change the index accordingly. In addition to using critical sections when retrieving an element from the circular buffer and when inserting a new one, consumers need also to wait until at least one element is available in the buffer, and producers have to wait until the buffer is not full. This kind of synchronization is different from mutual exclusion because it requires waiting for a given condition to occur. This is achieved by pthread *condition* variables acting as monitors. Once a condition variable has been declared and initialized, the following operations can be performed: *wait* and *signal*. The former will suspend the calling thread until some other thread executes a signal operation for that condition variable. The signal operation will have no effect if no thread is waiting for that condition variable; otherwise, it will wake only one waiting thread. In the producer/consumer program, two condition variables will be defined: one to signal the fact that the circular buffer is not full, and the other to signal that the circular buffer is not empty. The producer performs a wait operation over the first condition variable whenever it finds the buffer full, and the consumer will execute a signal operation over that condition variable after consuming one element of the buffer. A similar sequence occurs when the consumer finds the buffer empty.

The prototypes of the `pthread` routines for initializing, waiting, signaling, and destroying condition variables are respectively:

```
int pthread_cond_init(pthread_cond_t *condVar,
    pthread_condattr_t *attr)
int pthread_cond_wait(pthread_cond_t *cond ,
    pthread_mutex_t *mutex)
```

```
int pthread_cond_signal(pthread_cond_t *cond)
int pthread_cond_destroy(pthread_cond_t *cond)
```

The attr argument passed to pthread_cond_init() will specify whether the condition variable can be shared also among threads belonging to different processes. When NULL is passed as second argument, the condition variable is shared only by threads belonging to the same process. The first argument of pthread_cond_wait() and pthread_cond_signal() is the condition variable, and the second argument of pthread_cond_wait() is a mutex variable that must be locked at the time pthread_cond_wait() is called. This argument may seem somewhat confusing, but it reflects the normal way condition variables are used. Consider the producer/consumer example, and in particular, the moment in which the consumer waits, in a critical section, for the condition variable indicating that the circular buffer is not empty. If the mutex used for the critical section were not released prior to issuing a wait operation, the program would deadlock since no other thread could enter that critical section. If it were released prior to calling pthread_cond_wait(), it may happen that, just after finding the circular buffer empty and before issuing the wait operation, another producer adds an element to the buffer and issues a signal operation on that condition variable, which does nothing since no thread is still waiting for it. Soon after, the consumer issues a wait request, suspending itself *even if the buffer is not empty*. It is therefore necessary to issue the wait *at the same time* the mutex is unlocked, and this is the reason for the second argument of pthread_cond_wait(), which will atomically unlock the mutex and suspend the thread, and will lock again the mutex just before returning to the caller program when the thread is awakened.

The following program shows the usage of mutexes and condition variables when a producer thread puts integer data in a shared circular buffer, which are then read by a set of consumer threads. The number of consumer threads is passed as an argument to the program. A mutex is defined to protect insertion and removal of elements into/from the circular buffer, and two condition variables are used to signal the availability of data and room in the circular buffer.

```
#include <pthread.h>
#include <stdio.h>
#include <stdlib.h>
#include <sys/time.h>

/* The mutex used to protect shared data */
pthread_mutex_t mutex;
/* Condition variables to signal availability
   of room and data in the buffer */
pthread_cond_t roomAvailable, dataAvailable;

#define BUFFER_SIZE 128
/* Shared data */
int buffer[BUFFER_SIZE];
/* readIdx is the index in the buffer of the next item to be retrieved */
int readIdx = 0;
/* writeIdx is the index in the buffer of the next item to be inserted */
int writeIdx = 0;
/* Buffer empty condition corresponds to readIdx == writeIdx. Buffer full
```

```
      condition corresponds to (writeIdx + 1)%BUFFER_SIZE == readIdx */

/* Consumer Code: the passed argument is not used */
static void *consumer(void *arg)
{
  int item;
  while(1)
  {
/* Enter critical section */
    pthread_mutex_lock(&mutex);
/* If the buffer is empty, wait for new data */
    while(readIdx == writeIdx)
    {
      pthread_cond_wait(&dataAvailable, &mutex);
    }
/* At this point data are available
   Get the item from the buffer */
    item = buffer[readIdx];
    readIdx = (readIdx + 1)%BUFFER_SIZE;
/* Signal availability of room in the buffer */
    pthread_cond_signal(&roomAvailable);
/* Exit critical section */
    pthread_mutex_unlock(&mutex);

 /* Consume the item and take actions (e.g. return)*/
    ...
  }
  return NULL;
}
/* Producer code. Passed argument is not used */
static void *producer(void *arg)
{
  int item = 0;
  while(1)
  {
/* Produce a new item and take actions (e.g. return) */
    ...
/* Enter critical section */
    pthread_mutex_lock(&mutex);
/* Wait for room availability */
    while((writeIdx + 1)%BUFFER_SIZE == readIdx)
    {
      pthread_cond_wait(&roomAvailable, &mutex)
    }
/* At this point room is available
   Put the item in the buffer */
    buffer[writeIdx] = item;
    writeIdx = (writeIdx + 1)%BUFFER_SIZE;
/* Signal data avilability */
    pthread_cond_signal(&dataAvailable)
/* Exit critical section */
    pthread_mutex_unlock(&mutex);
  }
  return NULL;
}

int main(int argc, char *args[])
{
  pthread_t threads[MAX_THREADS];
  int nConsumers;
  int i;
/* The number of consumer is passed as argument */
  if(argc != 2)
  {
    printf("Usage: prod_cons <numConsumers>\n");
    exit(0);
```

```
}
  sscanf(args[1], "%d", &nConsumers);

/* Initialize mutex and condition variables */
  pthread_mutex_init(&mutex, NULL)
  pthread_cond_init(&dataAvailable, NULL)
  pthread_cond_init(&roomAvailable, NULL)

/* Create producer thread */
  pthread_create(&threads[0], NULL, producer, NULL);
/* Create consumer threads */
  for(i = 0; i < nConsumers; i++)
    pthread_create(&threads[i+1], NULL, consumer, NULL);

/* Wait termination of all threads */
  for(i = 0; i < nConsumers + 1; i++)
  {
    pthread_join(threads[i], NULL);
  }
  return 0;
}
```

No check on the returned status of pthread routines is carried out in the above program to reduce the length of the listed code. Be conscious, however, that a good programming practice is to check every time the status of the called functions, and this is true in particular for the system routines used to synchronize threads and processes. A trivial error, such as passing a wrong argument making the routine fail synchronization, may not produce an evident symptom in program execution, but potentially raises race conditions that are very difficult to diagnose.

In the above program, both the consumers and the producer, once entered in the critical section, check the availability of data and room, respectively, possibly issuing a wait operation on the corresponding condition variable. Observe that, in the code, the check is repeated once `pthread_cond_wait()` returns, being the check within a `while` loop. This is the correct way of using `pthread_cond_wait()` because pthread library does not guarantee that the waiting process cannot be awakened by spurious events, requiring therefore the repeat of the check for the condition before proceeding. Even if spurious events were not generated, using an `if` statement in place of the `while` statement, that is, not checking the condition after exiting the wait operation, leads to a race condition in the above program when the following sequence occurs: (1) a consumer finds the buffer empty and waits; (2) a producer puts a new data item and signals the condition variable; (3) another consumer thread enters the critical section and consumes the data item before the first consumer gains processor ownership; (4) the first consumer awakes and reads the data item *when the buffer is empty*.

Mutexes and condition variables are provided by pthread library for thread synchronization and cover, in principle, all the required synchronization mechanisms in practice. We shall see in the next section that there are several other synchronization primitives to be used for processes that can be used for threads as well. Nevertheless, it is good programming practice to use pthread primitives when programming with threads. Library pthreads is, in fact, implemented not only in Linux but also in other operating systems, so a program using only pthreads primitive is more easily portable across different platforms than a program using Linux-specific synchronization primitives.

7.3   INTERPROCESS COMMUNICATION AMONG PROCESSES

7.3.1   SEMAPHORES

Linux semaphores are counting semaphores and are widely used to synchronize processes. When a semaphore has been created and an initial value assigned, two operations can be performed on it: `sem_wait()` and `sem_post()`. Operation `sem_wait()` will decrement the value of the semaphore: if the semaphore's value is greater than zero, then the decrement proceeds and the function returns immediately. If the semaphore currently has the value zero, then the call blocks until it becomes possible to perform the decrement, that is, the semaphore value rises above zero. Operation `sem_post()` increments the semaphore. If the semaphore's value consequently becomes greater than zero, then another process or thread may be blocked in a `sem_wait()` call. In this case, it will be woken up and will proceed in decrementing the semaphore's value. Semaphores can be used to achieve the same functionality of `pthread` mutexes and condition variables. To protect a critical section, it suffices to initialize a semaphore with an initial value equal to one: `sem_wait()` and `sem_post()` will be called by each process just before entering and exiting the critical section, respectively. To achieve the signaling mechanism carried out by condition variables, the semaphore will be created with a value equal to zero. When `sem_wait()` is called the first time prior to `sem_post()`, the calling process will suspend until another process will call `sem_post()`. There is, however, a subtle difference between posting a semaphore and signaling a condition variable: when the latter is signaled, if no thread is waiting for it, nothing happens, and if a thread calls `pthread_cond_wait()` for that condition variable soon after, it will suspend anyway. Conversely, posting a semaphore will permanently increase its value until one process will perform a wait operation on it. So, if no process is waiting for the semaphore at the time it is posted, the first process that waits on it afterward will not be stopped.

There are two kinds of semaphores in Linux: *named semaphores* and *unnamed semaphores*. Named semaphores, as the name suggests, are associated with a name (character string) and are created by the following routine:

```
sem_t *sem_open(const char *name, int oflag, mode_t mode,
    unsigned int value)
```

where the first argument specifies the semaphore's name. The second argument defines associated flags that specify, among other information, if the semaphore has to be created if not yet existing. The third argument specifies the associated access protection (as seen for shared memory), and the last argument specifies the initial value of the semaphore in the case where this has been created. `sem_open()` will return the address of a `sem_t` structure to be passed to `sem_wait()` and `sem_post()`. Named semaphores are used when they are shared by different processes, using then their associated name to identify the right semaphores. When the communicating processes are all children of the same process, unnamed semaphores are preferable because it is not necessary to define names that may collide with other semaphores used by different processes. Unnamed semaphores are created by the following routine:

```
int sem_init(sem_t *sem, int pshared, unsigned int value)
```

sem_init() will always create a new semaphore whose data structure will be allocated in the sem_t variable passed as first argument. The second argument specifies whether the semaphore will be shared by different processes and will be set to 0 only if the semaphore is to be accessed by threads belonging to the same process. If the semaphore is shared among processes, the sem_t variable to host the semaphore data structures must be allocated in shared memory. Lastly, the third argument specifies the initial value of the semaphore.

The following example is an implementation of our well-known producer/consumer application where the producer and the consumers execute on different processes and use unnamed semaphores to manage the critical section and to handle producer/consumer synchronization. In particular, the initial value of the semaphore (mutexSem) used to manage the critical section is set to one, thus ensuring that only one process at a time can enter the critical section by issuing first a P() (sem_wait()) and then a V() (sem_post()) operation. The other two semaphores (dataAvailableSem and roomAvailableSem)will contain the current number of available data slots and free ones, respectively. Initially there will be no data slots and BUFFER_SIZE free slots and therefore the initial values of dataAvailableSem and roomAvailableSem will be 0 and BUFFER_SIZE, respectively.

```
#include <stdio.h>
#include <stdlib.h>
#include <sys/ipc.h>
#include <sys/shm.h>
#include <sys/wait.h>
#include <semaphore.h>

#define MAX_PROCESSES 256
#define BUFFER_SIZE 128
/* Shared Buffer, indexes and semaphores are held in shared memory
   readIdx is the index in the buffer of the next item to be retrieved
   writeIdx is the index in the buffer of the next item to be inserted
   Buffer empty condition corresponds to readIdx == writeIdx
   Buffer full condition corresponds to
   (writeIdx + 1)%BUFFER_SIZE == readIdx)
   Semaphores used for synchronization:
   mutexSem is used to protect the critical section
   dataAvailableSem is used to wait for data availability
   roomAvailableSem is used to wait for room abailable in the buffer */

struct BufferData {
  int readIdx;
  int writeIdx;
  int buffer[BUFFER_SIZE];
  sem_t mutexSem;
  sem_t dataAvailableSem;
  sem_t roomAvailableSem;
};

struct BufferData *sharedBuf;

/* Consumer routine */
static void consumer()
```

```
{
  int item;
  while(1)
  {
/* Wait for availability of at least one data slot */
    sem_wait(&sharedBuf->dataAvailableSem);
/* Enter critical section */
    sem_wait(&sharedBuf->mutexSem);
/* Get data item */
    item = sharedBuf->buffer[sharedBuf->readIdx];
/* Update read index */
    sharedBuf->readIdx = (sharedBuf->readIdx + 1)%BUFFER_SIZE;
/* Signal that a new empty slot is available */
    sem_post(&sharedBuf->roomAvailableSem);
/* Exit critical section */
    sem_post(&sharedBuf->mutexSem);
/* Consume data item and take actions (e.g return)*/
    ...
  }
}
/* producer routine */
static void producer()
{
  int item = 0;
  while(1)
  {
/* Produce data item and take actions (e.g. return)*/
    ...
/* Wait for availability of at least one empty slot */
    sem_wait(&sharedBuf->roomAvailableSem);
/* Enter critical section */
    sem_wait(&sharedBuf->mutexSem);
/* Write data item */
    sharedBuf->buffer[sharedBuf->writeIdx] = item;
/* Update write index */
    sharedBuf->writeIdx = (sharedBuf->writeIdx + 1)%BUFFER_SIZE;
/* Signal that a new data slot is available */
    sem_post(&sharedBuf->dataAvailableSem);
/* Exit critical section */
    sem_post(&sharedBuf->mutexSem);
  }
}
/* Main program: the passed argument specifies the number
   of consumers */
int main(int argc, char *args[])
{
  int memId;
  int i, nConsumers;
  pid_t pids[MAX_PROCESSES];
  if(argc != 2)
  {
    printf("Usage: prodcons  <numProcesses>\n");
    exit(0);
  }
  sscanf(args[1], "%d", &nConsumers);
/* Set-up shared memory */
  memId = shmget(IPC_PRIVATE, sizeof(struct BufferData), SHM_R | SHM_W);
  if(memId == -1)
  {
    perror("Error in shmget");
    exit(0);
  }
  sharedBuf = shmat(memId, NULL, 0);
  if(sharedBuf == (void *)-1)
  {
    perror("Error in shmat");
```

```
      exit(0);
   }
/* Initialize buffer indexes */
   sharedBuf->readIdx = 0;
   sharedBuf->writeIdx = 0;
/* Initialize semaphores. Initial value is 1 for mutexSem,
    0 for dataAvailableSem (no filled slots initially available)
    and BUFFER_SIZE for roomAvailableSem (all slots are
    initially free). The second argument specifies
    that the semaphore is shared among processes */
   sem_init(&sharedBuf->mutexSem, 1, 1);
   sem_init(&sharedBuf->dataAvailableSem, 1, 0);
   sem_init(&sharedBuf->roomAvailableSem, 1, BUFFER_SIZE);

/* Launch producer process */
   pids[0] = fork();
   if(pids[0] == 0)
   {
/* Child process */
      producer();
      exit(0);
   }
/* Launch consumer processes */
   for(i = 0; i < nConsumers; i++)
   {
      pids[i+1] = fork();
      if(pids[i+1] == 0)
      {
         consumer();
         exit(0);
      }
   }
/* Wait process termination */
   for(i = 0; i <= nConsumers; i++)
   {
      waitpid(pids[i], NULL, 0);
   }
   return 0;
}
```

Observe that, in the above example, there is no check performed on read and write indexes to state whether data or free room are available. This check is, in fact, implicit in the P (semWait()) and V (semPost()) operations carried out on dataAvailableSem and roomAvailableSem semaphores.

7.3.2 MESSAGE QUEUES

In the previous section, the exchange of information between the producer and consumers has been managed using shared memory and semaphores. In POSIX it is possible to use another IPC mechanism: *message queues*. Message queues allow different processes to exchange information by inserting and extracting data elements into and from FIFO queues that are managed by the operating system. A message queue is created in a very similar way as shared memory segments are created by shmget(), that is, either passing a unique identifier so that different processes can connect to the same message queue, or by defining the IPC_PRIVATE option in the case where the message queue is to be shared among the parent and children processes. In fact, when a child process is created by fork(), it inherits the message queue references of the parent process. A new message queue is created by routine:

```
int msgget(key_t key, int msgflg)
```

whose first argument, if not IPC_PRIVATE, is the message queue unique identifier, and the second argument specifies, among others, the access protection to the message queue, specified as a bitmask as for the shared memory. The returned value is the message queue identifier to be used in the following routines. New data items are inserted in the message queue by the following routine:

```
int msgsnd(int msqid,
    const void *msgp, size_t msgsz, int msgflg)
```

where the first argument is the message queue identifier. The second argument is a pointer to the data structure to be passed, whose length is specified in the third argument. Such a structure defines, as its first long element, a user-provided message type that can be used to select the messages to be received. The last argument may define several options, such as specifying whether the process is put in wait state in the case the message queue is full, or if the routine returns immediately with an error in this case.

Message reception is performed by the following routine:

```
ssize_t msgrcv(int msqid, void *msgp, size_t msgsz, long msgtyp,
    int msgflg);
```

whose arguments are the same for of the previous routine, except for msgtyp, which, if different from 0, specifies the type of message to be received. Unless differently specified, msgrcv() will put the process in wait state if a message of the specified type is not present in the queue.

The following example uses message queues to exchange data items between a producer and a set of consumers processes.

```
#include <stdio.h>
#include <stdlib.h>
#include <sys/ipc.h>
#include <sys/wait.h>
#include <sys/msg.h>
#define MAX_PROCESSES 256
/* The type of message */
#define PRODCONS_TYPE 1
/* Message structure definition */
struct msgbuf {
  long mtype;
  int item;
};
/* Message queue id */
int msgId;

/* Consumer routine */
static void consumer()
{
  int retSize;
  struct msgbuf msg;
  int item;
  while(1)
  {
```

```
/* Receive the message. msgrcv returns the size of the received message */
   retSize = msgrcv(msgId, &msg, sizeof(int), PRODCONS_TYPE, 0);
   if(retSize == -1) //If Message reception failed
   {
     perror("error msgrcv");
     exit(0);
   }
   item = msg.item;
/* Consume data item */
   ...
  }
}
/* Consumer routine */
static void producer()
{
  int item = 0;
  struct msgbuf msg;
  msg.mtype = PRODCONS_TYPE;
  while(1)
  {
/* produce data item */
   ...
   msg.item = item;
   msgsnd(msgId, &msg, sizeof(int), 0);
  }
}
/* Main program. The number of consumer
   is passed as argument */
int main(int argc, char *args[])
{
  int i, nConsumers;
  pid_t pids[MAX_PROCESSES];
  if(argc != 2)
  {
    printf("Usage: prodcons  <nConsumers>\n");
    exit(0);
  }
  sscanf(args[1], "%d", &nConsumers);
/* Initialize message queue */
  msgId = msgget(IPC_PRIVATE, 0666);
  if(msgId == -1)
  {
    perror("msgget");
    exit(0);
  }
/* Launch producer process */
  pids[0] = fork();
  if(pids[0] == 0)
  {
/* Child process */
    producer();
    exit(0);
  }
/* Launch consumer processes */
  for(i = 0; i < nConsumers; i++)
  {
    pids[i+1] = fork();
    if(pids[i+1] == 0)
    {
      consumer();
      exit(0);
    }
  }
/* Wait process termination */
  for(i = 0; i <= nConsumers; i++)
  {
```

```
   waitpid(pids[i], NULL, 0);
  }
  return 0;
}
```

The above program is much simpler than the previous ones because there is no need to worry about synchronization: everything is managed by the operating system! Several factors however limit in practice the applicability of message queues, among which is the fact that they consume more system resources than simpler mechanisms such as semaphores.

Routines `msgget()`, `msgsnd()`, and `msgrcv()`, now in the POSIX standard, originally belonged to the System V interface. POSIX defines also a different interface for named message queues, that is, routines `mq_open()` to create a message queue, and `mq_send()` and `mq_receive()` to send and receive messages over a message queue, respectively. As for the shared memory object creation, the definition of the message queue name is more immediate: the name is directly passed to `mq_open()`, without the need for using `ftok()` to create the identifier to be passed to the message queue creation routine. On the other side, `msgget()` (as well as `shmget()`) allows creating unnamed message queues, that are shared by the process and its children with no risk of conflicts with other similar resources with the same name.

7.3.3 SIGNALS

The synchronization mechanisms we have seen so far provide the necessary components, which, if correctly used, allow building concurrent and distributed systems. However sometime it is necessary to handle the occurrence of *signals*, that is, asynchronous event requiring some kind of action in response. In POSIX and ANSI, a set of signals is defined, summarized by table 7.2, and the corresponding action can be specified using the following routine:

```
signal(int signum, void (*handler)(int))
```

where the first argument is the event number, and the second one is the address of the event handler routine, which will be executed asynchronously when an event of the specified type is sent to the process.

A typical use of routine `signal()` is for "trapping" the SIG_INT event that is generated by the <ctrl> C key. In this case, instead of an abrupt program termination, it is possible to let a cleanup routine be executed, for example closing the files which have been opened by the process and making sure that their content is not corrupted. Another possible utilization of event handlers is in association with timers, as explained in the next section. Care is necessary in programming event handlers since they are executed asynchronously. Since events may occur at any time during the execution of the process, no assumption can be made on the current status of the data structures managed by programs at the time a signal is received. For the same reason, it is necessary that event handlers call only "safe" system routines, that is, system routines that are guaranteed to execute correctly regardless of the current system state (luckily, most pthread and Linux system routines are safe).

Table 7.2
Some signal events defined in Linux

Signal Name and Number		Description
SIGHUP	1	Hangup (POSIX)
SIGINT	2	Terminal interrupt (ANSI)
SIGQUIT	3	Terminal quit (POSIX)
SIGILL	4	Illegal instruction (ANSI)
SIGTRAP	5	Trace trap (POSIX)
SIGFPE	8	Floating point exception (ANSI)
SIGKILL	9	Kill (can't be caught or ignored) (POSIX)
SIGUSR1	10	User-defined signal 1 (POSIX)
SIGSEGV	11	Invalid memory segment access (ANSI)
SIGUSR2	12	User-defined signal 2 (POSIX)
SIGPIPE	13	Write on a pipe with no reader, Broken pipe (POSIX)
SIGALRM	14	Alarm clock (POSIX)
SIGTERM	15	Termination (ANSI)
SIGSTKFLT	16	Stack fault
SIGCHLD	17	Child process has stopped or exited, changed (POSIX)
SIGCONT	18	Continue executing, if stopped (POSIX)
SIGSTOP	19	Stop executing (can't be caught or ignored) (POSIX)
SIGTSTP	20	Terminal stop signal (POSIX)
SIGTTIN	21	Background process trying to read, from TTY (POSIX)
SIGTTOU	22	Background process trying to write, to TTY (POSIX)

7.4   CLOCKS AND TIMERS

Sometimes it is necessary to manage time in program. For example, a control cycle in an embedded system may be repeated every time period, or an action has to finish within a given timeout. Two classes of routines are available for handling time: wait routines and timers. Wait routines, when called, force the suspension of the calling thread or process for the specified time. Traditionally, programmers have used routine `sleep()` whose argument specifies the number of seconds the caller has to wait before resuming execution. More accurate wait time definition is achieved by routine

```
int nanosleep(const struct timespec *req, struct timespec *rem);
```

The first argument defines the wait time, and its type is specified as follows:

```
struct timespec {
    time_t tv_sec;      /* seconds */
    long   tv_nsec;     /* nanoseconds */
};
```

The second argument, if not NULL, is a pointer to a `struct timespec` argument that will report the remaining time in the case `nanosleep()` returned (with an error) before the specified time. This happens in the case where a signal has been delivered to the process that issued `nanosleep()`. Observe that, even if it is possible to specify the wait time with nanosecond precision, the actual wait time will be rounded to a much larger period, normally ranging from 10 to 20 ms. The reason lies in the mechanism used by the operating system to manage wait operations: when a process or a thread calls `nanosleep()`, it is put in wait state, thus losing control of the processor, and a new descriptor will be added to a linked list of descriptors of processes/threads, recording, among other information, the wake time. At every *tick*, that is, every time the processor is interrupted by the system clock, such list is checked, and the processes/threads for which the wait period has expired are awakened. So, the granularity in the wait period is dictated by the clock interrupt rate, which is normally around 50–60 Hz. Observe that, even if it is not possible to let processes/threads wait with microsecond precision, it is possible to get the current time with much more precision because every computer has internal counters that are updated at very high frequencies, often at the processor clock frequency. For this reason, the time returned by routine `gettimeofday()`, used in Chapter 2, is a very accurate measurement of the current time.

The other class of time-related routines creates *timers*, which allows an action to be executed after a given amount of time. Linux routine `timer_create()` will set up an internal timer, and upon the expiration of the timer, a signal will be sent to the calling process. An event handler will then be associated with the event via routine `signal()` in order to execute the required actions upon the timer expiration.

7.5 THREADS OR PROCESSES?

We have seen that in Linux there are two classes of entities able to execute programs. Processes have been implemented first in the history of UNIX and Linux. Threads have been introduced later as a lightweight version of processes, and are preferable to processes for two main reasons

1. *Efficiency*: context switch in threads belonging to the same process is fast when compared with context switch between processes, mainly because the page table information needs not to be updated because all threads in a process share the same memory space;
2. *Simplified programming model*: sharing memory among threads is trivial, it suffices to use static variables. Mutexes and Condition Variables then provide all the required synchronization mechanisms.

At this point the reader may wonder why any more processes need to be used when developing a concurrent application. After all, a single big process, hosting all the threads that cooperate in carrying out the required functionality, may definitely appear as the best choice. Indeed, very often this is the case, but threads have a weak aspect that sometimes cannot be acceptable, that is, *the lack of protection*. As threads share the same memory, except for stacks, a wrong memory access performed by one thread may corrupt the data structure of other threads. We have already seen that this fact would be impossible among processes since their memories are guaranteed to be insulated by the operating system, which builds a "fence" around them by properly setting the processes' page tables. Therefore, if some code to be executed is not trusted, that is, there is any likelihood that errors could arise during execution, the protection provided by the process model is mandatory. An example is given by Web Servers, which are typically concurrent programs because they must be able to serve multiple clients at the same time. Serving an HTTP connection may, however, also imply the execution of external code (i.e., not belonging to the Web Server application), for example, when CGI scripts are activated. If the Web Server were implemented using threads, the failure of a CGI script potentially crashes the whole server. Conversely, if the Web Server is implemented as a multiprocess application, failure of a CGI script will abort the client connection, but the other connections remain unaffected.

7.6 SUMMARY

This chapter has presented the Linux implementation of the concepts introduced in Chapters 3 and 5. Firstly, the difference between Linux processes and threads has been described, leading to a different memory model and two different sets of interprocess communication primitives. The main difference between threads and processes lies in the way memory is managed; since threads live in the context of a process, they share the same address space of the hosting process, duplicating only the stack segment containing local variables and the call frames. This means in practice that static variables, that are not located in the stack, are shared by all the threads created by a given process (or thread). Conversely, shared memory segments must be created in order to exchange memory data among different processes.

Threads can be programmed using library `pthread`, and they represent a complete model for concurrency, defining mutexes and condition variables for synchronization. Interprocess communication is carried out by semaphores and message queues, which can be used for threads, too.

Several implementations of the producer-consumer example, introduced in Chapter 3, have been presented using different synchronization primitives.

7.7 EXERCISES

The solutions of the proposed exercises are available in the GitHub repository at `https://github.com/minimap-xl/RTOS_Book`.

EXERCISE 1

In this chapter we have seen how it is possible to parallelize the computation of the sum of all the elements in a given large matrix. The complexity of this operation is $O(N^2)$, where N is the dimension of the matrix. If you replicate on your computer the presented program, you can barely appreciate the gain in performance, unless using a very large matrix. You can however appreciate the gain in performance due to parallelization if you considers instead matrix multiplication, whose complexity is $O(N^3)$.

Hint: as every element of the resulting matrix is obtained by a row-column multiplication performed over the two input matrices, computation of each element can be performed in parallel without the need of any locking as only read operations are performed over input matrices and no other element of the output matrix is affected. Observe that parallel execution of matrix multiplications represents the hearth of the computation carried out in Artificial Intelligence, where computation is normally performed in parallel by the many computation elements of General Purpose Graphical Processing Units (GPGPU). In this case, however, it is necessary to take into account also the time required to transfer matrices from the processor memory to the different levels of GPGPU memory, complicating a bit the algorithm. The interested reader can find a good introduction to GPGPU algorithms in the CUDA Programming Guide from nvidia at `https://docs.nvidia.com/cuda/cuda-c-programming-guide`.

EXERCISE 2

In the chapter several examples of producer-consumer applications have been presented using threads, condition variables and mutexes. A good software engineering practice is to present such functionality via a set of classes that hide the mechanism used to achieve parallelism and avoid race conditions.

Hint: the following C++ classes may be defined to implement message queueing and implementing the required mutexes and condition variables:

```
#include <pthread.h>
//Support class for Queue
class QueueItem
{
        public:
        char *buf;
        int size;
        QueueItem *nxt;
        QueueItem(char *buf, int size);
};
class Queue
{
        pthread_mutex_t lock;
        pthread_cond_t empty, full;
        QueueItem *first, *last;
        int numItems;
        int maxItems;
public:
        Queue(int maxItems = 1000);
        ~Queue();
        void addItem(char *buf, int size);
        char *getItem(int &size);
        int getNumItems();
};
```

Queue and Queue Item classes handle the enqueueing and dequeuing of messages exchanged by producers and consumers, represented in a generic way as a byte buffer of given size. The management of a single producer and multiple consumers can be presented by the following class definition

```cpp
#include "queue.h"
#include <pthread.h>
#define MAX_THREADS 100
class ProducerConsumer {
        Queue queue;
        pthread_t threadPool[MAX_THREADS];
        int numThreads;
        bool started;
public:
        ProducerConsumer(int numThreads, int maxItems = 100)
        ~ProducerConsumer()
        Queue *getQueue();
        void start();
        void stop();
        void produce(char *buf, int size);
        int getQueueLen();
        virtual void consume(char *buf, int size) = 0;
};
```

A multithreaded C++ producer consumer application will not directly use the ProducerConsumer class, but will first define a specific class inheriting from ProducerConsumer and defining the specific virtual consume() method. A program using this organization, just producing integer numbers and letting consumer print the received number can be the following:

```cpp
#include "prod_con.h"
#include <stdio.h>
#include <sys/types.h>
#include <unistd.h>
#include <stdlib.h>
#include <sys/syscall.h>

class SimpleProducerConsumer: public ProducerConsumer
{
        public:
                SimpleProducerConsumer(int numThreads):
            ProducerConsumer(numThreads)
                {}

        virtual void consume(char *buf, int size)
        {
                pid_t tid = syscall(SYS_gettid);
                printf("Thread %d: value: %d\n", (int)tid, *(int *)buf);
        }
};

int main(int argc, char *argv[])
{
        int numConsumers;
        if(argc != 2)
        {
                printf("Usage: simple_prod_con <Number of Consumers>\n");
                exit(0);
        }
        sscanf(argv[1], "%d", &numConsumers);
        SimpleProducerConsumer spd(numConsumers);
        spd.start();
        //produce 10 items
        for (int i = 0; i < 10; i++)
        {
                int *currInt = new int[1];
                *currInt = i;
```

```
            spd.produce((char *)currInt, sizeof(int));
    }
    spd.stop();
}
```

As you can see all the involved machinery is hidden by the C++ classes and the application developer can concentrate on the consumer functionality. The number of threads is passed as argument to the class constructor but no pthread function call is made visible outside such classes. The implementation of `ProducerConsumer::start()` will create the specified number of threads storing the returned thread identifiers in array `threadPool`. Method `ProducerConsumer::stop()` will terminate all the created threads and wait for their termination via `pthread_join()` calls for every active thread in `threadPool`. A problem arises here because it is not possible to explicitly kill a running thread. A trick here is to define a special message with size = 0 that, once received by the thread code forces its termination. So, in order to force the termination of N threads, it suffices sending N special messages over the queue, and then simply waiting for their termination.

EXERCISE 3

The C++ class organization presented in the previous exercise hid the specific implementation using POSIX threads, mutexes and condition variables. You may now provide a completely different implementation using Linux messages and Processes in place of Queue items allocated in memory. The same main program presented in the previous exercise can then be reused as it is.

Hint: even if the interface is the same, the implementation of the `ProducerConsumer` class will be completely different, not using classes `Queue` and `QueueItem` but relying on the queueing mechanism provided by Linux messages. A new mechanism must be defined here in to ensure that the length of the queue will never exceed a given maximum number of items. As suggested in the chapter, a possibility is to define a separate queue of slot placeholders. A producer must first get a placeholder to be allowed to send a message. For this purpose you may define two different message classes in the same Linux message queue. Indeed, when a message is requested via `msgget()` or sent via `msgsnd()`, the class of the message must be given among the other arguments.

EXERCISE 4

Using the same abstraction presented in exercise 2 you may now provide a smarter implementation that control the number of enqueued messages between the producer and the consumers. The message queue length depends on the rate new items are produced and on the rate these items are consumed. In a real-world application this may change over time and the pre-established number of consumer threads may be become insufficient to sustain the production rate. A mechanism that adapts the number of consumer threads trying to keep the queue length at a given value would in this case make the system adaptable to varying workloads.

Hint: The requested control can be implemented by a new special thread that periodically (say, every second) check the current queue length, comparing it with the target queue length and computes the new number of active threads (`newNumTThreads`) starting from the current number of threads (`prevNumThreads`) following a simple control law:

```
delta =   queueLen - targetQueueLen;
deltaSum += delta;
newNumThreads = prevNumThreads + kp * delta + ki * deltaSum;
```

This implementation must allow dynamic thread disposal and creation. We can use the same trick of exercise 2 to dispose N running threads, i.e. sending N special messages, while thread creation is simply performed via `pthread_create()` call. These is however an additional problem in respect of the implementation of exercise 2, where a `threadPool` array was used to keep track of the thread identifiers to be used in the synchronization performed by `stop()` method. Here we don't know which thread is disposed when a special message is sent in order to decrease the number of consumer threads, and the content of `threadPool` soon becomes inconsistent. The solution is to use a shared vector (e.g. `std::vector` of the C++ Standard Template Library (STL)) and let the consumer threads themselves add and remove their thread identifiers when they start and finish their execution, respectively. Of course the vector must be protected with a new mutex as it can be updated concurrently.

8 Interprocess Communication Primitives in FreeRTOS

The previous chapter discussed the interprocess communication primitives available to the user of a "full-fledged" operating system that supports the POSIX standard at the Application Programming Interface (API) level. Since those operating systems were initially intended for general-purpose computers, one of their main goals is to put at the user's disposal a rich set of high-level primitives, meant to be powerful and convenient to use.

At the same time, they must also deal with multiple applications, maybe pertaining to different users, being executed at the same time. Protecting those applications from each other and preventing, for example, an error in one application from corrupting the memory space of another, is a daunting task that requires the adoption of complex process and memory models, and must be suitably supported by sophisticated hardware components, such as a Memory Management Unit (MMU).

Unfortunately, all these features come at a cost, in terms of operating system code size, memory requirements, execution overhead, and complexity of the underlying hardware. When designing a small-scale embedded system, it may be impossible to afford such a cost, and therefore, the developer is compelled to settle on a cheaper architecture in which both the operating system and its interface are much simpler. Besides the obvious disadvantages, making the operating system simpler and smaller has several advantages as well. For instance, it becomes easier to ensure that the operating system behavior is correct for what concerns real-time execution. A smaller code base usually leads to a more efficient and reliable system, too.

Up to a certain extent, it is possible to reach this goal within the POSIX framework: the IEEE Standard 1003.13 [45], also recognized as an ANSI standard, defines several POSIX subsets, or profiles, oriented to real-time and embedded application environments. Each profile represents a different trade-off between complexity and features.

However, in some cases, it may still be convenient to resort to an even simpler set of features with a streamlined, custom programming interface. This is the case with FreeRTOS [13], a small open-source operating system focusing on high portability, very limited footprint and, of course, real-time capabilities. It is licensed under a variant of the GNU General Public License (GPL). The most important difference is an exception to the GPL that allows application developers to distribute a combined work that includes FreeRTOS without being obliged to provide the source code for any proprietary components.

DOI: 10.1201/9781003593416-8

180

Table 8.1

Summary of the task-related primitives of FreeRTOS

Function	Purpose	Optional
vTaskStartScheduler	Start the scheduler	-
vTaskEndScheduler	Stop the scheduler	-
xTaskCreate	Create a new task	-
vTaskDelete	Delete a task given its handle	*
uxTaskPriorityGet	Get the priority of a task	*
vTaskPrioritySet	Set the priority of a task	*
vTaskSuspend	Suspend a specific task	*
vTaskResume	Resume a specific task	*
xTaskResumeFromISR	Resume a specific task from an ISR	*
xTaskIsTaskSuspended	Check whether a task is suspended	*
vTaskSuspendAll	Suspend all tasks but the running one	-
xTaskResumeAll	Resume all tasks	-
uxTaskGetNumberOfTasks	Return current number of tasks	-

This chapter is meant as an overview of the FreeRTOS API, to highlight the differences that are usually found when comparing a small real-time operating system with a POSIX system and to introduce the reader to small-scale, embedded software development. The FreeRTOS manual [14] provides in-depth information on this topic, along with a number of real-world code examples.

8.1   FREERTOS THREADS AND PROCESSES

FreeRTOS supports only a single process, and multiprogramming is achieved by means of multiple threads of execution, all sharing the same address space. They are called *tasks* by FreeRTOS, as it happens in most other real-time operating systems. The main FreeRTOS primitives related to task management and scheduling are summarized in Table 8.1. To invoke them, it is necessary to include the main FreeRTOS header file, FreeRTOS.h, followed by task.h.

With the FreeRTOS approach, quite common also in many other small real-time operating systems, the address space used by tasks is also shared with the operating system itself. The operating system is, in fact, a library of object modules, and the application program is linked against it when the application's executable image is built, exactly as any other library. The application and the operating system modules are therefore bundled together in the resulting executable image.

This approach keeps the operating system as simple as possible and makes any shared memory-based interprocess communication mechanism extremely easy and efficient to use because all tasks can share memory with no effort. On the other hand, tasks cannot be protected from each other with respect to illegal memory accesses, but it should be noted that many microcontrollers intended for embedded application lack any hardware support for this protection anyway. For some processor architectures, FreeRTOS is able to use a Memory Protection Unit (MPU), when available, to implement a limited form of data access protection among tasks.

Usually, the executable image is stored in a nonvolatile memory within the target system and is invoked either directly or through a minimal boot loader when the system is turned on. Therefore, unlike for Linux, the image must also include an appropriate startup code, which takes care of initializing the target hardware and is invoked before calling the `main()` entry point of the application. Another important difference with respect to Linux is that, when `main()` gets executed, the operating system scheduler is *not yet* active and must be explicitly started by means of the following function call:

```
void vTaskStartScheduler(void);
```

It should be noted that this function reports errors back in an unusual way. When successful, it does not return to the caller. Instead, the execution proceeds with the FreeRTOS tasks that have been created before starting the scheduler, according to their priorities. On the other hand, `vTaskStartScheduler` may return to the caller for two distinct reasons:

1. An error occurred during scheduler initialization, and so it was impossible to start it successfully.
2. The scheduler was successfully started, but one of the tasks executed afterward stopped the operating system by invoking

   ```
   void vTaskEndScheduler(void);
   ```

The two scenarios are clearly very different because, in the first case, the return is immediate and the application tasks are never actually executed, whereas in the second the return is delayed and usually occurs when the application is shut down in an orderly manner, for instance, at the user's request. However, since `vTaskStartScheduler` has no return value, there is no immediate way to distinguish between them.

If the distinction is important for the application being developed, then the programmer must make the necessary information available on his or her own, for example, by setting a shared flag after a full and successful application startup so that it can be checked by the code that follows `vTaskStartScheduler`.

It is possible to create a new FreeRTOS task either before or after starting the scheduler, by calling the `xTaskCreate` function:

```
portBASE_TYPE xTaskCreate(
    pdTASK_CODE pvTaskCode,
    const char * const pcName,
    unsigned short usStackDepth,
    void *pvParameters,
    unsigned portBASE_TYPE uxPriority,
    xTaskHandle *pvCreatedTask);
```

where:

- `pvTaskCode` is a pointer to a function returning void and with one void
  ∗ argument. It represents the entry point of the new task, that is, the function that the task will start executing from. This function must be designed to *never* return to the caller because this operation has undefined results in FreeRTOS. It is very important to remember this because, in a POSIX-compliant operating system, a thread is indeed allowed to return from its starting function; when it does so, it is implicitly terminated with no adverse consequences. On the contrary, returning from the start function of a FreeRTOS task may be quite unfortunate because, on most platforms, it leads to the execution of code residing at an unpredictable address.
- `pcName`, a constant string of characters, represents the human-readable name of the task being created. The operating system simply stores this name along with the other task information it keeps track of, without interpretation, but it is useful when inspecting the operating system data structures, for example, during debugging. The maximum length of the name actually stored by the operating system is limited by a configuration parameter; longer names are silently truncated.
- `usStackDepth` indicates how many stack *words* must be reserved for the task stack. The stack word size depends on the underlying hardware architecture and is configured when the operating system is being ported onto it. If necessary, the actual size of a stack word can be calculated by looking at the `portSTACK_TYPE` data type, defined in an architecture-dependent header file that is automatically included by the main FreeRTOS header file.
- `pvParameters` is a void ∗ pointer that will be passed to the task entry point upon execution without any interpretation by the operating system. It is most commonly used to point at a shared memory structure that holds the task parameters and, possibly, return values.
- `uxPriority` represents the initial, or baseline, priority of the new task, expressed as a positive integer. The symbolic constant `tskIDLE_PRIORITY`, defined in the operating system's header files, gives the priority of the idle task, that is, the lowest priority in the system, and higher priority values correspond to higher priorities. The total number of priority levels available is set in the operating system configuration depending on the application requirements because the size of several operating system data structures depend on it. The currently configured value is available in the symbolic constant `configMAX_PRIORITIES`. Hence, the legal range of priorities in the system goes from `tskIDLE_PRIORITY` to `tskIDLE_PRIORITY`+`configMAX_PRIORITIES`−1, extremes included.
- `pvCreatedTask` points to the task *handle*, which will be filled upon successful completion of this function. The handle must be used to refer to the new task in the future and is taken as a parameter by all operating system functions that operate on, or refer to, a task.

The return value of xTaskCreate is a status code. If its value is pdPASS, the function was successful in creating the new task, whereas any other value means that an error occurred. For example, the value errCOULD_NOT_ALLOCATE_REQUIRED_MEMORY denotes that it was impossible to create the new task because not enough memory was available. The FreeRTOS header projdefs.h, automatically included by the main FreeRTOS header file, contains the full list of error codes that may be returned by the operating system functions.

After creation, a task can be deleted by means of the function

```
void vTaskDelete(xTaskHandle pxTaskToDelete);
```

Its only argument, pxTaskToDelete, is the handle of the task to be deleted. It is possible to delete the currently running task, and the effect in that case is that vTaskDelete will never return to the caller.

For technical reasons, the memory dynamically allocated to the task by the operating system (for instance, to store its stack) cannot be freed immediately during the execution of vTaskDelete itself; this duty is instead delegated to the idle task. If the application makes use of vTaskDelete, it is important to ensure that a portion of the processor time is available to the idle task, as otherwise the system may run out of memory not because there is not enough but because the idle task was unable to free it fast enough before reuse.

As many other operating system functions, the availability of vTaskDelete depends on the operating system configuration so that it can be excluded from systems in which it is not used in order to save code and data memory. In Table 8.1, as well as in all the ensuing ones, those functions are marked as optional.

Another important difference with respect to a POSIX-compliant operating system is that FreeRTOS—like most other small, real-time operating systems—does not provide anything comparable to the POSIX *thread cancellation* mechanism. This mechanism is rather complex and allows POSIX threads to decline or postpone deletion requests, or *cancellation* requests as they are called in POSIX, directed to them. This is useful in ensuring that these requests are honored only when it is safe to do so.

In addition, POSIX threads can also register a set of functions, called *cleanup handlers*, which will be invoked automatically by the system while a cancellation request is being honored, before the target thread is actually deleted. Cleanup handlers, as their name says, provide therefore a good opportunity for POSIX threads to execute any last-second cleanup action they may need to make sure that they leave the application in a safe and consistent state upon termination.

On the contrary, task deletion is immediate in FreeRTOS, that is, it can neither be refused nor delayed by the target task. As a consequence, the target task may be deleted and cease execution at any time and location in the code, and it will not have the possibility of executing any cleanup handler before terminating. From the point of view of concurrent programming, it means that, for example, if a task is deleted when it is within a critical region controlled by a mutual exclusion semaphore, the semaphore will never be unlocked.

The high-level effect of the deletion is therefore the same as the terminated task never having exited from the critical region: no other tasks will ever be allowed to enter a critical region controlled by the same semaphore in the future. Since this usually corresponds to a complete breakdown of any concurrent program, the direct invocation of vTaskDelete should usually be avoided, and it should be replaced by a more sophisticated deletion mechanism.

One simple solution, mimicking the POSIX approach, is to send a deletion request to the target task by some other means—for instance, one of the interprocess communication mechanisms described in Sections 8.2 and 8.3, and design the target task so that it responds to the request by terminating itself at a well-known location in the target task's code and after any required cleanup operation has been carried out.

After creation, it is possible to retrieve the priority of a task and change it by means of the functions

```
unsigned portBASE_TYPE uxTaskPriorityGet(xTaskHandle pxTask);
```

```
void vTaskPrioritySet(xTaskHandle pxTask,
    unsigned portBASE_TYPE uxNewPriority);
```

Both functions are optional, that is, they can be excluded from the operating system to reduce its code and data space requirements. They both take a task handle, pxTask, as their first argument. The special value NULL can be used as a shortcut to refer to the calling task.

The function vTaskPrioritySet modifies the priority of a task after it has been created, and uxTaskPriorityGet returns the current priority of the task. It should, however, be noted that both the priority given at task creation and the priority set by vTaskPrioritySet represent the *baseline* priority of the task.

Instead, uxTaskPriorityGet returns its *active* priority, which may differ from the baseline priority when one of the mechanisms to prevent unbounded priority inversion, to be discussed in Chapter 14, is in effect. More specifically, FreeRTOS implements the priority inheritance protocol for mutual exclusion semaphores. See also Section 8.3 for more information.

The pair of optional functions vTaskSuspend and vTaskResume take an argument of type xTaskHandle according to the following prototypes:

```
void vTaskSuspend(xTaskHandle pxTaskToSuspend);
```

```
void vTaskResume(xTaskHandle pxTaskToResume);
```

They are used to suspend and resume the execution of the task identified by the argument. For vTaskSuspend, the special value NULL can be used to suspend the invoking task, whereas, obviously, it is impossible for a task to resume executing of its own initiative.

Like vTaskDelete, vTaskSuspend also may suspend the execution of a task at an arbitrary point. Therefore, it must be used with care when the task to be suspended contains critical sections—or, more generally, can get mutually exclusive access to one or more shared resources—because those resources are not implicitly released while the task is suspended.

FreeRTOS, like most other monolithic operating systems, does not hold a full task context for interrupt handlers, and hence, they are not full-fledged tasks. One of the consequences of this design choice is that interrupt handlers cannot block or suspend themselves (informally speaking, there is no dedicated space within the operating system to save their context into), and hence, calling vTaskSuspend(NULL) from an interrupt handler makes no sense. For related reasons, interrupt handlers are also not allowed to suspend regular tasks by invoking vTaskSuspend with a valid xTaskHandle as argument.

The function

```
portBASE_TYPE xTaskResumeFromISR(xTaskHandle pxTaskToResume);
```

is a variant of vTaskResume that must be used to resume a task from an interrupt handler, also known as Interrupt Service Routine (ISR) in the FreeRTOS jargon.

Since, as said above, interrupt handlers do not have a full-fledged, dedicated task context in FreeRTOS, xTaskResumeFromISR cannot perform a full context switch between tasks when needed as its regular counterpart would do. A context switch would be necessary, for example, when a low-priority task is interrupted and the interrupt handler wakes up a different task, with a higher priority.

On the contrary, xTaskResumeFromISR merely returns a nonzero value in this case in order to make the invoking interrupt handler aware of the situation. In response to this indication, the interrupt handler will eventually invoke the FreeRTOS scheduling algorithm so that the higher-priority task just resumed will get executed upon its exit instead of the interrupted one. This is accomplished by invoking a primitive such as, for example, vPortYieldFromISR() for the ARM Cortex-M3 port of FreeRTOS. Although the implementation of the primitive is port-dependent, its name is the same across most recent ports of FreeRTOS.

This course of action has the additional advantage that the scheduling algorithm—a quite expensive algorithm to be run in an interrupt context—will be triggered only when strictly necessary to avoid priority inversion.

The optional function

```
portBASE_TYPE xTaskIsTaskSuspended(xTaskHandle xTask)
```

can be used to tell whether a certain task, identified by xTask, is currently suspended or not. Its return value will be nonzero if the task is suspended, and zero otherwise.

The function

```
void vTaskSuspendAll(void);
```

suspends all tasks but the calling one. Interrupt handling is not suspended and is still performed as usual.

Symmetrically, the function

```
portBASE_TYPE xTaskResumeAll(void);
```

resumes all tasks suspended by vTaskSuspendAll. In turn, for example, when the priority of one of the resumed tasks is higher than the priority of the invoking task, this may require a context switch. In this case, the context switch is performed immediately within xTaskResumeAll itself, and therefore, this function cannot be called from an interrupt handler. The invoking task is later notified that it lost the processor for this reason because it will get a nonzero return value from xTaskResumeAll.

Contrary to what could be expected, both vTaskSuspendAll and xTaskResumeAll are extremely efficient on single-processor systems, such those targeted by FreeRTOS. In fact, these functions are not implemented by suspending and resuming all tasks one by one but by temporarily disabling the operating system scheduler, and the latter operation requires little more work than updating a shared counter.

Hence, they can be used to implement critical regions without using any semaphore and without fear of unbounded priority inversion simply by using them as a pair of brackets around the critical code. In fact, in a single-processor system, vTaskSuspendAll opens a mutual exclusion region because the first task to successfully execute it will effectively prevent all other tasks from being executed until it invokes xTaskResumeAll.

Moreover, they also realize an extremely aggressive form of immediate priority ceiling, as will be discussed in Chapter 14, because any task executing between vTaskSuspendAll and xTaskResumeAll implicitly gets the highest possible priority in the system, too, except interrupt handlers. That said, the method just described has two main shortcomings:

1. Any FreeRTOS primitive that might block the caller for any reason and even temporarily, or might require a context switch, must not be used within this kind of critical region. This is because blocking the only task allowed to run would completely lock up the system, and it is impossible to perform a context switch with the scheduler disabled.
2. Protecting critical regions with a sizable execution time in this way would probably be unacceptable in many applications because it leads to a large amount of unnecessary blocking. This is especially true for high-priority tasks, because if one of them becomes ready for execution while a low-priority task is engaged in a critical region of this kind, it will not run immediately, but only at the end of the critical region itself. See Chapter 14 for additional information on how to compute the worst-case blocking time a task will suffer, depending on the method used to address the unbounded priority inversion problem.

The last function related to task management simply returns the number of tasks currently present in the system, regardless of their state:

```
unsigned portBASE_TYPE uxTaskGetNumberOfTasks(void);
```

Therefore, the count also includes the calling task and blocked tasks. Moreover, it may also include some tasks that have been deleted by vTaskDelete. This is a side effect of the delayed dismissal of the operating system's data structures associated with a task upon deletion previously mentioned.

8.2 MESSAGE QUEUES

Message queues are the main Interprocess Communication (IPC) mechanism provided by FreeRTOS. They are also the basic block on which the additional IPC mechanisms discussed in the next sections are built. For this reason, none of the primitives that operate on message queues, summarized in Table 8.2, can be excluded from the operating system configuration. To use them, it is necessary to include the main FreeRTOS header FreeRTOS.h, followed by queue.h. There are also some additional primitives, intended either for internal operating system's use or to facilitate debugging, that will not be discussed here.

With respect to the nomenclature presented in Chapter 6, FreeRTOS adopts a symmetric, indirect naming scheme for message queues because the sender task does not name the intended receiver task directly. Rather, both the sender and the receiver make reference to an intermediate entity, that is, the message queue itself.

The synchronization model is asynchronous with finite buffering because the sender always proceeds as soon as the message has been stored into the message queue without waiting for the message to be received at destination. The amount of buffering is fixed and known in advance for each message queue. It is set when the queue is created and cannot be modified afterward.

Table 8.2

Summary of the main message-queue related primitives of FreeRTOS

Function	Purpose	Optional
xQueueCreate	Create a message queue	-
vQueueDelete	Delete a message queue	-
xQueueSendToBack	Send a message	-
xQueueSendToFront	Send a high-priority message	-
xQueueSendToBackFromISR	. . . from an interrupt handler	-
xQueueSendToFrontFromISR	. . . from an interrupt handler	-
xQueueReceive	Receive a message	-
xQueueReceiveFromISR	. . . from an interrupt handler	-
xQueuePeek	Nondestructive receive	-
uxQueueMessagesWaiting	Query current queue length	-
uxQueueMessagesWaitingFromISR	. . . from an interrupt handler	-
xQueueIsQueueEmptyFromISR	Check if a queue is empty	-
xQueueIsQueueFullFromISR	Check if a queue is full	-

No functions are provided for data serialization, but this does not have serious consequences because FreeRTOS message passing is restricted anyway to take place between tasks belonging to a single process. All these tasks are therefore necessarily executed on the same machine and share the same address space.

The function

```
xQueueHandle xQueueCreate(
    unsigned portBASE_TYPE uxQueueLength,
    unsigned portBASE_TYPE uxItemSize);
```

creates a new message queue, given the maximum number of elements it can contain, uxQueueLength, and the size of each element, uxItemSize, expressed in bytes. Upon successful completion, the function returns a valid message queue handle to the caller, which must be used for any subsequent operation on the queue just created. When an error occurs, the function returns a NULL pointer instead.

When a message queue is no longer needed, it is advisable to delete it, in order to reclaim its memory for future use, by means of the function

```
void vQueueDelete(xQueueHandle xQueue);
```

It should be noted that the deletion of a FreeRTOS message queue takes place immediately and is never delayed even if some tasks are waiting on it. The fate of the waiting tasks then depends on whether they specified a time limit for the wait or not:

- if they did specify a time limit for the message queue operation, they will receive an error indication when the operation times out;
- otherwise, they will be blocked forever.

After a message queue has been successfully created and its xQueue handle is available for use, it is possible to send a message to it by means of the functions

```
portBASE_TYPE xQueueSendToBack(
    xQueueHandle xQueue,
    const void *pvItemToQueue,
    portTickType xTicksToWait);

portBASE_TYPE xQueueSendToFront(
    xQueueHandle xQueue,
    const void *pvItemToQueue,
    portTickType xTicksToWait);

portBASE_TYPE xQueueSendToBackFromISR(
    xQueueHandle xQueue,
    const void *pvItemToQueue,
    portBASE_TYPE *pxHigherPriorityTaskWoken);
```

```
portBASE_TYPE xQueueSendToFrontFromISR(
    xQueueHandle xQueue,
    const void *pvItemToQueue,
    portBASE_TYPE *pxHigherPriorityTaskWoken);
```

The first function, xQueueSendToBack, sends a message to the back of a message queue. The message to be sent is pointed by the pvItemToQueue argument, whereas its size is implicitly assumed to be equal to the size of a message queue item, as declared when the queue was created.

The last argument, xTicksToWait, specifies the maximum amount of time allotted to the operation. In particular,

- If the value is 0 (zero), the function returns an error indication to the caller when the operation cannot be performed immediately because the message queue is completely full at the moment.
- If the value is portMAX_DELAY (a symbolic constant defined in a port-dependent header file that is automatically included by the main FreeRTOS header file), when the message queue is completely full, the function blocks indefinitely until the space it needs becomes available. For this option to be available, the operating system must be configured to support task suspend and resume, as described in Section 8.1.
- Any other value is interpreted as the maximum amount of time the function will wait, expressed as an integral number of clock *ticks*. See Section 8.4 for more information about ticks.

The return value of xQueueSendToBack will be pdPASS if the function was successful; any other value means than an error occurred. In particular, the error code errQUEUE_FULL means that the function was unable to send the message within the maximum amount of time specified by xTicksToWait because the queue was full.

Unlike in POSIX, FreeRTOS messages do not have a full-fledged priority associated with them, and hence, they are normally sent and received in First-In, First-Out (FIFO) order. However, a high-priority message can be sent using the xQueueSendToFront function instead of xQueueSendToBack. The only difference between those two functions is that xQueueSendToFront sends the message to the *front* of the message queue so that it passes over the other messages stored in the queue and will be received before them.

Neither xQueueSendToBack nor xQueueSendToFront can be called from an interrupt handler. Instead, either xQueueSendToBackFromISR or xQueueSendToFrontFromISR must be used. The only differences with respect to their regular counterparts are

- They cannot block the caller, and hence, they do not have a xTicksToWait argument and always behave as if the timeout were 0, that is, they return an error indication to the caller if the operation cannot be concluded immediately.

- The argument pxHigherPriorityTaskWoken points to a portBASE_TYPE variable. The function will set the referenced variable to either pdTRUE or pdFALSE, depending on whether or not it awakened a task with a priority higher than the task which was running when the interrupt handler started.
  The interrupt handler should use this information, as discussed in Section 8.1, to determine if it should invoke the FreeRTOS scheduling algorithm before exiting.

The functions xQueueSend and xQueueSendFromISR are just synonyms of xQueueSendToBack and xQueueSendToBackFromISR, respectively, They have been retained for backward compatibility with previous versions of FreeRTOS, which did not have the ability to send a message to the front of a message queue.

Messages are always received from the front of a message queue by means of the following functions:

```
portBASE_TYPE xQueueReceive(
    xQueueHandle xQueue,
    void *pvBuffer,
    portTickType xTicksToWait);

portBASE_TYPE xQueueReceiveFromISR(
    xQueueHandle xQueue,
    void *pvBuffer,
    portBASE_TYPE *pxHigherPriorityTaskWoken);

portBASE_TYPE xQueuePeek(
    xQueueHandle xQueue,
    void *pvBuffer,
    portTickType xTicksToWait);
```

All these functions take a message queue handle, xQueue, as their first argument; this is the message queue they will work upon. The second argument, pvBuffer, is a pointer to a memory buffer into which the function will store the message just received. It must be large enough to hold the message, that is, at least as large as a message queue item as declared when the queue was created.

In the case of xQueueReceive, the last argument, xTicksToWait, specifies how much time the function should wait for a message to become available if the message queue was completely empty when it was invoked. The valid values of xTicksToWait are the same already mentioned when discussing xQueueSendToBack.

The return value of xQueueReceive will be pdPASS if the function was successful; any other value means than an error occurred. In particular, the error code errQUEUE_EMPTY means that the function was unable to receive a message within the maximum amount of time specified by xTicksToWait because the queue was empty. In this case, the buffer pointed by `pvBuffer` will not contain any valid message after `xQueueReceive` returns.

The function `xQueueReceive`, when successful, *removes* the message it just received from the message queue so that each message sent to the queue is received exactly once. On the contrary, the function `xQueuePeek` simply *copies* the message into the memory buffer indicated by the caller without removing it for the queue. It takes the same arguments as `xQueueReceive`.

The function `xQueueReceiveFromISR` is the variant of `xQueueReceive` that must be used within an interrupt handler. It never blocks, but it returns to the caller in the variable pointed by `pxHigherPriorityTaskWoken`, an indication on whether it awakened a higher priority task or not.

The last group of functions,

```
unsigned portBASE_TYPE
    uxQueueMessagesWaiting(const xQueueHandle xQueue);
```

```
unsigned portBASE_TYPE
    uxQueueMessagesWaitingFromISR(const xQueueHandle xQueue);
```

```
portBASE_TYPE
    xQueueIsQueueEmptyFromISR(const xQueueHandle xQueue);
```

```
portBASE_TYPE
    xQueueIsQueueFullFromISR(const xQueueHandle xQueue);
```

queries various aspects of a message queue status. In particular,

- `uxQueueMessagesWaiting` and `uxQueueMessagesWaitingFromISR` return the number of items currently stored in the message queue xQueue. The latter variant must be used when the invoker is an interrupt handler.
- `xQueueIsQueueEmptyFromISR` and `xQueueIsQueueFullFromISR` return a Boolean value that will be pdTRUE if the message queue xQueue is empty (or full, respectively) and pdFALSE otherwise. Both can be invoked safely from an interrupt handler.

These functions should be used with caution because, although the information they return is certainly correct and valid at the time of the call, the scope of its validity is somewhat limited. It is worth mentioning, for example, that the information may *no longer* be valid and should not be relied upon when any subsequent message queue operation is attempted because other tasks may have changed the queue status in the meantime.

For example, the preventive execution of `uxQueueMessageWaiting` by a task, with a result less than the total length of the message queue, is not enough to guarantee that the same task will be able to immediately conclude a `xQueueSendToBack` in the near future: other tasks, or interrupt handlers, may have sent additional items into the queue and filled it completely in the meantime.

The following program shows how the producers/consumers problem can be solved using a FreeRTOS message queue.

```c
/* Producers/Consumers problem solved with a FreeRTOS message queue */

#include <stdio.h>      /* For printf() */
#include <stdlib.h>
#include <FreeRTOS.h>  /* Main RTOS header */
#include <task.h>       /* Task and time functions */
#include <queue.h>      /* Message queue functions */

#define  N          10  /* Buffer size (# of items) */

#define  NP         3   /* Number of producer tasks */
#define  NC         2   /* Number of consumer tasks */

/* The minimal task stack size specified in the FreeRTOS configuration
   does not support printf().  Make it larger.
*/
#define  STACK_SIZE      (configMINIMAL_STACK_SIZE+512)

#define  PRODUCER_DELAY 500   /* Delays in producer/consumer tasks */
#define  CONSUMER_DELAY 300

/* Data type for task arguments, for both producers and consumers */
struct task_args_s {
  int n;               /* Task number */
  xQueueHandle q;      /* Message queue to use */
};

void producer_code(void *argv)
{
  /* Cast the argument pointer, argv, to the right data type */
  struct task_args_s *args = (struct task_args_s *)argv;
  int c = 0;
  int item;

  while(1)
  {
    /* A real producer would put together an actual data item.
       Here, we block for a while and then make up a fake item.
    */
    vTaskDelay(PRODUCER_DELAY);
    item = args->n*1000 + c;
    c++;

    printf("Producer #%d - sending  item %6d\n", args->n, item);

    /* Send the data item to the back of the queue, waiting if the
       queue is full.  portMAX_DELAY means that there is no upper
       bound on the amount of wait.
    */
    if(xQueueSendToBack(args->q, &item, portMAX_DELAY) != pdPASS)
      printf("*  Producer %d unable to send\n", args->n);
  }
}

void consumer_code(void *argv)
{
  struct task_args_s *args = (struct task_args_s *)argv;
  int item;

  while(1)
  {
    /* Receive a data item from the front of the queue, waiting if
       the queue is empty.  portMAX_DELAY means that there is no
```

```
      upper bound on the amount of wait.
  */
  if(xQueueReceive(args->q, &item, portMAX_DELAY) != pdPASS)
    printf("* Consumer #%d unable to receive\n", args->n);
  else
    printf("Consumer #%d - received item %6d\n", args->n, item);

  /* A real consumer would do something meaningful with the data item.
     Here, we simply block for a while
  */
  vTaskDelay(CONSUMER_DELAY);
  }
}

int main(int argc, char *argv[])
{
  xQueueHandle q;                      /* Message queue handle */
  struct task_args_s prod_args[NP];   /* Task arguments for producers */
  struct task_args_s cons_args[NC];   /* Task arguments for consumers */
  xTaskHandle dummy;
  int i;

  /* Create the message queue */
  if((q = xQueueCreate(N, sizeof(int))) == NULL)
    printf("* Cannot create message queue of %d elements\n", N);

  else
  {
    /* Create NP producer tasks */
    for(i=0; i<NP; i++)
    {
      prod_args[i].n = i;  /* Prepare the arguments */
      prod_args[i].q = q;

      /* The task handles are not used in the following,
         so a dummy variable is used for them
      */
      if(xTaskCreate(producer_code, "PROD", STACK_SIZE,
                  &(prod_args[i]), tskIDLE_PRIORITY, &dummy) != pdPASS)
        printf("* Cannot create producer #%d\n", i);
    }

    /* Create NC consumer tasks */
    for(i=0; i<NC; i++)
    {
      cons_args[i].n = i;
      cons_args[i].q = q;

      if(xTaskCreate(consumer_code, "CONS", STACK_SIZE,
                  &(cons_args[i]), tskIDLE_PRIORITY, &dummy) != pdPASS)
        printf("* Cannot create consumer #%d\n", i);
    }

    vTaskStartScheduler();
    printf("* vTaskStartScheduler() failed\n");
  }

  /* Since this is just an example, always return a success
     indication, even this might not be true.
  */
  return EXIT_SUCCESS;
}
```

The main program first creates the message queue that will be used for interprocess communication, and then a few producer and consumer tasks. For the sake of the example, the number of tasks to be created is controlled by the macros NP and NC, respectively.

The producers will all execute the same code, that is, the function `producer_code`. Each of them receives as argument a pointer to a `struct task_args_s` that holds two fields:

1. a task number (n field) used to distinguish one task from another in the debugging printouts, and
2. the message queue (q field) that the task will use to send data.

In this way, all tasks can work together by only looking at their arguments and without sharing any variable, as foreseen by the message-passing paradigm. Symmetrically, the consumers all execute the function `consumer_code`, which has a very similar structure.

8.3 COUNTING, BINARY, AND MUTUAL EXCLUSION SEMAPHORES

FreeRTOS provides four different kinds of semaphores, representing different trade-offs between features and efficiency:

1. **Counting semaphores** are the most general kind of semaphore provided by FreeRTOS. They are also the only kind of semaphore actually able to hold a count, as do the abstract semaphores discussed in Chapter 5. The main difference is that, in the FreeRTOS case, the maximum value the counter may assume must be declared when the semaphore is first created.
2. **Binary semaphores** have a maximum value as well as an initial value of one. As a consequence, their value can only be either one or zero, but they can still be used for either mutual exclusion or task synchronization.
3. **Mutex semaphores** are similar to binary semaphores, with the additional restriction that they must *only* be used as mutual exclusion semaphores, that is, P() and V() must always appear in pairs and must be placed as brackets around critical regions. Hence, mutex semaphores cannot be used for task synchronization. In exchange for this, mutex semaphores implement priority inheritance, which as discussed in Chapter 14, is especially useful to address the unbounded priority inversion problem.
4. **Recursive mutex semaphores** have all the features ordinary mutex semaphores have, and also optionally support the so-called "recursive" locks and unlocks in which a process is allowed to contain more than one nested critical region, all controlled by the same semaphore and delimited by their own P()/V() brackets, without deadlocking. In this case, the semaphore is automatically locked and unlocked only at the outermost region boundary, as it should.

The four different kinds of semaphores are created by means of distinct functions, all listed in Table 8.3. In order to call any semaphore-related function, it is necessary to include the main FreeRTOS header `FreeRTOS.h`, followed by `semphr.h`.

Table 8.3

Summary of the semaphore creation/deletion primitives of FreeRTOS

Function	Purpose	Optional
xSemaphoreCreateCounting	Create a counting semaphore	*
vSemaphoreCreateBinary	Create a binary semaphore	-
xSemaphoreCreateMutex	Create a mutex semaphore	*
xSemaphoreCreateRecursiveMutex	Create a recursive mutex	*
vQueueDelete	Delete a semaphore of any kind	-

The function

```
xSemaphoreHandle xSemaphoreCreateCounting(
    unsigned portBASE_TYPE uxMaxCount,
    unsigned portBASE_TYPE uxInitialCount);
```

creates a counting semaphore with a given maximum (uxMaxCount) and initial (uxInitialCount) value. When successful, it returns to the caller a valid semaphore handle; otherwise, it returns a NULL pointer. To create a binary semaphore, use the macro xSemaphoreCreateBinary instead:

```
void vSemaphoreCreateBinary(xSemaphoreHandle xSemaphore);
```

It should be noted that

- Unlike xSemaphoreCreateCounting, xSemaphoreCreateBinary has no return value. Being a macro rather than a function, it directly manipulates xSemaphore instead. Upon success, xSemaphore will be set to the semaphore handle just created; otherwise, it will be set to NULL.
- Both the maximum and initial value of a binary semaphore are constrained to be 1, and hence, they are not explicitly indicated.
- Binary semaphores are the only kind of semaphore that is always available for use in FreeRTOS, regardless of its configuration. All the others are optional.

Mutual exclusion semaphores are created by means of two different functions, depending on whether the recursive lock and unlock feature is desired or not:

```
xSemaphoreHandle xSemaphoreCreateMutex(void);
xSemaphoreHandle xSemaphoreCreateRecursiveMutex(void);
```

In both cases, the creation function returns either a semaphore handle upon successful completion, or NULL. All mutual exclusion semaphores are unlocked when they are first created, and priority inheritance is always enabled for them.

Since FreeRTOS semaphores of all kinds are built on top of a message queue, they can be deleted by means of the function vQueueDelete, already discussed in Section 8.2. Also in this case, the semaphore is destroyed immediately even if there are some tasks waiting on it.

After being created, all kinds of semaphores except recursive, mutual exclusion semaphores are acted upon by means of the functions xSemaphoreTake and xSemaphoreGive, the FreeRTOS counterpart of P() and V(), respectively. Both take a semaphore handle xSemaphore as their first argument:

```
portBASE_TYPE xSemaphoreTake(xSemaphoreHandle xSemaphore,
    portTickType xBlockTime);


portBASE_TYPE xSemaphoreGive(xSemaphoreHandle xSemaphore);
```

In addition, xSemaphoreTake also takes a second argument, xBlockTime, that specifies the maximum amount of time allotted to the operation. The interpretation and valid values that this argument can assume are the same as for message queues.

The function xSemaphoreTake returns pdTRUE if it was successful, that is, it was able to conclude the semaphore operation before the specified amount of time elapsed. Otherwise, it returns pdFALSE. Similarly, xSemaphoreGive returns pdTRUE when successful, and pdFALSE if an error occurred.

The function xSemaphoreGiveFromISR is the variant of xSemaphoreGive that must be used within an interrupt handler:

```
portBASE_TYPE xSemaphoreGiveFromISR(xSemaphoreHandle xSemaphore,
    portBASE_TYPE *pxHigherPriorityTaskWoken);
```

Like many other FreeRTOS primitives that can be invoked from an interrupt handler, this function returns to the caller, in the variable pointed by pxHigherPriorityTaskWoken, an indication on whether or not it awakened a task with a priority higher than the task which was running when the interrupt handler started.

The interrupt handler should use this information, as discussed in Section 8.1, to determine if it should invoke the FreeRTOS scheduling algorithm before exiting. The function also returns either pdTRUE or pdFALSE, depending on whether it was successful or not.

The last pair of functions to be discussed here are the counterpart of xSemaphoreTake and xSemaphoreGive, to be used with recursive mutual exclusion semaphores:

```
portBASE_TYPE xSemaphoreTakeRecursive(xSemaphoreHandle xMutex,
    portTickType xBlockTime);


portBASE_TYPE xSemaphoreGiveRecursive(xSemaphoreHandle xMutex);
```

Table 8.4

Summary of the semaphore manipulation primitives of FreeRTOS

Function	Purpose	Optional
xSemaphoreTake	Perform a P() on a semaphore	-
xSemaphoreGive	Perform a V() on a semaphore	-
xSemaphoreGiveFromISR	...from an interrupt handler	-
xSemaphoreTakeRecursive	P() on a recursive mutex	*
xSemaphoreGiveRecursive	V() on a recursive mutex	*

Both their arguments and return values are the same as xSemaphoreTake and xSemaphoreGive, respectively. Table 8.4 summarizes the FreeRTOS functions that work on semaphores.

The following program shows how the producers–consumers problem can be solved using a shared buffer and FreeRTOS semaphores.

```
/* Producers/Consumers problem solved with FreeRTOS semaphores */

#include <stdio.h>     /* For printf() */
#include <stdlib.h>
#include <FreeRTOS.h>  /* Main RTOS header */
#include <task.h>      /* Task and time functions */
#include <semphr.h>    /* Semaphore functions */

#define  N        10   /* Buffer size (# of items) */

#define  NP       3    /* Number of producer tasks */
#define  NC       2    /* Number of consumer tasks */

/* The minimal task stack size specified in the FreeRTOS configuration
   does not support printf(). Make it larger.
*/
#define  STACK_SIZE      (configMINIMAL_STACK_SIZE+512)

#define  PRODUCER_DELAY 500   /* Delays in producer/consumer tasks */
#define  CONSUMER_DELAY 300

/* Data type for task arguments, for both producers and consumers */
struct task_args_s {
  int n;                 /* Task number */
};

/* Shared variables and semaphores.  They implement the shared
   buffer, as well as mutual exclusion and task synchronization
*/

int buf[N];
int in = 0, out = 0;
xSemaphoreHandle empty, full;
xSemaphoreHandle mutex;


void producer_code(void *argv)
{
  /* Cast the argument pointer, argv, to the right data type */
  struct task_args_s *args = (struct task_args_s *)argv;
```

```
  int c = 0;
  int item;

  while(1)
  {
    /* A real producer would put together an actual data item.
       Here, we block for a while and then make up a fake item.
    */
    vTaskDelay(PRODUCER_DELAY);
    item = args->n*1000 + c;
    c++;

    printf("Producer #%d - sending  item %6d\n", args->n, item);

    /* Synchronize with consumers */
    if(xSemaphoreTake(empty, portMAX_DELAY) != pdTRUE)
      printf("*  Producer %d unable to take 'empty'\n", args->n);

    /* Mutual exclusion for buffer access */
    else if(xSemaphoreTake(mutex, portMAX_DELAY) != pdTRUE)
      printf("*  Producer %d unable to take 'mutex'\n", args->n);

    else
    {
      /* Store data item into 'buf', update 'in' index */
      buf[in] = item;
      in = (in + 1) % N;

      /* Release mutex */
      if(xSemaphoreGive(mutex) != pdTRUE)
        printf("*  Producer %d unable to give 'mutex'\n", args->n);

      /* Synchronize with consumers */
      if(xSemaphoreGive(full) != pdTRUE)
        printf("*  Producer %d unable to give 'full'\n", args->n);
    }
  }
}

void consumer_code(void *argv)
{
  struct task_args_s *args = (struct task_args_s *)argv;
  int item;

  while(1)
  {
    /* Synchronize with producers */
    if(xSemaphoreTake(full, portMAX_DELAY) != pdTRUE)
      printf("*  Consumer %d unable to take 'full'\n", args->n);

    /* Mutual exclusion for buffer access */
    else if(xSemaphoreTake(mutex, portMAX_DELAY) != pdTRUE)
      printf("*  Consumer %d unable to take 'mutex'\n", args->n);

    else
    {
      /* Get data item from 'buf', update 'out' index */
      item = buf[out];
      out = (out + 1) % N;

      /* Release mutex */
      if(xSemaphoreGive(mutex) != pdTRUE)
        printf("*  Consumer %d unable to give 'mutex'\n", args->n);

      /* Synchronize with producers */
      if(xSemaphoreGive(empty) != pdTRUE)
```

```c
      printf("*   Consumer %d unable to give 'full'\n", args->n);

    /* A real consumer would do something meaningful with the data item.
       Here, we simply print it out and block for a while
    */
    printf("Consumer #%d - received item %6d\n", args->n, item);
    vTaskDelay(CONSUMER_DELAY);
    }
  }
}

int main(int argc, char *argv[])
{
  struct task_args_s prod_args[NP];   /* Task arguments for producers */
  struct task_args_s cons_args[NC];   /* Task arguments for consumers */
  xTaskHandle dummy;
  int i;

  /* Create the two synchronization semaphores, empty and full.
     They both have a maximum value of N (first argument), and
     an initial value of N and 0, respectively
  */
  if((empty = xSemaphoreCreateCounting(N, N)) == NULL
     || (full = xSemaphoreCreateCounting(N, 0)) == NULL)
    printf("* Cannot create counting semaphores\n");

  /* Create the mutual exclusion semaphore */
  else if((mutex = xSemaphoreCreateMutex()) == NULL)
    printf("* Cannot create mutex\n");

  else
  {
    /* Create NP producer tasks */
    for(i=0; i<NP; i++)
    {
      prod_args[i].n = i;   /* Prepare the argument */

      /* The task handles are not used in the following,
         so a dummy variable is used for them
      */
      if(xTaskCreate(producer_code, "PROD", STACK_SIZE,
                &(prod_args[i]), tskIDLE_PRIORITY, &dummy) != pdPASS)
        printf("* Cannot create producer #%d\n", i);
    }

    /* Create NC consumer tasks */
    for(i=0; i<NC; i++)
    {
      cons_args[i].n = i;

      if(xTaskCreate(consumer_code, "CONS", STACK_SIZE,
                &(cons_args[i]), tskIDLE_PRIORITY, &dummy) != pdPASS)
        printf("* Cannot create consumer #%d\n", i);
    }

    vTaskStartScheduler();
    printf("* vTaskStartScheduler() failed\n");
  }

  /* Since this is just an example, always return a success
     indication, even this might not be true.
  */
  return EXIT_SUCCESS;
}
```

As before, the main program takes care of initializing the shared synchronization and mutual exclusion semaphores needed by the application, creates several producers and consumers, and then starts the scheduler. Even if the general parameter passing strategy adopted in the previous example has been maintained, the only argument passed to the tasks is their identification number because the semaphores, as well as the data buffer itself, are shared and globally accessible.

With respect to the solution based on message queues, the most important difference to be remarked is that, in this case, the data buffer shared between the producers and consumers must be allocated and handled explicitly by the application code, instead of being hidden behind the operating system's implementation of message queues. In the example, it has been implemented by means of the (circular) buffer buf [], assisted by the input and output indexes in and out.

8.4   CLOCKS AND TIMERS

Many of the activities performed in a real-time system ought to be correlated, quite intuitively, with time. Accordingly, all FreeRTOS primitives that may potentially block the caller, such as those discussed in Sections 8.2 and 8.3, allow the caller to specify an upper bound to the blocking time.

Moreover, FreeRTOS provides a small set of primitives, listed in Table 8.5, to *keep track* of the elapsed time and *synchronize* a task with it by delaying its execution. Since they do not have their own dedicated header file, it is necessary to include the main FreeRTOS header file, FreeRTOS.h, followed by task.h, in order to use them.

In FreeRTOS, the same data type, portTickType, is used to represent both the current time and a time interval. The current time is simply represented by the number of clock *ticks* elapsed from when the operating system scheduler was first started. The length of a tick depends on the operating system configuration and, to some extent, on hardware capabilities.

The configuration macro configTICK_RATE_HZ represents the tick frequency in Hertz. In addition, most porting layers define the macro portTICK_RATE_MS as the fraction 1000/configTICK_RATE_HZ so that is represents the tick period, expressed in milliseconds. Both of them can be useful to convert back and forth between the usual time measurement units and clock ticks.

Table 8.5
Summary of the time-related primitives of FreeRTOS

Function	Purpose	Optional
xTaskGetTickCount	Get current time, in ticks	-
vTaskDelay	Relative time delay	*
vTaskDelayUntil	Absolute time delay	*

The function

```
portTickType xTaskGetTickCount(void);
```

returns the current time, expressed in ticks. It is quite important to keep in mind that the returned value comes from a tick counter of type `portTickType` maintained by FreeRTOS. Barring some details, the operating system resets the counter to zero when the scheduler is first started, and increments it with the help of a periodic interrupt source, running at a frequency of `configTICK_RATE_HZ`.

In most microcontrollers, the tick counter data type is either a 16- or 32-bit unsigned integer, depending on the configuration. Therefore, it will sooner or later wrap around and resume counting from zero. For instance, an unsigned, 32-bit counter incremented at 1000 Hz—a common configuration choice for FreeRTOS—will wrap around after about 1193 hours, that is, a bit more than 49 days.

It is therefore crucial that any application planning to "stay alive" for a longer time, as many real-time applications must do, is aware of the wrap-around and handles it appropriately if it manipulates time values directly. If this is not the case, the application will be confronted with time values that suddenly "jump into the past" when a wraparound occurs, with imaginable consequences. The delay functions, to be discussed next, already handle time wraparound automatically, and hence, no special care is needed to use them.

Two distinct delay functions are available, depending on whether the delay should be *relative*, that is, measured with respect to the instant in which the delay function is invoked, or *absolute*, that is, until a certain instant in the future, measured as a number of ticks, from when the scheduler has been started:

```
void vTaskDelay(portTickType xTicksToDelay);
void vTaskDelayUntil(portTickType *pxPreviousWakeTime,
  portTickType xTimeIncrement);
```

The function `vTaskDelay` implements a relative time delay: it blocks the calling task for `xTicksToDelay` ticks, then returns. As shown in Figure 8.1, the time interval is relative to the time of the call and the amount of delay is fixed, that is, `xTicksToDelay`.

Instead, the function `vTaskDelayUntil` implements an absolute time delay: referring again to Figure 8.1, the next wake-up time is calculated as the previous one, `*pxPreviousWakeTime`, plus the time interval `xTimeIncrement`. Hence, the amount of delay varies from call to call, and the function might not block at all if the prescribed wake-up time is already in the past. Just before returning, the function also increments `*pxPreviousWakeTime` by `xTimeIncrement` so that it is ready for the next call.

The right kind of delay to be used depends on its purpose. A relative delay may be useful, for instance, if an I/O device must be allowed (at least) a certain amount of time to react to a command. In this case, the delay must be measured from when the command has actually been sent to the device, and a relative delay makes sense.

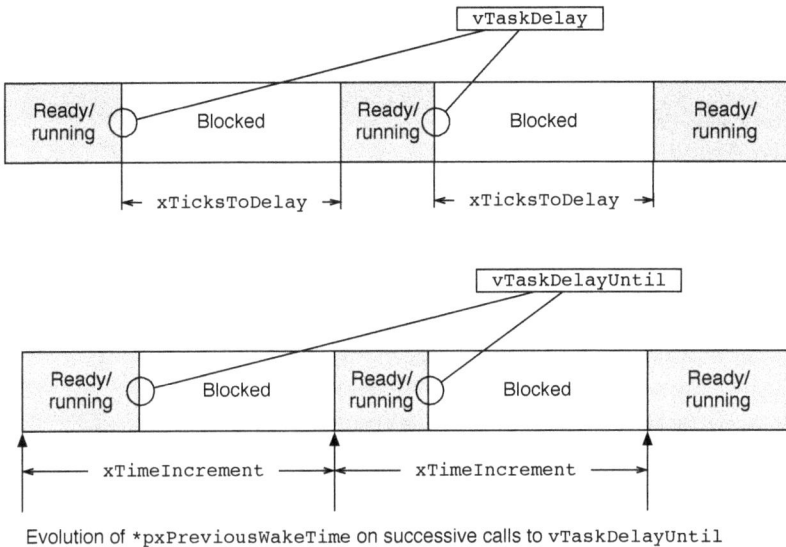

Evolution of *pxPreviousWakeTime on successive calls to vTaskDelayUntil

Figure 8.1 Comparison between relative and absolute time delays, as implemented by vTaskDelay and vTaskDelayUntil.

On the other hand, an absolute delay is better when a task has to carry out an operation periodically because it guarantees that the period will stay constant even if the response time of the task—the grey rectangles in Figure 8.1—varies from one instance to another.

The last example program, listed below, shows how absolute and relative delays can be used in an actual piece of code. When run for a long time, the example is also useful in better highlighting the difference between those two kinds of delay. In fact, it can be seen that the wake-up time of task rel_delay (that uses a relative delay) not only drifts forward but is also irregular because the variations in its response time are not accounted for when determining the delay before its next activation. On the contrary, the wake-up time of task abs_delay (that uses an absolute delay) does not drift, and it strictly periodic.

```
/* FreeRTOS vTaskDelay versus vTaskDelayUntil */

#include <stdio.h>    /* For printf() */
#include <stdlib.h>
#include <FreeRTOS.h> /* Main RTOS header */
#include <task.h>     /* Task and time functions */

/* The minimal task stack size specified in the FreeRTOS configuration
   does not support printf().  Make it larger.

   Period is the nominal period of the tasks to be created
*/
#define  STACK_SIZE     (configMINIMAL_STACK_SIZE+512)
#define  PERIOD         ((portTickType)100)

/* Periodic task with relative delay (vTaskDelay) */
```

```
void rel_delay(void *dummy)
{
  while(1)
  {
    /* Block for PERIOD ticks, measured from 'now' */
    vTaskDelay(PERIOD);

    printf("rel_delay active at --------- %9u ticks\n",
           (unsigned int)xTaskGetTickCount());
  }
}

/* Periodi task with absolute delay (vTaskDelayUntil) */
void abs_delay(void *dummy)
{
  portTickType last_wakeup = xTaskGetTickCount();

  while(1)
  {
    /* Block until the instant last_wakeup + PERIOD,
       then update last_wakeup and return
    */
    vTaskDelayUntil(&last_wakeup, PERIOD);

    printf("abs_delay active at %9u --------- ticks\n",
           (unsigned int)xTaskGetTickCount());
  }
}

int main(int argc, char *argv[])
{
  xTaskHandle dummy;

  /* Create the two tasks to be compared, with no arguments.
     The task handles are not used, hence they are discarded
  */
  if(xTaskCreate(rel_delay, "REL", STACK_SIZE, NULL,
                 tskIDLE_PRIORITY, &dummy) != pdPASS)
    printf("* Cannot create task rel_delay\n");

  else if(xTaskCreate(abs_delay, "ABS", STACK_SIZE, NULL,
                      tskIDLE_PRIORITY, &dummy) != pdPASS)
    printf("* Cannot create task abs_delay\n");

  else
  {
    vTaskStartScheduler();
    printf("* vTaskStartScheduler() failed\n");
  }

  /* Since this is just an example, always return a success
     indication, even this might not be true.
  */
  return EXIT_SUCCESS;
}
```

8.5   SUMMARY

In this chapter, we filled the gap between the abstract concepts of multiprogramming and IPC, presented in Chapters 3 through 6, and what real-time operating systems actually offer programmers when the resources at their disposal are severely constrained.

FreeRTOS, an open-source, real-time operating system targeted to small embedded systems has been considered as a case study. This is in sharp contrast to what was shown in Chapter 7, which deals instead with a full-fledged, POSIX-compliant operating system like Linux. This also gives the readers the opportunity of comparing several real-world code examples, written in C for these two very dissimilar execution environments.

The first important difference is about how FreeRTOS implements multiprocessing. In fact, to cope with hardware limitations and to simplify the implementation, FreeRTOS does not support multiple processes but only threads, or *tasks*, all living within the same address space. With respect to a POSIX system, task creation and deletion are much simpler and less sophisticated, too.

For what concerns IPC, the primitives provided by FreeRTOS are rather established, and do not depart significantly from any of the abstract concepts discussed earlier in the book. The most important aspect that is worth noting is that FreeRTOS sometimes maps a single abstract concept into several distinct, concrete objects.

For example, the abstract *semaphore* corresponds to four different "flavors" of semaphore in FreeRTOS, each representing a different trade-off between the flexibility and power of the object and the efficiency of its implementation. This is exactly the reason why this approach is rather common and is also taken by most other, real-world operating systems.

Another noteworthy difference is that, quite unsurprisingly, time plays a central role in a real-time operating system. For this reason, all abstract primitives that may block a process (such as a P() on a semaphore) have been extended to support a *timeout* mechanism. In this way, the caller can specify a maximum amount of time it is willing to block for any given primitive and do not run the risk of being blocked forever if something goes wrong.

Last but not least, FreeRTOS also provides a couple of primitives to synchronize a task with the elapsed time. Even if those primitives were not discussed in abstract terms, they are especially important anyway in a real-time system because they lay out the foundation for executing any kind of periodic activity in the system. Moreover, they also provide a convenient way to insert a controlled delay in a task without wasting processing power for it.

8.6   EXERCISES

EXERCISE 1

Write the code for Exercise 2, Chapter 5 using two FreeRTOS counting semaphores. Why are mutual exclusion semaphores ill-suited in this particular case?

EXERCISE 2

Write the code for Exercise 1, Chapter 6 using a FreeRTOS queue of $n = 5$ elements. The function that generates an event must report an error if unable to do so because the queue is full.

EXERCISE 3

Improve the function that waits for an event, developed in the previous exercise, so that it informs the calling process about the loss of events due to the message queue being full. The event loss indication must be provided as closely as possible to the time at which the lost events would have been received. It may be assumed that only one event, or a group of consecutively generated events, may have been lost in the time span covered by the events in the queue.

9 Network Communication

We have seen in chapters 7 and 8 how processes and threads can communicate within the *same* computer. This chapter will introduce the concepts and interfaces for achieving communication among *different* computers to implement distributed applications. Distributed applications involving network communication are used in embedded systems for a variety of reasons, among which are

- *Computing Power*: Whenever the computing power needed by the application cannot be provided by a single computer, it is necessary to distribute the application among different machines, each carrying out a part of the required computation and coordinating with the others via the network.
- *Distributed Data*: Often, an embedded system is required to acquire and elaborate data coming from different locations in the controlled plant. In this case, one or more computers will be dedicated to data acquisition and first-data processing. They will then send preprocessed data to other computers that will complete the computation required for the control loop.
- *Single Point of failure*: For some safety-critical applications, such as aircraft control, it is important that the system does not exhibit a single point of failure, that is, the failure of a single computer cannot bring the system down. In this case, it is necessary to distribute the computing load among separate machines so that, in case of failure of one of them, another one can resume the activity of the failed component.

Here we shall concentrate on the most widespread programming interface for network communication based on the concept of *socket*. Before describing the programming interface, we shall briefly review some basic concepts in network communication with an eye on Ethernet, a network protocol widely used in local area networks (LANs).

9.1 THE ETHERNET PROTOCOL

Every time different computers are connected for exchanging information, it is necessary that they strictly adhere to a *communication protocol*. A communication protocol defines a set of rules that allow different computers, possibly from different vendors, to communicate over a network link. Such rules are specified in several *layers*, usually according to the Open Systems Interconnection (OSI) ISO/IEC standard [51]. At the lowest abstraction level is the *Physical Layer*, which specifies how bits are transferred over the physical communication media. The *Data Link Layer* specifies how data is transferred between network entities. The *Network Layer* specifies the functional and procedural means to route data among different networks, and the *Transport Layer* provides transparent transfer of data between end users, providing reliable data transfer services.

DOI: 10.1201/9781003593416-9

The definition of the Ethernet protocol is restricted to the physical layer and the data link layer. The physical layer defines the electrical characteristics of the communication media, including:

- Number of communication lines;
- Impedance for input and output electronics;
- Electrical levels and timing characteristics for high and low levels;
- The coding schema used to transmit ones and zeroes;
- Rules for ensuring that the communication links are not contended.

The physical layer specification often reflects the state of the art of electronic technology and therefore rapidly evolves over time. As an example, the physical layer of Ethernet evolved in the past years through several main stages, all discussed in IEEE Standard 802.3 [47]:

- 10 Mbit/s connection over a coaxial cable. In this, a single coaxial cable was shared by all the partners in communication using Manchester Coding for the transmission of the logical ones and zeroes. Carrier sense multiple access with collision detection (CSMA/CD) was defined to avoid the communication line being driven by more than one transmitter.
- 100BASE-T (Fast Ethernet), which runs over two wire-pairs, normally one pair of twisted wires in each direction, using 4B5B coding and providing 100 Mbit/s of throughput in each direction (full-duplex). Each network segment can have a maximum distance of 100 meters and can be shared only by two communication partners. Ethernet hubs and switches provide the required connectivity among multiple partners.
- 1000BASE-T and 1000BASE-X (Gigabit Ethernet), ensuring a communication speed of 1 Gbit/s over twisted pair cable or optical fiber, respectively.

The physical layer is continuously evolving, and 10 Gigabit Ethernet is currently entering the mainstream market. Conversely, the data link layer of Ethernet is more stable. This layer defines how information is coded by using the transmitted logical zeroes and ones (how logical zeroes and ones are transmitted is defined by the physical layer). Due to its complexity, the data link layer is often split into two or more sublayers to make it more manageable. For Ethernet, the lower sublayer is called *Media Access Control* (MAC) and is discussed in Reference [47], along with the physical layer. The upper sub-layer is the *Logical Link Control* (LLC) and is specified in Reference [50]. Data exchanged over Ethernet is grouped in *Frames*, and every frame is a packet of binary data that contains the following fields:

- **Preamble and Start Frame Identifier** (8 octets): Formed by a sequence of identical octets with a predefined value (an octet in network communication terminology corresponds to a byte), followed by a single octet whose value differs only for the least significant bit. The preamble and the start frame identifier are used to detect the beginning of the frame in the received bit stream.

- **MAC Destination** (6 octets): the Media Access Control (MAC) address of the designated receiver for the frame.
- **MAC Source** (6 octets): The MAC address of the sender of the frame.
- **Packet Length** (2 octets): Coding either the length of the data frame or other special information about the packet type.
- **Payload** (46–1500 octets): Frame data.
- **CRC** (4 octets): Cyclic Redundancy Check (CRC) used to detect possible communication errors. This field is obtained from the frame content at the time the frame is sent, and the same algorithm is performed when the packet is received. If the new CRC value is different form the CRC field, the packet is discarded because there has indeed been a transmission error.
- **Interframe Gap** (12 octets): Minimum number of bytes between different frames.

An Ethernet frame can therefore be seen as an envelope containing some data (the payload). The additional fields are only required for the proper management of the packet, such as the definition of the sender and receiver addresses and checksum fields. The envelope is normally processed by the network board firmware, and the payload is returned to the upper software layers when a packet is received.

9.2   TCP/IP AND UDP

It would be possible to develop distributed application directly interfacing to the data link layer of Ethernet, but in this way, in order to ensure proper and reliable communication, the program should also handle the following facts:

- **Address resolution**: The Ethernet addresses are unique for every Hardware Board, and they must be known to the program. Changing a computer, or even a single Ethernet board, would require a change in the program.
- **Frame Splitting**: The maximum payload in Ethernet is 1500 bytes, and therefore, if a larger amount of data must be transmitted, it has to be split in two or more packets. The original data must then be reassembled upon the reception of the packets.
- **Transmission Error Management**: The network board firmware discards those packets for which a communication error has been detected using the CRC field. So the program must take into account the possibility that packets could be lost in the transmission and therefore must be able to detect this fact and take corrective actions, such as request for a new data packet.

It is clear that programming network communication at this level would be a nightmare: the programmer would be requested to handle a variety of problems that would overwhelm the application requirements. For this reason, further communication layers are defined and can be used to achieve effective and reliable network communication. Many different network communication layers are defined for different network protocols, and every layer, normally built on top of one or more other layers, provides some added functionality in respect of that provided by the layers below. Here we shall consider the *Internet Protocol* (IP), which addresses the Network Layer, and the *Transmission Control Protocol* (TCP), addressing the Transport Layer. IP and TCP together implement the well-known TCP/IP protocol. Both protocols are specified and discussed in detail in a number of Request for Comments (RFC), a series of informational and standardization documents about Internet. In their most basic form, IP version 4 and TCP are presented in References [78], and [79], respectively.

The IP defines the functionality needed for handling the transmission of packets along one or more networks and performs two basic functions:

- **Addressing**: It defines a hierarchical addressing system using IP addresses represented by a 4-byte integer.
- **Routing**: It defines the rules for achieving communication among different networks, that is, getting packets of data from source to destination by sending them from network to network. Based on the destination IP address, the data packet will be sent over the local networks from router to router up to the final destination.

The Internet Protocol can be built on top of other data link layers for different communication protocols. For LANs, the IP is typically built over Ethernet. Observe that, in both layers, data are sent over packets, and the IP packet defines, among other information, the source and destination IP addresses. However, when sent over an Ethernet segment, these fields cannot be recognized by the Ethernet board, which is only able to recognize Ethernet addresses. This apparent contradiction is explained by the fact the Internet layer data packet is represented by the payload of the Ethernet packet, as shown in Figure 9.1. So, when an Ethernet packet is received, the

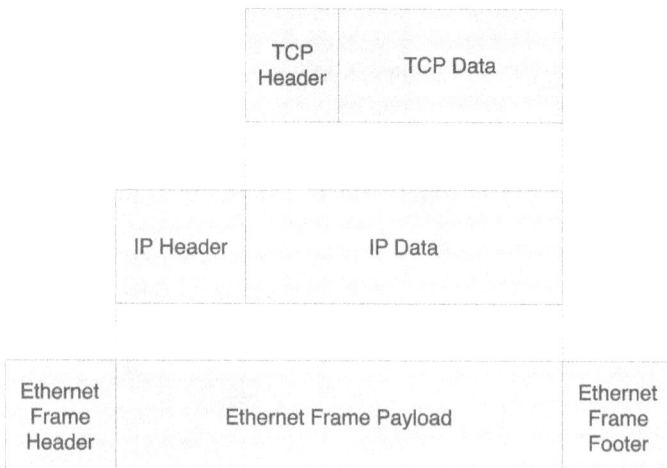

Figure 9.1   Network frames: Ethernet, IP, and TCP/IP.

lowest communication layers (normally carried out by the board firmware) will use the Ethernet header, to acquire the payload and pass it to the upper layer. The upper Internet layer will interpret this chunk of bytes as an Internet Packet and will retrieve its content. For sending an Internet Packet over an Ethernet network, the packet will be encapsulated into an Ethernet packet and then sent over the Ethernet link. Observe that the system must know how to map IP addresses with Ethernet addresses: such information will be maintained in routing tables, as specified by the routing rules of the IP. If the resolution for a given IP address is not currently known by the system, it is necessary to discover it. The Address Resolution Protocol (ARP) [76] is intended for this purpose and defines how the association between IP addresses and Ethernet MAC addresses is exchanged over Ethernet. Basically, the machine needing this information sends a broadcast message (i.e., a message that is received by all the receivers for that network segment), bringing the MAC address of the requester and the IP address for which the translation is required. The receiver that recognizes the IP address sends a reply with its MAC address so that the client can update its routing tables with the new information.

Even if the IP solves the important problem of routing the network packets so that the world network can be seen as a whole, communication is still packet based, and reliability is not ensured because packets can be lost when transmission errors occur. These limits are removed by TCP, built over IP. This layer provides a connection-oriented view of the network transmissions: the partners in the communication first establish a connection and then exchange data. When the connection has been established, a stream of data can be exchanged between the connected entities. This layer removes the data packet view and ensures that data arrive with no errors, no duplications, and in order. In addition, this layer introduces the concept of port, that is, a unique integer identifier of the communicating entity within a single computer (associated with a given IP address). So, in TCP/IP, the address of the sender and the receiver will be identified by the pair (IP address, port), allowing communication among different entities even if sharing the same IP address. The operations defined by the TCP layer are complex and include management of the detection and retransmission of lost packets, proper sequencing of the packets, check for duplicated data assembling and de-assembling of data packets, and traffic congestion control. TCP defines its own data packet format, which brings, in addition to data themselves, all the required information for reliable and stream-oriented communication, including Source/Destination port definitions and Packet Sequence and Acknowledge numbers used to detect lost packets and handle retransmission.

Being the TCP layer built on top of the Internet layer, the latter cannot know anything about the structure of the TCP data packet, which is contained in the data part of the Internet packet, as shown in Figure 9.1. So, when a data packet is received by the Internet layer (possibly contained in the payload of an Ethernet data packet), the specific header fields will be used by the Internet layer, which will pass the data content of the packet to the above TCP layer, which in turn will interpret this as a TCP packet.

The abstraction provided by the TCP layer represents an effective way to achieve network communication and, for this reason, TCP/IP communication is widely used in applications. In the next section we shall present the programming model of TCP/IP and illustrate it in a sample client/server application. This is, however, not the end of the story: many other protocols are built over TCP/IP, such as File Transfer Protocol (FTP), and the ubiquitous Hypertext Transfer Protocol (HTTP) used in web communication.

Even if the connection-oriented communication provided by TCP/IP is widely used in practice, there are situations in which a connectionless model is required instead. Consider, for example, a program that must communicate asynchronous events to a set of listener entities over the network, possibly without knowing which are the recipients. This would not be possible using TCP/IP because a connection should be established with every listener, and therefore, its address must be known in advance. The User Datagram Protocol (UDP) [77], which is built over the Internet layer, lets computer applications send messages, in this case referred to as `datagrams`, to other hosts on an Internet network without the need of establishing point-to-point connections. In addition, UDP provides multicast capability, that is, it allows sending of datagrams to sets of recipients without even knowing their IP addresses. On the other side, the communication model offered by UDP is less sophisticated than that of TCP/IP, and data reliability is not provided. Later in this chapter, the programming interface of UDP will be presented, together with a sample application using UDP multicast communication.

9.3 SOCKETS

The programming interface for TCP/IP and UDP is centered around the concept of *socket*, which represents the endpoint of a bidirectional interprocess communication flow. The creation of a socket is therefore the first step in the procedure for setting up and managing network communication. The prototype of the socket creation routine is

```
int socket(int domain, int type, int protocol)
```

where `domain` selects the protocol family that will be used for communication. In the case of the Internet, the communication domain is `AF_INET`. `type` specifies the communication semantics, which can be `SOCK_STREAM` or `SOCK_DGRAM` for TCP/IP or UDP communication, respectively. The last argument, `protocol`, specifies a particular protocol within the communication domain to be used with the socket. Normally, only a single protocol exists and, therefore, the argument is usually specified as 0.

The creation of a socket represents the only common step when managing TCP/IP and UDP communication. In the following we shall first describe TCP/IP programming using a simple client–server application. Then UDP communication will be described by presenting a program for multicast notification.

9.3.1   TCP/IP SOCKETS

We have seen that TCP/IP communication requires the establishment of a connection before transmission. This implies a client–server organization: the client will request a connection to the server. The server may be accepting multiple clients' connections in order to carry out a given service. In the program shown below, the server accepts character strings, representing some sort of command, from clients and returns other character strings representing the answer to the commands. It is worth noting that a high-level protocol for information exchange must be handled by the program: TCP/IP sockets, in fact, provide full duplex point-to-point communication where the communication partners can send and transmit bytes, but it is up to the application to handle transmission and reception to avoid, for example, situations in which the two communication partners both hang waiting to receive some data from the other. The protocol defined in the program below is a simple one and can be summarized as follows:

- The client initiates the transaction by sending a command to be executed. To do this, it first sends the length (4 bytes) of the command string, followed by the command characters. Sending the string length first allows the server to receive the correct number of bytes afterward.
- The server, after receiving the command string, executes the command getting and answer string, which is sent back to the client. Again, first the length of the string is sent, followed by the answer string characters. The transaction is then terminated and a new one can be initiated by the client.

Observe that, in the protocol used in the example, numbers and single-byte characters are exchanged between the client and the server. When exchanging numbers that are represented by two, four, or more bytes, the programmer must take into account the possible difference in byte ordering between the client and the server machine. Getting weird numbers from a network connection is one of the main source of headache to novel network programmers. Luckily, there is no need to find out exotic ways of discovering whether the client and the server use a different byte order and to shuffle bytes manually, but it suffices to use a few routines available in the network API that convert short and integer numbers to and from the network byte order, which is, by convention, big endian.

Another possible source of frustration for network programmers is due to the fact that the `recv()` routine for receiving a given number of bytes from the socket does not necessarily return after the specified number of bytes has been read, but it may end when a lower number of bytes has been received, returning the actual number of bytes read. This occurs very seldom in practice and typically not when the program is tested, since it is related to the level of congestion of the network. Consequently, when not properly managed, this fact generates random communication errors that are very hard to reproduce. In order to receive a given number of bytes, it is therefore necessary to check the number of bytes returned by `recv()`, possibly issuing again the read operation until all the expected bytes are read, as done by routine `receive()` in this program.

The client program is listed below:

```
#include <stdio.h>
#include <sys/types.h>
#include <sys/socket.h>
#include <netinet/in.h>
#include <netdb.h>
#include <string.h>
#include <stdlib.h>
#define FALSE 0
#define TRUE 1


/* Receive routine: use recv to receive from socket and manage
   the fact that recv may return after having read less bytes than
   the passed buffer size
   In most cases recv will read ALL requested bytes, and the loop body
   will be executed once. This is not however guaranteed and must
   be handled by the user program. The routine returns 0 upon
   successful completion, -1 otherwise */
static int receive(int sd, char *retBuf, int size)
{
  int totSize, currSize;
  totSize = 0;
  while(totSize < size)
  {
    currSize = recv(sd, &retBuf[totSize], size - totSize, 0);
    if(currSize <= 0)
/* An error occurred */
      return -1;
    totSize += currSize;
  }
  return 0;
}


/* Main client program. The IP address and the port number of
   the server are passed in the command line. After establishing
   a connection, the program will read commands from the terminal
   and send them to the server. The returned answer string is
   then printed. */
main(int argc, char **argv)
{
  char hostname[100];
  char command[256];
  char *answer;
  int  sd;
  int port;
  int stopped = FALSE;
  int len;
  unsigned int netLen;
  struct sockaddr_in sin;
  struct hostent *hp;
/* Check number of arguments and get IP address and port */
  if (argc < 3)
  {
    printf("Usage: client <hostname> <port>\n");
    exit(0);
  }
  sscanf(argv[1], "%s", hostname);
  sscanf(argv[2], "%d", &port);


/* Resolve the passed name and store the resulting long representation
   in the struct hostent variable */
  if ((hp = gethostbyname(hostname)) == 0)
  {
    perror("gethostbyname");
    exit(0);
  }
```

```
/* fill in the socket structure with host information */
  memset(&sin, 0, sizeof(sin));
  sin.sin_family = AF_INET;
  sin.sin_addr.s_addr = ((struct in_addr *)(hp->h_addr))->s_addr;
  sin.sin_port = htons(port);
/* create a new socket */
  if ((sd = socket(AF_INET, SOCK_STREAM, 0)) == -1)
  {
    perror("socket");
    exit(0);
  }
/* connect the socket to the port and host
   specified in struct sockaddr in */
  if (connect(sd,(struct sockaddr *)&sin, sizeof(sin)) == -1)
  {
    perror("connect");
    exit(0);
  }
  while(!stopped)
  {
/* Get a string command from terminal */
    printf("Enter command: ");
    scanf("%s", command);
    if(!strcmp(command, "quit"))
      break;
/* Send first the number of characters in the command and then
   the command itself */
    len = strlen(command);
/* Convert the integer number into network byte order */
    netLen = htonl(len);
/* Send number of characters */
    if(send(sd, &netLen, sizeof(netLen), 0) == -1)
    {
      perror("send");
      exit(0);
    }
/* Send the command */
    if (send(sd, command, len, 0) == -1)
    {
      perror("send");
      exit(0);
    }
/* Receive the answer: first the number of characters
   and then the answer itself */
    if(receive(sd, (char *)&netLen, sizeof(netLen)))
    {
      perror("recv");
      exit(0);
    }
/* Convert from Network byte order */
    len = ntohl(netLen);
/* Allocate and receive the answer */
    answer = malloc(len + 1);
    if(receive(sd, answer, len))
    {
      perror("send");
      exit(1);
    }
    answer[len] = 0;
    printf("%s\n", answer);
    free(answer);
  }
/* Close the socket */
  close(sd);
}
```

The above program first creates a socket and connects it to the server whose IP Address and port are passed in the command string. Socket connection is performed by routine connect(), and the server address is specified in a variable of type struct sockaddr_in, which is defined as follows

```
struct sockaddr_in {
  short           sin_family; // Address family e.g. AF_INET
  unsigned short  sin_port; // Port number in Network Byte order
  struct in_addr  sin_addr; // see struct in_addr, below
  char            sin_zero[8]; //Padding zeroes
};


struct in_addr {
    unsigned long s_addr;   //4 byte IP address
};
```

The Internet Address is internally specified as a 4-byte integer but is presented to users in the usual dot notation. The conversion from human readable notation and the integer address is carried out by routine gethostbyname(), which fills a struct hostent variable with several address-related information. We are interested here (and in almost all the applications in practice) in field h_addr, which contains the resolved IP address and which is copied in the corresponding field of variable sin. When connect() returns successfully, the connection with the server is established, and data can be exchanged. Here, the exchanged information is represented by character strings: command string are sent to the server and, for every command, an answer string is received. The length of the string is sent first, converted in network byte order by routine htonl(), followed by the string characters. Afterward, the answer is obtained by reading first its length and converting from network byte order via routine ntohl(), and then reading the expected number of characters.

The server code is listed below, and differs in several points from the client one. First of all, the server does not have to know the address of the clients: after creating a socket and binding it to the port number (i.e., the port number clients will specify to connect to the server), and specifying the maximum length of pending clients via listen() routine, the server suspends itself in a call to routine accept(). This routine will return a new socket to be used to communicate with the client that just established the connection.

```
#include <stdio.h>
#include <sys/types.h>
#include <sys/socket.h>
#include <netinet/in.h>
#include <arpa/inet.h>
#include <netdb.h>
#include <string.h>
#include <stdlib.h>

/* Handle an established   connection
   routine receive is listed in the previous example */
static void handleConnection(int currSd)
{
```

```
   unsigned int netLen;
   int len;
   char *command, *answer;
   for(;;)
   {
/* Get the command string length
   If receive fails, the client most likely exited */
     if(receive(currSd, (char *)&netLen, sizeof(netLen)))
       break;
/* Convert from network byte order */
     len = ntohl(netLen);
     command = malloc(len+1);
/* Get the command and write terminator */
     receive(currSd, command, len);
     command[len] = 0;
/* Execute the command and get the answer character string */
     ...
/* Send the answer back */
     len = strlen(answer);
/* Convert to network byte order */
     netLen = htonl(len);
/* Send answer character length */
     if (send(currSd, &netLen, sizeof(netLen), 0) == -1)
       break;
/* Send answer characters */
     if (send(currSd, answer, len, 0) == -1)
       break;
   }
/* The loop is most likely exited when the connection is terminated */
   printf("Connection terminated\n");
   close(currSd);
}

/* Main Program */
main(int argc, char *argv[])
{
   int     sd, currSd;
   int     sAddrLen;
   int     port;
   int     len;
   unsigned int netLen;
   char *command, *answer;
   struct  sockaddr_in sin, retSin;
/* The port number is passed as command argument */
   if(argc < 2)
   {
     printf("Usage: server <port>\n");
     exit(0);
   }
   sscanf(argv[1], "%d", &port);
/* Create a new socket */
   if ((sd = socket(AF_INET, SOCK_STREAM, 0)) == -1)
   {
     perror("socket");
     exit(1);
   }
/* Initialize the address (struct sokaddr`in) fields */
   memset(&sin, 0, sizeof(sin));
   sin.sin_family = AF_INET;
   sin.sin_addr.s_addr = INADDR_ANY;
   sin.sin_port = htons(port);

/* Bind the socket to the specified port number */
   if (bind(sd, (struct sockaddr *) &sin, sizeof(sin)) == -1)
   {
     perror("bind");
```

```
      exit(1);
  }
/* Set the maximum queue length for clients requesting connection to 5 */
  if (listen(sd, 5) == -1)
  {
    perror("listen");
    exit(1);
  }
  sAddrLen = sizeof(retSin);
/* Accept and serve all incoming connections in a loop */
  for(;;)
  {
    if ((currSd =
         accept(sd, (struct sockaddr *) &retSin, &sAddrLen)) == -1)
    {
      perror("accept");
      exit(1);
    }
/* When execution reaches this point a client established the connection.
   The returned socket (currSd) is used to communicate with the client */
    printf("Connection received from %s\n", inet_ntoa(retSin.sin_addr));
    handleConnection(currSd);
  }
}
```

In the above example, the server program has two nested loops: the external loop waits for incoming connections, and the internal one, defined in routine handleConnection(), handles the connection just established until the client exits. Observe that the way the connection is terminated in the example is rather harsh: the inner loop breaks whenever an error is issued when either reading or writing the socket, under the assumption that the error is because the client exited. A more polite management of the termination of the connection would have been to foresee in the client–server protocol an explicit command for closing the communication. This would also allow discriminating between possible errors in the communication and the natural termination of the connection.

Another consequence of the nested loop approach in the above program is that the server, while serving one connection, is not able to accept any other connection request. This fact may pose severe limitations to the functionality of a network server: imagine a web server that is able to serve only one connection at a time! Fortunately, there is a ready solution to this problem: let a separate thread (or process) handle the connection established, allowing the main process accepting other connection requests. This is also the reason for the apparently strange fact why routine accept() returns a new socket to be used in the following communication with the client. The returned socket, in fact, is specific to the communication with that client, while the original socket can still be used to issue accept() again.

The above program can be turned into a multithreaded server just replacing the external loop accepting incoming connections as follows:

```
/* Thread routine. It calls routine handleConnection()
   defined in the previous program. */
static void *connectionHandler(void *arg)
{
    int currSock = *(int *)arg;
    handleConnection(currSock);
    free(arg);
    pthread_exit(0);
```

```
      return NULL;
}
...
/* Replacement of the external (accept) loop of the previous program */
  for (;;)
  {
/* Allocate the current socket.
   It will be freed just before thread termination. */
    currSd = malloc(sizeof(int));
    if ((*currSd =
        accept(sd, (struct sockaddr *) &retSin, &sAddrLen)) == -1)
    {
      perror("accept");
      exit(1);
    }
    printf("Connection received from %s\n", inet_ntoa(retSin.sin_addr));
/* Connection received, start a new thread serving the connection */
    pthread_create(&handler, NULL, connectionHandler, currSd);
  }
```

In the new version of the program, routine handleConnection() for the communication with the client is wrapped into a thread. The only small change in the program is due to the address of the socket being passed because the thread routine accepts a pointer argument. The new server can now accept and serve any incoming connection in parallel.

9.4 UDP SOCKETS

We have seen in the previous section how the communication model provided by TCP/IP ensures reliable connection between a client and a server application. TCP is built over IP and provides the functionality necessary to achieve communication reliability over unreliable packet-based communication layer, such as IP is. This is obtained using several techniques for timestamping messages in order to detect missing, duplicate, or out-of-order message reception and to handle retransmission in case of lost packets. As a consequence, although TCP/IP is ubiquitous and is the base protocol for a variety of other protocols, it may be not optimal for real-time communication. In real-time communication, in fact, it is often preferable not to receive a data packet at all rather than receive it out of time. Consider, for example, a feedback system where a controller receives from the network data from sensors and computes the actual reference values for actuators. Control computation must be performed on the most recent samples. Suppose that a reliable protocol such as TCP/IP is used to transfer sensor data, and that a data packet bringing current sensor values is lost: in this case, the protocol would handle the retransmission of the packet, which is eventually received correctly. However, at the time this packet has been received, it brings out-of-date sensor values, and the following sensor samples will likely arrive delayed as well, at least until the transmission stabilizes. From the control point of view, this situation is often worse than not receiving the input sample at all, and it is preferable that the input values are not changed in the next control cycle, corresponding to the assumption that sensor data did not change during that period. For this reason a faster protocol is often preferred for real-time application, relaxing the reliability requirement, and the UDP is normally adopted. UDP is a protocol built above IP that allows that applications send and receive messages, called *datagrams*, over an IP network. Unlike TCP/IP, the communication does not require prior communication to set up client–server connection, and for this reason, it is called *connectionless*. UDP provides an unreliable service, and datagrams may arrive out of order, duplicated, or lost, and these conditions must be handled in the user application. Conversely, faster communication can be achieved in respect of other reliable protocols because UDP introduces less overhead. As for TCP/IP, message senders and receivers are uniquely identified by the pair (IP Address, port). No connection is established prior to communication, and datagrams sent and received by routines `sendto()` and `revfrom()`, respectively, can be sent and received to/from any other partner in communication. The following example lists a simple point–to–point communication between a sender and a receiver using UDP. First the sender code:

```c
#include <stdio.h>
#include <stdlib.h>
#include <string.h>
#include <unistd.h>
#include <arpa/inet.h>
#include <sys/socket.h>

/* As usual, actors are identified by an IP address
   and a port number */
#define RECEIVER_IP "192.168.1.100"
#define RECEIVER_PORT 8888
#define MESSAGE "HELLO"
int main() {
  int sockfd;
  struct sockaddr_in receiver_addr;

/* UDP socket creation */
  if ((sockfd = socket(AF_INET, SOCK_DGRAM, 0)) < 0) {
    perror("Socket creation failed");
    exit(EXIT_FAILURE);
  }
/* Set up the receiver address */
  memset(&receiver_addr, 0, sizeof(receiver_addr));
  receiver_addr.sin_family = AF_INET;
  receiver_addr.sin_port = htons(RECEIVER_PORT);
  if (inet_pton(AF_INET, RECEIVER_IP, &receiver_addr.sin_addr) <= 0) {
    perror("Invalid address");
    close(sockfd);
    exit(EXIT_FAILURE);
  }
/* Send message to the receiver */
  if (sendto(sockfd, MESSAGE, strlen(MESSAGE), 0,
      (struct sockaddr*)&receiver_addr, sizeof(receiver_addr)) < 0) {
    perror("Send failed");
    close(sockfd);
    exit(EXIT_FAILURE);
  }
  close(sockfd);
  return 0;
}
```

Then the receiver code:

```c
#include <stdio.h>
#include <stdlib.h>
#include <string.h>
#include <unistd.h>
#include <arpa/inet.h>
```

```c
#include <sys/socket.h>

#define SERVER_PORT 8888
#define MAX_MESSAGE_SIZE 1024

int main() {
  int sockfd;
  struct sockaddr_in server_addr, client_addr;
/* Make sure there is enough room for the received message
   In any case, unless using Jumbo frames, it will be 16k max */
  char buffer[BUFFER_SIZE];
  socklen_t addr_len = sizeof(client_addr);

/* Create UDP socket */
  if ((sockfd = socket(AF_INET, SOCK_DGRAM, 0)) < 0) {
    perror("Socket creation failed");
    exit(EXIT_FAILURE);
  }

/* Set up the receiver address.  Using INADDR`ANY ensures
   that the receiver will receive messages from every
   client on the network */
  memset(&server_addr, 0, sizeof(server_addr));
  server_addr.sin_family = AF_INET;
/* Port numbers (4 bytes) must be converted to network endianity */
  server_addr.sin_port = htons(SERVER_PORT);
  server_addr.sin_addr.s_addr = INADDR_ANY;

/* Bind the socket to listen at incoming messages */
  if (bind(sockfd, (struct sockaddr*)&server_addr,
    sizeof(server_addr)) < 0) {
  perror("Bind failed");
  close(sockfd);
  exit(EXIT_FAILURE);
  }
/* Receive message */
  size_t bytes_received = recvfrom(sockfd, buffer, BUFFER_SIZE - 1,
       0, (struct sockaddr*)&client_addr, &addr_len);
  if (bytes_received < 0) {
    perror("Receive failed");
  }
  else
/* The number of successfully received bytes always corresponds
   to the size of the sentt datagram    */
  {
/* We are sending the "HELLO" stirng. Add terminator */
    buffer[bytes_received] = '\0';
    printf("Received message from %s:%d: %s\n",
      inet_ntoa(client_addr.sin_addr),
      ntohs(client_addr.sin_port), buffer);
  }
  close(sockfd);
  return 0;
}
```

In addition to specifying a datagram recipient in the form (IP Address, port), as we did in the sender example above, the UDP protocol allows *broadcast* communication, that is, sending the datagram to all the recipients in the network, and *multicast*, that is, sending the datagram to a set of recipients.

Broadcast communication is used when a given UDP message has to be received by *all* the nodes belonging to the same local network. For the most common broadcast configuration, that is *Directed Broadcast*, it suffices to change the destination IP specifying all address bits equal to one in the host address part of the IP address.

Normally, the eight least significant bits of the IP address specify the host address within the network whose address is defined by the remaining 24 most significant bits (corresponding to 255.255.255.0 netmask). In this case the only modification in the above example to achieve broadcast is the replacement of

```
#define RECEIVER_IP "192.168.1.100"
```

with

```
#define RECEIVER_IP "192.168.1.255"
```

Multicast communication is useful in distributed embedded applications because it is often required that data are exchanged among groups of communicating actors. The approach taken in UDP multicast is called *publish–subscribe*, and the set of IP addresses ranging from 224.0.0.0 to 239.255.255.255 is reserved for multicast communication. When an address is chosen for multicast communication, it is used by the sender, and receivers must register themselves for receiving datagrams sent to such address. So, the sender is not aware of the actual receivers, which may change over time.

The use of UDP multicast communication is explained by the following sender and receiver programs: the sender sends a string message to the multicast address 225.0.0.37, and the message is received by every receiver that subscribed to that multicast address.

```
#include <sys/types.h>
#include <sys/socket.h>
#include <netinet/in.h>
#include <arpa/inet.h>
#include <string.h>
#include <stdio.h>
#include <stdlib.h>
/* Port number used in the application */
#define PORT 4444
/* Multicast address */
#define GROUP "225.0.0.37"
/* Sender main program: get the string from the command argument */
main(int argc, char *argv[])
{
  struct sockaddr_in addr;
  int sd;
  char *message;
/* Get message string */
  if(argc < 2)
  {
    printf("Usage: sendUdp <message>\n");
    exit(0);
  }
  message = argv[1];
/* Create the socket. The second argument specifies that
   this is an UDP socket */
  if ((sd = socket(AF_INET,SOCK_DGRAM,0)) < 0)
  {
    perror("socket");
    exit(0);
  }
/* Set up destination address: same as TCP/IP example */
  memset(&addr,0,sizeof(addr));
```

```
   addr.sin_family = AF_INET;
   addr.sin_addr.s_addr = inet_addr(GROUP);
   addr.sin_port=htons(PORT);
/* Send the message */
   if (sendto(sd,message,strlen(message),0,
                (struct sockaddr *) &addr, sizeof(addr)) < 0)
   {
     perror("sendto");
     exit(0);
   }
/* Close the socket */
   close(sd);
}
```

In the above program, the translation of the multicast address 225.0.0.37 from the dot notation into its internal integer representation is carried out by routine inet_addr(). This routine is a simplified version of gethostbyname() used in the TCP/IP socket example. The latter, in fact, provides the resolution of names based on the current information maintained by the IP, possibly communicating with other computers using a specific protocol to retrieve the appropriate mapping. On the Internet, this is usually attained by means of the Domain Name System (DNS) infrastructure and protocol. The general ideas behind DNS are discussed in Reference [68], while Reference [67] contains the full specification.

The UDP sender program is simpler than in the TCP/IP connection because there is no need to call connect() first, and the recipient address is passed directly to the send routine. Even simpler is the receiver program because it is no more necessary to handle the establishment of the connection. In this case, however, the routine for receiving datagrams must also return the address of the sender since different clients can send datagrams to the receiver. The receiver program is listed below:

```
#include <sys/types.h>
#include <sys/socket.h>
#include <netinet/in.h>
#include <arpa/inet.h>
#include <time.h>
#include <string.h>
#include <stdio.h>
#include <stdlib.h>

#define PORT 4444
#define GROUP "225.0.0.37"
/* Maximum dimension of the receiver buffer */
#define BUFSIZE 256
/* Receiver main program. No arguments are passed in the command line. */
main(int argc, char *argv[])
{
   struct sockaddr_in addr;
   int sd, nbytes,addrLen;
   struct ip_mreq mreq;
   char msgBuf[BUFSIZE];

/* Create a UDP socket */
   if ((sd=socket(AF_INET,SOCK_DGRAM,0)) < 0)
   {
     perror("socket");
     exit(0);
   }
/* Set up receiver address. Same as in the TCP/IP example. */
   memset(&addr,0,sizeof(addr));
   addr.sin_family = AF_INET;
```

```
    addr.sin_addr.s_addr = INADDR_ANY;
    addr.sin_port = htons(PORT);
/* Bind to receiver address */
    if (bind(sd,(struct sockaddr *) &addr,sizeof(addr)) < 0)
    {
      perror("bind");
      exit(0);
    }
/* Use setsockopt() to request that the receiver join a multicast group */
    mreq.imr_multiaddr.s_addr = inet_addr(GROUP);
    mreq.imr_interface.s_addr = INADDR_ANY;
    if (setsockopt(sd,IPPROTO_IP,IP_ADD_MEMBERSHIP,&mreq,sizeof(mreq)) < 0)
    {
      perror("setsockopt");
      exit(0);
    }
/* Now the receiver belongs to the multicast group:
   start accepting datagrams in a loop */
    for(;;)
    {
      addrLen = sizeof(addr);
/* Receive the datagram. The sender address is returned in addr */
      if ((nbytes = recvfrom(sd, msgBuf, BUFSIZE, 0,
       (struct sockaddr *) &addr,&addrLen)) < 0)
      {
        perror("recvfrom");
        exit(0);
      }
/* Insert terminator */
      msgBuf[nBytes] = 0;
      printf("%s\n", msgBuf);
    }
}
```

After creating the UDP socket, the required steps for the receiver are

1. Bind to the receiver port, as for TCP/IP.
2. Join the multicast group. This is achieved via the generic setsockopt()
   routine for defining the socket properties (similar in concept to ioctl())
   where the IP_ADD_MEMBERSHIP operation is specified and the multicast
   address is specified in a variable of type struct ip_mreq.
3. Collect incoming datagrams using routine recvfrom(). In addition to the
   received buffer containing datagram data, the address of the sender is returned.

Observe that, in this example, there is no need to send the size of the character strings.
In fact, sender and receivers agree on communicating the characters (terminator excluded) in the exchanged datagram whose size will depend on the number of characters in the transferred string: it will be set by the sender and detected by the receiver.

9.5   SUMMARY

This chapter has presented the programming interface of TCP/IP and UDP, which are
widely used in computer systems and embedded applications. Even if the examples
presented here refer to Linux, the same interface is exported in other operating systems, either natively as in Windows and VxWorks or by separate modules, and so it
can be considered a multiplatform communication standard. For example, lwIP [29]

is a lightweight, open-source protocol stack that can easily be layered on top of FreeRTOS [13] and other small operating systems. It exports a subset of the socket interface to the users.

TCP/IP provides reliable communication and represents the base protocol for a variety of other protocols such as HTTP, FTP and Secure Shell (SSH).

UDP is a lighter protocol and is often used in embedded systems, especially for real-time applications, because it introduces less overhead. Using UDP, user programs need to handle the possible loss, duplication and out-of-order reception of datagrams. Such a management is not as complicated as it might appear, provided the detected loss of data packets is acceptable. In this case, it suffices to add a timestamp to each message: the sender increases the timestamp for every sent message, and the timestamp is checked by the receiver. If the timestamp of the received message is the previous received timestamp plus one, the message has been correctly received, and no datagram has been lost since the last reception. If the timestamp is greater than the previous one plus one, at least another datagram has been lost or will arrive out of order. Finally, if the timestamp is less or equal the previous one, the message is a duplicated one or arrived out of order, and will be discarded.

The choice between TCP/IP and UDP in an embedded system depends on the requirements: whenever fast communication is required, and the occasional loss of some data packet is tolerable, UDP is a good candidate. There are, however, other applications in which the loss of information is not tolerable: imagine what would happen if UDP were used for communicating alarms in a nuclear plant! So, in practice, both protocols are used, often in the same application, where TCP/IP is used for offline communication (no real-time requirements) and whenever reliability is an issue. The combined use of TCP/IP and UDP is common in many applications. For example, the H.323 protocol [55], used to provide audiovisual communication sessions on any packet network, prescribes the use of UDP for voice and image transmission, and TCP/IP for communication control and management. In fact, the loss of datapacket introduces degradation in the quality of communication, which can be acceptable to a certain extent. Conversely, failure in management information exchange may definitely abort a videoconference session.

Even if this chapter concentrated on Ethernet, TCP/IP, and UDP, which represent the most widespread communication protocols in many fields of application, it is worth noting that several other protocols exist, especially in industrial applications. For example, EtherCAT [48] is an Ethernet-based protocol oriented toward high-performance communication. This is achieved by minimizing the number of exchanged data packets, and the protocol is used in industrial machine controls such as assembly systems, printing machines, and robotics. Other widespread communication protocols in industrial application are not based on Ethernet and define their own physical and data link layers. For example, the Controller Area Network (CAN) [53, 54] bus represents a message-based protocol designed specifically for automotive applications, and Process Field Bus (PROFIBUS) [44] is a standard for field bus communication in automation technology.

As a final remark, recall that one of the main reasons for distributed computing is the need for a quantity of computing power that cannot be provided by a single machine. Farms of cheap personal computers have been widely used for applications that would have otherwise required very expensive solutions based on super-computers. The current trend in computer technology, however, reduces the need of distributed systems for achieving more computing power because modern multicore servers allow distribution of computing power among the processor cores hosted in the same machine, with the advantages that communication among computing units is much faster since it is carried out in memory and not over a network link.

9.6   EXERCISES

EXERCISE 1

In this chapter we have seen how using the data structures and the routines provided by the socket library to handle TCP and UDP communication. We can easily convince ourselves that exposing the data structures and the routines for communication and management in larger distributed program is not a good software engineering practice. A better approach is to embed the TCP and UPD socket functionality into two different C++ classes, say, TCPSocket and UDPSocket.

Hint: even is similar in functionality, the socket interfaces for two protocols are not identical, and therefore the exported methods will be slightly different. Indeed, while in UDP communication messages can be freely exchanged among UDP actors, to establish TCP communication, a rendezvous is first required in order to establish a pair of communication actors. In this case a client-server interface is provided, where the servers accepts an incoming communication to establish peer-to-peer communication. For this reason a further C++ class, TCPSocketServer is required, whose accept() method returns when a client connects to this server. The returned value is the pointer to a dynamically allocated TCPSocket instance, connected to the client. Aa possible class interface for TCP and UDP sockets is listed below. The implementation is left as exercise.

```
#include <stdio.h>
#include <stdlib.h>
#include <string.h>
#include <unistd.h>
#include <arpa/inet.h>
#include <sys/socket.h>
#include <sys/types.h>
#include <string>


class UDPSocket:public Socket
{
public:
        /* Public constructor. It will create the socket and bind
        it to the passed port number */
    UDPSocket(int port);
        /* Public desctrctor. It will close the socket */
    ~UDPSocket();
        /* send a message to the actor specified by IP address and port */
    int sendMsgTo(char *buf, int size, std::string ip, int port);
        /* Receive any message set to this IP address and the port
```

```
        specified in the costructor */
    int receiveMsg(char *buf, int size);
        /* For multicast communication */
    void subscribeTo(std::string ip);
    void unsubscribeFrom(std::string multicastIp);
};



class TCPSocket:
{
        /* Private data */
    bool connected;
    int sock;
    struct sockaddr_in ipAddr;
        /* constructor used by TCPSocketServer */Used by TCPSocketServer
    TCPSocket(int sock, struct sockaddr_in ipAddr);
public:
        /* public constructor. It will create and configure the socket */
    TCPSocket();
        /* Public desctrctor. It will close the socket */
    ~TCPSocket();
        /* connectTo() method is called by TCP cleints connecting
        to a server */
    void connectTo(std::string ip, int port);
        /* return the peer IP address */
    std::string getConnectedIp();
        /* return the peer port number */
    int getConnectedPort();
        /*sendMsg and receiveMsg are called when the connection
        has been established */
    int sendMsg(char *buf, int size);
    int receiveMsg(char *buf, int size);
        /*class TCPSocket server will call a private TCPSocket constructor */
    friend class TCPSocketServer;
};

class TCPSocketServer
{
        /* Private data */
    int serverSock;
public:
        /* public constructor. It will bind the socket server
        to the passed port */
    TCPSocketServer(int port);
        /* accept incoming connection */
    TCPSocket *acceptConnection();
};
```

EXERCISE 2

Using the TCPSocket class presented in the previous exercise, implement a publish-subscribe application. Actors will subscribe for a given item name, passed as argument to the subscribe method, and publish messages, specifying the item name so that every actor that subscribed for that item will receive the message.

Hint: to achieve the desired functionality it is necessary to develop a *broker* application that will keep track of what actors subscribed for a given item. When an actor publishes a message for a given item, it will send the message to the broker that, in turn, will deliver it to every actor that subscribed for that item. The broker will maintain a dictionary (you may use the C++ Standard Template Library (STL) `std::unordered_map` component) keeping for every item name a vector (again, you may use `std::vector`) of TCP connections (that is of `TCPSocket` instances) for all the actors that registered for that item. In this case the actor will first open a socket connection toward the broker and will also accept, from a different port, incoming TCP connections from the broker itself. When an actor subscribes for a given item, it will send a message to the broker specifying the item name and the actor specific port. In response, the broker will update the dictionary and will establish a new connection with the actor, using the port specified in the message.

Observe that in this implementation a number of different TCP socket connections is activated, i.e., a connection for every actor toward the broker, and a connection from the broker to every subscribed actor for every active item. Things become more complicated if actors can enter and exit the system because the broker must handle the broken TCP connections and update its dictionary accordingly.

EXERCISE 3

In the publish-subscribe implementation suggested in the previous exercise, a separate TCP connection was established between the broker and every subscribed actor for every subscription item, in addition to a connection between the broker and every client for handling the publish and subscribe requests. This leads to a potentially large number of peer-to-peer connections. If this is not a problem per se, but the management of the error conditions that arises when an actor either exits or crashes may complicate a bit the implementation. A possible solution proposed in this exercise is the combined usage of TCP and UDP sockets using UDP multicast to send at the same time a published message to all subscribers.

Hint: Part of the code developed in the previous exercise can be retained here, in particular most part of the management of the publish and subscribe requests. The use of a `std::unordered_map` dictionary can be retained but in this case, however, instead of a vector of `TCPSocket` instances, a multicast address is associated in this case with every item name. The actor will read in a separate thread via a `UDPSocket` instance incoming published messages. To subscribe to a given item, an actor will send, via its connected `TCPSocket` instance, a subscribe message bringing the item name. In response the broker will check its dictionary, picking a new multicast address in case this item has been declared the first time, and returning via the same `TCPSocket` the corresponding multicast address, that will be used by the requesting actor to subscribe its `UDPSocket` instance to this multicast address via method `subscribeTo()`. When an actor publishes a message, it will send it along with the item name to the broker via its connected `TCPSocket` instance. The broker will search the multicast address corresponding to that item in its dictionary and, if found, it will send the published message via its unique `UDPSocket` instance to that multicast address using method `sendMsgTo()` triggering the simultaneous reception of the message by means of all subscribed actors.

Section II

Real-Time Scheduling Analysis

10 Real-Time Scheduling Based on the Cyclic Executive

In any concurrent program, the exact order in which processes execute is not completely specified. According to the concurrent programming theory discussed in Chapter 3, the interprocess communication and synchronization primitives described in Chapters 5 and 6 are used to enforce as many ordering constraints as necessary to ensure that the result of a concurrent program is correct in all cases.

For example, a mutual exclusion semaphore can be used to ensure that only one process at a time is allowed to operate on shared data. Similarly, a message can be used to make one process wait for the result of a computation carried out by another process and pass the result along. Nevertheless, the program will still exhibit a significant amount of nondeterminism because its processes may *interleave* in different ways without violating any of those constraints. The concurrent program output will of course be the same in all cases, but its *timings* may still vary considerably from one execution to another.

Going back to our reference example, the producers–consumers problem, if many processes are concurrently producing data items, the final result does not depend on the exact order in which they are allowed to update the shared buffer because all data items will eventually be in the buffer. However, the amount of time spent by the processes to carry out their operations does depend on that order.

If one of the processes in a concurrent program has a tight deadline on its completion time, only *some* of the interleavings that are acceptable from the concurrent programming perspective will also be adequate from the real-time execution point of view. As a consequence, a real-time system must *further restrict* the nondeterminism found in a concurrent system because some interleavings that are acceptable with respect to the *results* of the computation may be unacceptable for what concerns *timings*. This is the main goal of *scheduling models*, the main topic of the second part of this book.

10.1 SCHEDULING AND PROCESS MODELS

The main goal of a scheduling model is to ensure that a concurrent program does not only produce the expected output in all cases but is also correct with respect to *timings*. In order to do this, a scheduling model must comprise two main elements:

1. A *scheduling algorithm*, consisting of a set of rules for ordering the use of system resources, in particular the processors.

DOI: 10.1201/9781003593416-10

2. An analytical means of analyzing the system and predicting its *worst-case behavior* with respect to timings when that scheduling algorithm is applied.

In a hard real-time scenario, the worst-case behavior is compared against the timing constraints the system must fulfill, to check whether it is acceptable or not. Those constraints are specified at system design time and are typically dictated by the physical equipment to be connected to, or controlled by, the system.

When choosing a scheduling algorithm, it is often necessary to look for a compromise between optimizing the mean performance of the system and its determinism and ability to certainly meet timing constraints. For this reason, general-purpose scheduling algorithms are often very different from their real-time counterparts.

Since they are less concerned with determinism, most general-purpose scheduling algorithms emphasize aspects such as, for instance,

Fairness In a general-purpose system, it is important to grant to each process a fair share of the available processor time and not to systematically put any process at a disadvantage with respect to the others. Dynamic priority assignments are often used for this purpose.

Efficiency The scheduling algorithm is invoked very often in an operating system, and applications perceive this as an overhead. After all, the system is not doing anything useful from the application's point of view while it is deciding what to execute next. For this reason, the complexity of most general-purpose scheduling algorithms is forced to be $\mathcal{O}(1)$. In particular, it must not depend on how many processes there are in the system.

Throughput Especially for batch systems, this is another important parameter to optimize because it represents the average number of jobs completed in a given time interval. As in the previous cases, the focus of the scheduling algorithm is on carrying out as much useful work as possible given a certain set of processes, rather than satisfying any time-related property of a specific process.

On the other hand, real-time scheduling algorithms must put the emphasis on the *timing* requirements of each individual process being executed, even if this entails a greater overhead and the mean performance of the system becomes *worse*. In order to do this, the scheduling algorithms can take advantage of the greater amount of information that, on most real-time systems, is available on the processes to be executed.

This is in sharp contrast with the scenario that general-purpose scheduling algorithms usually face: nothing is known in advance about the processes being executed, and their future characteristics must be inferred from their past behavior. Moreover, those characteristics, such as processor time demand, may vary widely with time. For instance, think about a web browser: the interval between execution bursts and the amount of processor time each of them requires both depend on what its human user is doing at the moment and on the contents of the web pages he or she is looking at.

Even in the context of real-time scheduling, it turns out that the analysis of an arbitrarily complex concurrent program, in order to predict its worst-case timing behavior, is very difficult. It is necessary to introduce a simplified *process model* that imposes some restrictions on the structure of real-time concurrent programs to be considered for analysis.

The simplest model, also known as the *basic* process model, has the following characteristics:

1. The concurrent program consists of a fixed number of processes, and that number is known in advance.
2. Processes are periodic, with known periods. Moreover, process periods do not change with time. For this reason, processes can be seen as an infinite sequence of *instances*. Process instances becomes ready for execution at regular time intervals at the beginning of each period.
3. Processes are completely independent of each other.
4. Timing constrains are expressed by means of *deadlines*. For a given process, a deadline represents the upper bound on the completion time of a process instance. All processes have *hard* deadlines, that is, they must obey their temporal constraints all the time, and the deadline of each process is equal to its period.
5. All processes have a fixed worst-case execution time that can be computed offline.
6. All system's overheads, for example, context switch times, are negligible.

The basic model just introduced has a number of shortcomings, and will be generalized to make it more suitable to describe real-world systems. In particular,

- Process independence must be understood in a very broad sense. It means that there are no synchronization constraints among processes at all, so no process must even wait for another. This rules out, for instance, mutual exclusion and synchronization semaphores and is somewhat contrary to the way concurrent systems are usually designed, in which processes *must* interact with one another.
- The deadline of a process is not always related to its period, and is often shorter than it.
- Some processes are *sporadic* rather than periodic. In other words, they are executed "on demand" when an external event, for example an alarm, occurs.
- For some applications and hardware architectures, scheduling and context switch times may not be negligible.
- The behavior of some nondeterministic hardware components, for example, caches, must sometimes be taken into account, and this makes it difficult to determine a reasonably tight upper bound on the process execution time.
- Real-time systems may sometimes be *overloaded*, a critical situation in which the computational demand exceeds the system capacity during a certain time interval. Clearly, not all processes will meet their deadline in this case, but some residual system properties may still be useful. For instance, it may be interesting to know what processes will miss their deadline first.

Table 10.1

Notation for real-time scheduling algorithms and analysis methods

Symbol	Meaning
τ_i	The i-th task
$\tau_{i,j}$	The j-th instance of the i-th task
T_i	The period of task τ_i
D_i	The relative deadline of task τ_i
C_i	The worst-case execution time of task τ_i
R_i	The worst-case response time of task τ_i
$r_{i,j}$	The release time of $\tau_{i,j}$
$f_{i,j}$	The response time of $\tau_{i,j}$
$d_{i,j}$	The absolute deadline of $\tau_{i,j}$

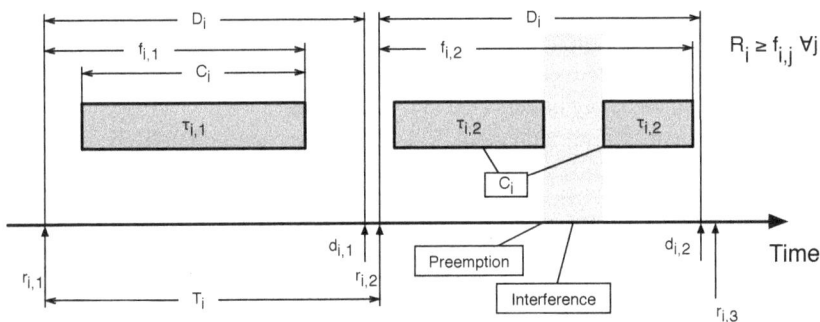

Figure 10.1 Notation for real-time scheduling algorithms and analysis methods.

The basic notation and nomenclature most commonly adopted to define scheduling algorithms and the related analysis methods are summarized in Table 10.1. It will be used throughout the second part of this book. Figure 10.1 contains a graphical depiction of the same terms and shows the execution of two instances, $\tau_{i,1}$ and $\tau_{i,2}$, of task τ_i.

As shown in the figure, it is important to distinguish among the worst-case execution time of task τ_i, denoted by C_i, the response time of its j-th instance $f_{i,j}$, and its worst-case response time, denoted by R_i. The worst-case execution time is the time required to complete the task without any interference from other activities, that is, if the task being considered were alone in the system.

The response time may (and usually will) be longer due to the effect of other tasks. As shown in the figure, a higher-priority task becoming ready during the execution of τ_i will lead to a preemption for most scheduling algorithms, so the execution of τ_i will be postponed and its completion delayed. Moreover, the execution of any tasks does not necessarily start as soon as they are released, that is, as soon as they become ready for execution.

It is also important to clearly distinguish between relative and absolute deadlines. The relative deadline D_i is defined for task τ_i as a whole and is the same for all instances. It indicates, for each instance, the distance between its release time and the deadline expiration. On the other hand, there is one distinct absolute deadline $d_{i,j}$ for each task instance $\tau_{i,j}$. Each of them denotes the instant in which the deadline expires for that particular instance.

10.2   THE CYCLIC EXECUTIVE

The *cyclic executive*, also known as *timeline scheduling* or *cyclic scheduling*, is one of the most ancient, but still widely used, real-time scheduling methods or algorithms. A full description of this scheduling model can be found in Reference [11].

In its most basic form, it is assumed that the basic model just introduced holds, that is, there is a fixed set of periodic tasks. The basic idea is to lay out *offline* a completely static schedule such that its repeated execution causes all tasks to run at their correct rate and finish within their deadline. The existence of such a schedule is also a proof "by construction" that all tasks will actually and always meet their deadline at runtime. Moreover, the sequence of tasks in the schedule is always the same so that it can be easily understood and visualized.

For what concerns its implementation, the schedule can essentially be thought of as a table of procedure calls, where each call represents (part of) the code of a task. During execution, a very simple software component, the cyclic executive, loops through the table and invokes the procedures it contains in sequence. To keep the executive in sync with the real elapsed time, the table also contains synchronization points in which the cyclic executive aligns the execution with a time reference usually generated by a hardware component.

In principle, a static schedule can be entirely crafted by hand but, in practice, it is desirable for it to adhere to a certain well-understood and agreed-upon structure, and most cyclic executives are designed according to the following principles. The complete table is also known as the *major cycle* and is typically split into a number of slices called *minor cycles*, of equal and fixed duration.

Minor cycle boundaries are also synchronization points: during execution, the cyclic executive switches from one minor cycle to the next after waiting for a periodic clock interrupt. As a consequence, the activation of the tasks at the beginning of each minor cycle is synchronized with the real elapsed time, whereas all the tasks belonging to the same minor cycle are simply activated in sequence. The minor cycle interrupt is also useful in detecting a critical error known as minor cycle *overrun*, in which the total execution time of the tasks belonging to a certain minor cycle exceeds the length of the cycle itself.

As an example, the set of tasks listed in Table 10.2 can be scheduled on a single-processor system as shown in the time diagram of Figure 10.2. If deadlines are assumed to be the same as periods for all tasks, from the figure it can easily be seen that all tasks are executed periodically, with the right period, and they all meet their deadlines.

Table 10.2

A simple task set to be executed by a cyclic executive

Task τ_i	Period T_i (ms)	Execution time C_i (ms)
τ_1	20	9
τ_2	40	8
τ_3	40	8
τ_4	80	2

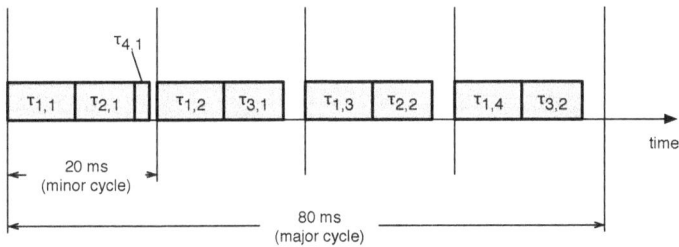

Figure 10.2   An example of how a cyclic executive can successfully schedule the task set of Table 10.2.

More in general, this kind of time diagram illustrates the job that each processor in the system is executing at any particular time. It is therefore useful to visualize and understand how a scheduling algorithm works in a certain scenario. For this reason, it is a useful tool not only for cyclic executives but for any scheduling algorithm.

To present a sample implementation, we will start from a couple of quite realistic assumptions:

- The underlying hardware has a programmable timer, and it can be used as an interrupt source. In particular, the abstract function void timer_setup(int p) can be used to set the timer up, start it, and ask for a periodic interrupt with period p milliseconds.
- The function void wait_for_interrupt(void) waits for the next timer interrupt and reports an error, for instance, by raising an exception, if the interrupt arrived before the function was invoked, denoting an overrun.
- The functions void task_1(void), ..., void task_4() contain the code of tasks $\tau_1, \ldots, \tau_4$, respectively.

The cyclic executive can then be implemented with the following program.

```
int main(int argc, void *argv[])
{
  ...
  timer_setup(20);

  while(1)
```

```
{
  wait_for_interrupt();
    task_1();
    task_2();
    task_4();

  wait_for_interrupt();
    task_1();
    task_3();

  wait_for_interrupt();
    task_1();
    task_2();

  wait_for_interrupt();
    task_1();
    task_3();
  }
}
```

In the sample program, the scheduling table is actually embedded into the main loop. The main program first sets up the timer with the right minor cycle period and then enters an endless loop. Within the loop, the function `wait_for_interrupt()` sets the boundary between one minor cycle and the next. In between, the functions corresponding to the task instances to be executed within the minor cycle are called in sequence.

Hence, for example, the first minor cycle contains a call to `task_1()`, `task_2()`, and `task_4()` because, as it can be seen on the left part of Figure 10.2, the first minor cycle must contain one instance of τ_1, τ_2, and τ_4.

With this implementation, no actual processes exist at run-time because the minor cycles are just a sequence of *procedure* calls. These procedures share a common address space, and hence, they implicitly share their global data. Moreover, on a single processor system, task bodies are always invoked *sequentially* one after another. Thus, shared data do not need to be protected in any way against concurrent access because concurrent access is simply not possible.

Once a suitable cyclic executive has been constructed, its implementation is straightforward and very *efficient* because no scheduling activity takes place at run-time and overheads are very low, without precluding the use of a very sophisticated (and computationally expensive) algorithm to construct the schedule. This is because scheduler construction is done completely offline.

On the downside, the cyclic executive "processes" cannot be *protected* from each other, as regular processes are, during execution. It is also difficult to incorporate nonperiodic activities efficiently into the system without changing the task sequence.

10.3   CHOICE OF MAJOR AND MINOR CYCLE LENGTH

The minor cycle is the smallest timing reference of the cyclic executive because task execution is synchronized with the real elapsed time only at minor cycle boundaries. As a consequence, all task periods must be an integer multiple of the minor cycle period. Otherwise, it would be impossible to execute them at their proper rate.

On the other hand, it is also desirable to keep the minor cycle length as large as possible. This is useful not only to reduce synchronization overheads but also to make it easier to accommodate tasks with a large execution time, as discussed in Section 10.4. It is easy to show that one simple way to satisfy both constraints is to set the minor cycle length to be equal to the *Greatest Common Divisor* (GCD) of the periods of the tasks to be scheduled. A more flexible and sophisticated way of selecting the minor cycle length is discussed in Reference [11].

If we call T_m the minor cycle length, for the example presented in the previous section we must choose

$$T_m = \text{gcd}(T_1, \ldots, T_4) = \text{gcd}(20, 40, 40, 80) = 20\,\text{ms} \tag{10.1}$$

The cyclic executive repeats the same schedule over and over at each major cycle. Therefore, the major cycle must be big enough to be an integer multiple of all task periods, but no larger than that to avoid making the scheduling table larger than necessary for no reason. A sensible choice is to let the major cycle length be the Least Common Multiple (LCM) of the task periods. Sometimes this is also called the *hyperperiod* of the task set. For example, if we call T_M the major cycle length, we have

$$T_M = \text{lcm}(T_1, \ldots, T_4) = \text{lcm}(20, 40, 40, 80) = 80\,\text{ms} \tag{10.2}$$

10.4 TASKS WITH LARGE PERIOD OR EXECUTION TIME

In the previous section, some general rules to choose the minor and major cycle length for a given task set were given. Although they are fine in theory, there may be some issues when trying to put them into practice. For instance, when the task periods are mutually prime, the major cycle length calculated according to (10.2) has its worst possible value, that is, the product of all periods. The cyclic executive scheduling table will be large as a consequence.

Although this issue clearly cannot be solved in general—except by adjusting task periods to make them more favorable—it turns out that, in many cases, the root cause of the problem is that only one or a few tasks have a period that is disproportionately large with respect to the others. For instance, in the task set shown in Table 10.3, the major cycle length is

$$T_M = \text{lcm}(T_1, \ldots, T_4) = \text{lcm}(20, 40, 40, 400) = 400\,\text{ms} \tag{10.3}$$

However, if we neglected τ_4 for a moment, the major cycle length would shrink by an order of magnitude

$$T_M' = \text{lcm}(T_1, \ldots, T_3) = \text{lcm}(20, 40, 40) = 40\,\text{ms} \tag{10.4}$$

If τ_4's period T_4 is a multiple of the new major cycle length T_M', that is, if

$$T_4/T_M' = k, \quad k \in \mathbb{N} \tag{10.5}$$

Table 10.3

A task set in which a since task, τ_4, leads to a large major cycle because its period is large

Task τ_i	Period T_i (ms)	Execution time C_i (ms)
τ_1	20	9
τ_2	40	8
τ_3	40	8
τ_4	400	2

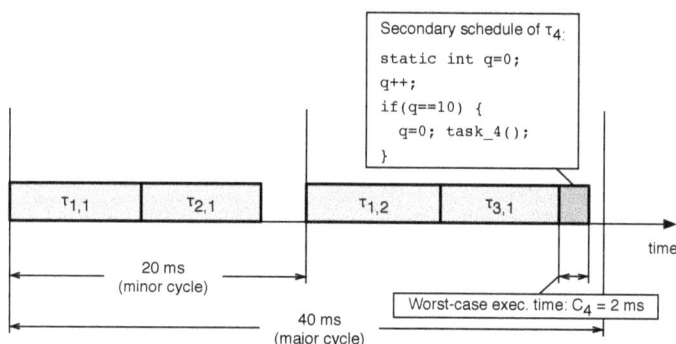

Figure 10.3   An example of how a simple secondary schedule can schedule the task set of Table 10.3 with a small major cycle.

the issue can be circumvented by designing the schedule as if τ_4 were not part of the system, and then using a so-called *secondary* schedule.

In its simplest form, a secondary schedule is simply a wrapper placed around the body of a task, `task_4()` in our case. The secondary schedule is invoked on every major cycle and, with the help of a private counter q that is incremented by one at every invocation, it checks if it has been invoked for the k-th time, with $k = 10$ in our example. If this is not the case, it does nothing; otherwise, it resets the counter and invokes `task_4()`. The code of the secondary schedule and the corresponding time diagram are depicted in Figure 10.3.

As shown in the figure, even if the time required to execute the wrapper itself is negligible, as is often the case, the worst-case execution time that must be considered during the cyclic executive design to accommodate the secondary schedule is still equal to the worst-case execution time of τ_4, that is, C_4. This is an extremely conservative approach because `task_4()` is actually invoked only on every k iterations of the schedule, and hence, the worst-case execution time of the wrapper is very different from its mean execution time.

A different issue may occur when one or more tasks have a large execution time. The most obvious case happens when the execution time of a task is greater than the

Table 10.4

Large execution times, of τ_3 in this case, may lead to problems when designing a cyclic executive

Task τ_i	Period T_i (ms)	Execution time C_i (ms)
τ_1	25	10
τ_2	50	8
τ_3	100	20

minor cycle length so that it simply cannot fit into the schedule. However, there may be subtler problems as well. For instance, for the task set shown in Table 10.4, the minor and major cycle length, chosen according to the rules given in Section 10.3, are

$$T_m = 25\,\text{ms} \tag{10.6}$$
$$T_M = 100\,\text{ms} \tag{10.7}$$

In this case, as shown in the upper portion of Figure 10.4, task instance $\tau_{3,1}$ could be executed entirely within a single minor cycle because $C_3 \le T_m$, but this choice would hamper the proper schedule of other tasks, especially τ_1. In fact, the first instance of τ_1 would not fit in the first minor cycle because $C_1 + C_3 > T_m$. Shifting $\tau_{3,1}$ into another minor cycle does not solve the problem either.

Hence, the only option is to split $\tau_{3,1}$ into two pieces: $\tau_{3,1a}$ and $\tau_{3,1b}$, and put them into two distinct minor cycles. For example, as shown in the lower part of Figure 10.4, we could split $\tau_{3,1}$ into two equal pieces with an execution time of 10 ms each and put them into the first and third minor cycle, respectively.

Although it is possible to work out a correct cyclic executive in this way, it should be remarked that splitting tasks into pieces may cut across the tasks in a way that has nothing to do with the structure of the code itself. In fact, the split is not made on the basis of some characteristics of the code but merely on the constraints the execution time of each piece must satisfy to fit into the schedule.

Moreover, task splits make shared data management much more complicated. As shown in the example of Figure 10.4—but this is also true in general—whenever a task instance is split into pieces, other task instances are executed between those pieces. In our case, two instances of task τ_1 and one instance of τ_2 are executed between $\tau_{3,1a}$ and $\tau_{3,1b}$. This fact has two important consequences:

1. If τ_1 and/or τ_2 share some data structures with τ_3, the code of $\tau_{3,1a}$ must be designed so that those data structures are left in a consistent state at the end of $\tau_{3,1a}$ itself. This requirement may increase the complexity of the code.
2. It is no longer completely true that shared data does not need to be protected against concurrent access. In this example, it is "as if" τ_3 were preempted

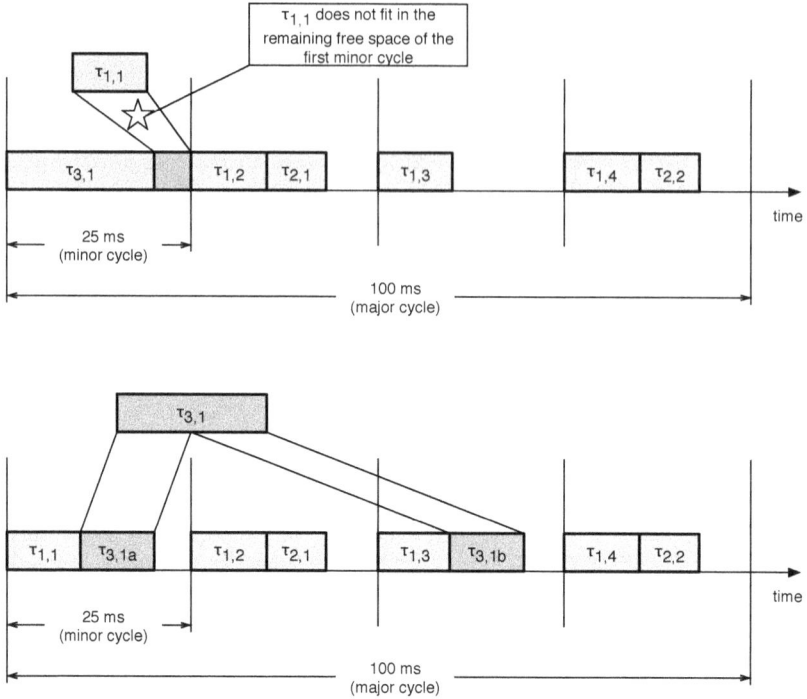

Figure 10.4 In some cases, such as for the task set of Table 10.4, it is necessary to split one or more tasks with a large execution time into pieces to fit them into a cyclic executive.

by τ_1 and τ_2 during execution. If τ_1 and/or τ_2 share some data structures with τ_3, this is equivalent to a concurrent access to the shared data. The only difference is that the preemption point is always the same (at the boundary between $\tau_{3,1a}$ and $\tau_{3,1b}$) and is known in advance.

Last but not least, building a cyclic executive is mathematically hard in itself. Moreover, the schedule is sensitive to any change in the task characteristics, above all their periods, which requires the entire scheduling sequence to be reconstructed from scratch when those characteristics change.

Even if the cyclic executive approach is a simple and effective tool in many cases, it may not be general enough to solve all kinds of real-time scheduling problems that can be found in practice. This reasoning led to the introduction of other, more sophisticated scheduling models, to be discussed in the following chapters. The relative advantages and disadvantages of cyclic executives with respect to other scheduling models have been subject to considerable debate. For example, Reference [65] contains an in-depth comparison between cyclic executives versus fixed-priority, task-based schedulers.

10.5   SUMMARY

To start discussing about real-time scheduling, it is first of all necessary to abstract away, at least at the beginning, from most of the complex and involved details of real concurrent systems and introduce a simple *process model*, more suitable for reasoning and analysis. In a similar way, an abstract *scheduling model* specifies a scheduling algorithm and its associated analysis methods without going into the fine details of its implementation.

In this chapter, one of the simplest process models, called the *basic process model*, has been introduced, along with the nomenclature associated with it. It is used throughout the book as a foundation to talk about the most widespread real-time scheduling algorithms and gain an insight into their properties. Since some of its underlying assumptions are quite unrealistic, it will also be progressively refined and extended to make it adhere better to what real-world processes look like.

Then we have gone on to specify how one of the simplest and most intuitive real-time scheduling methods, the *cyclic executive*, works. Its basic idea is to lay out a time diagram and place task instances into it so that all tasks are executed periodically at their proper time and they meet their timing constraints or *deadlines*.

The time diagram is completely built offline before the system is ever executed, and hence, it is possible to put into action sophisticated layout algorithms without incurring any significant overhead at runtime. The time diagram itself also provides intuitive and convincing evidence that the system really works as intended.

That said, the cyclic executive also has a number of disadvantages: it may be hard to build, especially for unfortunate combinations of task execution times and periods, it is quite inflexible, and may be difficult to properly maintain it when task characteristics are subject to change with time or the system complexity grows up. For this reason, we should go further ahead and examine other, more sophisticated, scheduling methods in the next chapters.

10.6   EXERCISES

EXERCISE 1

Design a cyclic executive for the following set of tasks:

Task τ_i	Period T_i (ms)	Execution time C_i (ms)
τ_1	60	2
τ_2	20	3
τ_3	30	3

EXERCISE 2

Extend the cyclic executive of the previous exercise to accommodate an additional task τ_4 with period $T_4 = 20\,$ms and execution time $C_4 = 4\,$ms *without* task splitting.

EXERCISE 3

How long would the major cycle of the cyclic executive designed in the previous exercise become if we added a fifth task τ_5 with period $T_5 = 600\,\text{ms}$ and $C_5 = 2\,\text{ms}$ without using a secondary schedule? Would a secondary schedule keep the major cycle the same as before?

EXERCISE 4

Further extend the cyclic executive by adding a sixth task τ_6 with period $T_6 = 600\,\text{ms}$ and execution time $C_6 = 10\,\text{ms}$. Tasks with such a long period and non-negligible executing time are typical, for example, of background network communication.

   Hint: A typical solution to this exercise requires a combination of task splitting and secondary scheduling.

11 Real-Time, Task-Based Scheduling

The previous chapter has introduced the basic model and terminology for real-time scheduling. The same notation will be used in this chapter as well as in the following ones, and therefore, it is briefly recalled here. A periodic real-time process is called a *task* and denoted by τ_i. A task models a periodic activity: at the j-th occurrence of the period T_i a *job* $\tau_{i,j}$ for a given *task* τ_i is released. The job is also called an *instance* of the task τ_i. The relative deadline D_i for a task τ_i represents the maximum time allowed between the release of any job $\tau_{i,j}$ and its termination, and, therefore, the absolute deadline $d_{i,j}$ for the job $\tau_{i,j}$ is equal to its release time plus the relative deadline. The worst-case execution time of task T_i represents the upper limit of the processor time required for the computation of any job for that task, while R_i indicates the worst-case response time of task T_i, that is, the maximum elapsed time between the release of any job for this task and its termination. The worst-case execution time (WCET) C_i is the time required to complete any job of the task τ_i without any interference from other activities. Finally, $f_{i,j}$ is the actual absolute response time (i.e., the time of its termination) for job $\tau_{i,j}$ of task τ_i.

While in the cyclic executive scheduling policy all jobs were executed in a predefined order, in this chapter we shall analyze a different situation where tasks correspond to processes or threads and are therefore scheduled by the operating system based on their current priority. Observe that, in the cyclic executive model, there is no need for a scheduler at all: the jobs are represented by routines that are invoked in a predefined order by a single program. Here we shall refer to a situation, which is more familiar to those who have read the first part of this book, where the operating system handles the concurrent execution of different units of execution. In the following, we shall indicate such units as *tasks*, being the distinction between processes and threads not relevant in this context. Depending on the practical requirements, tasks will be implemented either by processes or threads, and the results of the scheduling analysis are valid in both cases.

Many of the results presented in this chapter and in the following ones are due to the seminal work of Liu and Layland [63], published in 1973. Several proofs given in the original paper have later been refined and put in a more intuitive form by Buttazzo [19]. Interested readers are also referred to Reference [82] for more information about the evolution of real-time scheduling theory from a historical perspective. Moreover, References [19, 64] discuss real-time scheduling in much more formal terms than can be afforded here, and they will surely be of interest to readers with a stronger mathematical background.

DOI: 10.1201/9781003593416-11

11.1   FIXED AND VARIABLE TASK PRIORITY

We have seen in Chapter 3 that the priority associated with tasks is an indication of their "importance." Important tasks need to be executed first, and therefore, the scheduler, that is, the component of the operating system that supervises the assignment of the processors to tasks, selects the task with the highest priority among those that are currently ready (i.e., which are not in wait state, due, for example, to a pending I/O operation). Therefore, the policy that is adopted to assign priorities to tasks determines the behavior of the scheduling. In the following, we shall analyze different policies for assigning priorities to tasks and their impact in obtaining the desired real-time behavior, making sure that every job terminates within its assigned deadline.

11.1.1   PREEMPTION

Before discussing about priority assignment policies, we need to consider an important fact: what happens if, during the execution of a task at a given priority, another task with higher priority becomes ready? Most modern operating systems in this case reclaim the processor from the executing task and assign it to the task with higher priority by means of a context switch. This policy is called *preemption*, and it ensures that the most important task able to utilize the processor is always executing. Older operating systems, such as MS-DOS or the Mac OS versions prior to 10, did not support preemption, and therefore a task that took possession of the processor could not be forced to release it, unless it performed an I/O operation or invoked a system call. Preemption presents several advantages, such as making the system more reactive and preventing rogue tasks from monopolizing the processor. The other side of the coin is that preemption is responsible for most race conditions due to the possibly unforeseen interleaved execution of higher-priority tasks.

Since in a preemptive policy the scheduler must ensure that the task currently in execution is always at the highest priority among those that are ready, it is important to understand when a context switch is possibly required, that is, when a new higher priority task may request the processor. Let us assume first that the priorities assigned to tasks are fixed: a new task may reclaim the processor only when it becomes ready, and this may happen only when a pending I/O operation for that task terminates or a system call (e.g., waiting for a semaphore) is concluded. In all cases, such a change in the task scenario is carried out by the operating system, which can therefore effectively check current task priorities and ensure that the current task is *always* that with the highest priority among the ready ones. This fact holds also if we relax the fixed-priority assumption: the change in task priority would be in any case carried out by the operating system, which again is aware of any possible change in the priority distribution among ready tasks.

Within the preemptive organization, differences may arise in the management of multiple ready tasks with the same highest priority. In the following discussion, we shall assume that all the tasks have a different priority level, but such a situation represents somehow an abstraction, the number of available priority levels being limited in practice. We have already seen that POSIX threads allow two different management of multiple tasks at the same highest priority tasks:

1. The First In First Out (FIFO) management, where the task that acquires the processor will execute until it terminates or enters in wait state due to an I/O operation or a synchronization primitive, or a higher priority task becomes ready.
2. The Round Robin (RR) management where after some amount of time (often called time slice) the running task is preempted by the scheduler even if no I/O operation is performed and no higher-priority task is ready to let another task at the same priority gain processor usage.

11.1.2 VARIABLE PRIORITY IN GENERAL PURPOSE OPERATING SYSTEMS

Scheduling analysis refers to two broad categories in task priority assignment to tasks: *Fixed Priority* and *Variable Priority*. As the name suggests, in fixed priority scheduling, the priority assigned to tasks never changes during system execution. Conversely, in the variable-priority policy, the priority of tasks is dynamically changed during execution to improve system responsiveness or other parameters. Before comparing the two approaches, it is worth briefly describing what happens in general-purpose operating systems such as Linux and Windows. Such systems are intended for a variety of different applications, but interaction with a human user is a major use case and, in this case, the perceived responsiveness of the system is an important factor. When interacting with a computer via a user interface, in fact, getting a quick response to user events such as the click of the mouse is preferable over other performance aspects such as overall throughput in computation. For this reason, a task that spends most of its time doing I/O operations, including the response to user interface events, is considered more important than a task making intensive computation. Moreover, when the processor is assigned to a task making lengthy computation, if preemption were not supported by the scheduler, the user would experience delays in interaction due to the fact that the current task would not get a chance to release the processor if performing only computation and not starting any I/O. For these reasons, the scheduler in a general purpose operating system will assign a higher priority to I/O intensive tasks and will avoid that a computing-intensive task monopolize the processor, thus blocking interaction for an excessively long period. To achieve this, it is necessary to provide an answer to the following questions:

1. How to discriminate between I/O-intensive and computing-intensive tasks?
2. How to preempt the processor from a computing-intensive task that is not willing to relinquish it?
3. How to ensure enough fairness to avoid that a ready task is postponed forever or for a period of time that is too long?

The above problems are solved by the following mechanism for dynamic priority assignment, called *timesharing*, which relies on a clock device that periodically interrupts the processor and gives a chance to the operating system to get control, rearrange task priorities and possibly operate a context switch because a task with a higher priority than the current one is now available. Interestingly enough, many dynamic priority assignment schemes in use today still bear a strong resemblance to the scheduling algorithm designed by Corbató [22] back in 1962 for one of the first experimental timesharing systems.

A time slice (also called *quantum*) is assigned to every task when it acquires the processor, and at every clock period (called *tick*), the operating system decreases the quantum value of the running task in case no other task with a higher priority becomes ready. When the quantum reaches 0, and there is at least another task with equal or higher priority, the current task is preempted. In addition to the quantum mechanism, a variable is maintained by the operating system for each task: whenever the task is awakened, this variable is incremented; whenever the task is preempted or its quantum expires, the variable is decremented.

The actual value of the variable is used to compute a "priority bonus" that rewards I/O-intensive tasks that very seldom experience quantum expiration and preemption (an I/O-intensive task is likely to utilize the processor for a very short period of time before issuing a new I/O and entering in wait state), and which penalizes computing-intensive tasks that periodically experience quantum expiration and are preempted.

In this way, discrimination between I/O and computing-intensive tasks is carried out by the actual value of the associated task variable, the quantum expiration mechanism ensures that the processor is eventually preempted from tasks that would otherwise block the system for too long, and the dynamic priority mechanism lowers the priority of computing-intensive tasks that would otherwise block lower priority tasks waiting for the processor. Observe that I/O-intensive tasks will hardly cause any task starvation since typically they require the processor for very short periods of time.

Timesharing is an effective policy for interactive systems but cannot be considered in real-time applications because it is not possible to predict in advance the maximum response time for a given task since this depends on the behavior of the other tasks in the systems, affecting the priority of the task and therefore its response time. For this reason, real-time systems normally assign fixed priorities to tasks, and even general-purpose operating systems supporting timesharing reserve a range of higher priorities to be statically assigned to real-time tasks. These tasks will have a priority that is always higher than the priority of timesharing tasks and are therefore guaranteed to get the processor as soon as they become ready, provided no higher-priority real-time task is currently ready. The next section will present and analyze a widely adopted policy called *Rate Monotonic* in fixed-priority assignment. The reader may, at this point, wonder whether dynamic priority assignment can be of any help in real-time applications. The next section shows that this is the case and presents a dynamic priority assignment policy, called *Earliest Deadline First*, which not only ensures real-time behavior in a system of periodic tasks, but represents the "best" scheduling policy ever attainable for a given set of periodic tasks under certain conditions as described in the following pages.

Table 11.1

An example of Rate Monotonic priority assignment

Task	Period	Computation Time	Priority
τ_1	20	7	High
τ_2	50	13	Low
τ_3	25	6	Medium

11.2 RATE MONOTONIC

Rate monotonic is a policy for fixed-priority assignment in periodic tasks, which assigns a priority that is inversely proportional to the task period: the shorter the task period, the higher its priority. Consider, for example, the three tasks listed in Table 11.1: task τ_1, which has the smallest period, will have the highest priority, followed by tasks τ_3 and τ_2, in that order.

Observe that the priority assignment takes into account only the task period, and not the effective computation time. Priorities are often expressed by integers, but there is no general agreement on whether higher values represent higher priorities or the other way round. In most cases, lower numbers indicate higher priorities, but in any case, this is only an implementation issue and does not affect the following discussion.

In order to better understand how a scheduler (the rate monotonic scheduler in this case) works, and to draw some conclusions about its characteristics, we can simulate the behavior of the scheduler and build the corresponding scheduling diagram. To be meaningful, the simulation must be carried out for an amount of time that is "long enough" to cover all possible phase relations among the tasks. As for the cyclic executive, the right amount of time is the *Least Common Multiple* (LCM) of the task periods. After such period, if no overflow occurs (i.e., the scheduling does not fail), the same sequence will repeat, and therefore, no further information is obtained when simulation of the system behavior is performed for a longer period of time. Since we do not have any additional information about the tasks, we also assume that all tasks are simultaneously released at t = 0. We shall see shortly that this assumption is the most pessimistic one when considering the scheduling assignment, and therefore, if we prove that a given task priority assignment can be used for a system, that it will be feasible regardless of the actual initial release time (often called *phase*) of task jobs. A sample schedule is shown in Figure 11.1.

The periods of the three tasks are 20 ms, 25 ms, and 50 ms, respectively, and therefore the period to be considered in simulation is 100 ms, that is the Least Common Multiplier of 20, 25, and 50. At t = 0, all tasks are ready: the first one to be executed is τ_1 then, at its completion, τ_3. At t = 13 ms, τ_2 finally starts but, at t = 20 ms, τ_1 is released again. Hence, τ_2 is preempted in favor of τ_1. While τ_1 is executing, τ_3 is released, but this does not lead to a preemption: τ_3 is executed after τ_1

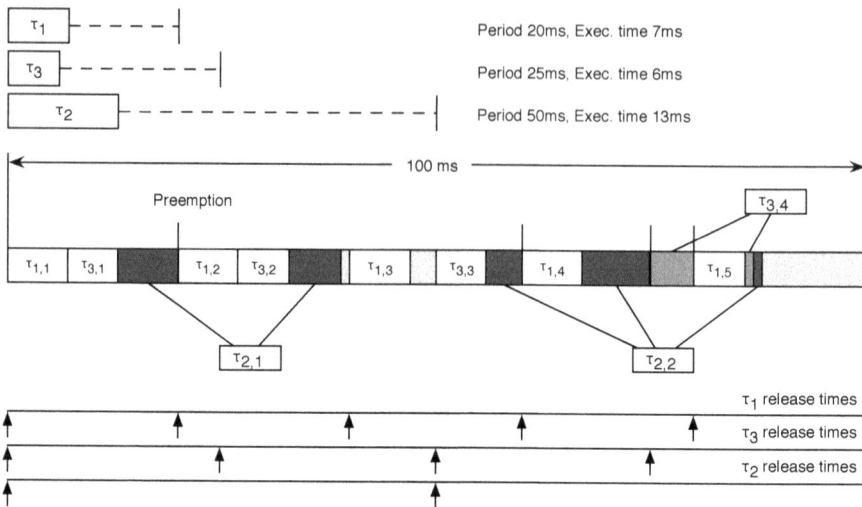

Figure 11.1   Scheduling sequence for tasks τ_1, τ_2, and τ_3.

has finished. Finally, τ_2 is resumed and then completed at t = 39 ms. At t = 40 ms, after 1 ms of idling, task τ_1 is released. Since it is the only ready task, it is executed immediately, and completes at t = 47 ms. At t = 50ms, both τ_3 and τ_2 become ready simultaneously. τ_3 is run first, then τ_2 starts and runs for 4 ms. However, at t = 60 ms, τ_1 is released again. As before, this leads to the preemption of τ_2 and τ_1 runs to completion. Then, τ_2 is resumed and runs for 8 ms, until τ_3 is released. τ_2 is preempted again to run τ_3. The latter runs for 5 ms but at, t = 80 ms, τ_1 is released for the fifth time. τ_3 is preempted, too, to run τ_1. After the completion of τ_1, both τ_3 and τ_2 are ready. τ_3 runs for 1 ms, then completes. Finally, τ_2 runs and completes its execution cycle by consuming 1 ms of CPU time. After that, the system stays idle until t = 100 ms, where the whole cycle starts again.

Intuitively Rate Monotonic makes sense: tasks with shorter period are expected to be executed before others because they have less time available. Conversely, a task with a long period can afford waiting for other more urgent tasks and finish its execution in time all the same. However intuition does not represent a mathematical proof, and we shall prove that Rate Monotonic is really the best scheduling policy among all the fixed priority scheduling policies. In other words, if every task job finishes execution within its deadline under any given fixed priority assignment policy, then the same system is feasible under Rate Monotonic priority assignment. The formal proof, which may be skipped by the less mathematically inclined reader, is given below.

11.2.1 PROOF OF RATE MONOTONIC OPTIMALITY

Proving the optimality of Rate Monotonic consists in showing that if a given set of periodic tasks with fixed priorities is schedulable in any way, then it will be schedulable using the Rate Monotonic policy. The proof will be carried out under the following assumptions:

1. Every task τ_i is periodic with period T_i.
2. The relative deadline D_i for every task τ_i is equal to its period T_i.
3. Tasks are scheduled preemptively and according to their priority.
4. There is only one processor.

We shall prove this in two steps. First we shall introduce the concept of "critical instant," that is, the "worst" situation that may occur when a set of periodic tasks with given periods and computation times is scheduled. Task jobs can, in fact, be released at arbitrary instants within their period, and the time between period occurrence and job release is called the *phase* of the task. We shall see that the worst situation will occur when all the jobs are initially released at the same time (i.e., when the phase of all the tasks is zero). The following proof will refer to such a situation: proving that the system under consideration is schedulable in such a bad situation means proving that it will be schedulable for every task phase.

We introduce first some considerations and definition:

- According to the simple process model, the relative deadline of a task is equal to its period, that is, $D_i = T_i \forall i$.
- Hence, for each task instance, the absolute deadline is the time of its next release, that is, $d_{i,j} = r_{i,j+1}$.
- We say that there is an overflow at time t if t is the deadline for a job that misses the deadline.
- A scheduling algorithm is *feasible* for a given set of task if they are scheduled so that no overflows ever occur.
- A *critical instant* for a task is an instant at which the release of the task will produce the largest response time.
- A *critical time zone* for a task is the interval between a critical instant and the end of the task response.

The following theorem, proved by Liu and Layland [63], identifies critical instants.

Theorem 11.1. *A critical instant for any task occurs whenever it is released simultaneously with the release of all higher-priority tasks.*

To prove the theorem, which is valid for every fixed-priority assignment, let τ_1, τ_2, ..., τ_m be a set of tasks, listed in order of decreasing priority, and consider the task with the lowest priority, τ_m. If τ_m is released at t_1, between t_1 and $t_1 + T_m$, that is, the time of the next release of τ_m, other tasks with a higher priority will possibly be released and interfere with the execution of τ_m because of preemption. Now, consider one of the interfering tasks, τ_i, with $i < m$ and suppose that, in the interval

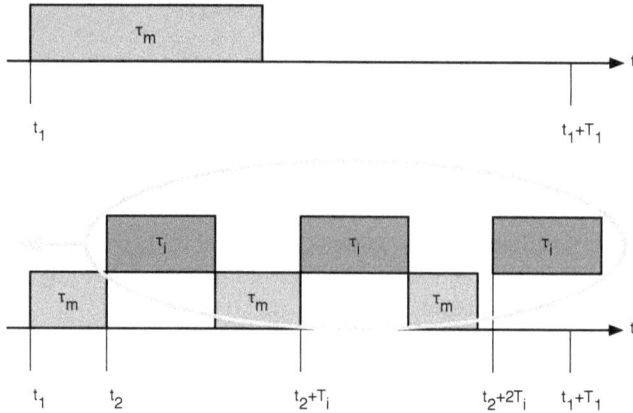

Figure 11.2 Interference to τ_m due to higher-priority tasks τ_i.

between t_1 and $t_1 + T_m$, it is released at t_2; $t_2 + T_i$, …; $t_2 + kT_i$, with $t_2 \geq t_1$. The preemption of τ_m by τ_i will cause a certain amount of delay in the completion of the instance of τ_m being considered, unless it has already been completed before t_2, as shown in Figure 11.2.

From the figure, it can be seen that the amount of delay depends on the relative placement of t_1 and t_2. However, moving t_2 toward t_1 will never decrease the completion time of τ_m. Hence, the completion time of τ_m will be either unchanged or further delayed, due to additional interference, by moving t_2 toward t_1. If t_2 is moved further, that is $t_2 < t_1$, the interference is not increased because the possibly added interference due to a new release of τ_i before the termination of the instance of τ_m is at least compensated by the reduction of the interference due the instance of τ_i released at t_1 (part of the work for the first instance of τ_i has already been carried out at t_2). The delay is therefore largest when $t_1 = t_2$, that is, when the tasks are *released simultaneously*.

The above argument can finally be repeated for all tasks τ_i; $1 \leq i < m$, thus proving the theorem.

Under the hypotheses of the theorem, it is possible to check whether or not a given priority assignment scheme will yield a feasible scheduling algorithm *without simulating* it for the LCM of the periods. If all tasks conclude their execution before the deadline—that is, they all fulfill their deadlines—when they are released simultaneously and therefore are at their critical instant, then the scheduling algorithm is feasible.

What we are going to prove is the optimality of Rate Monotonic in the worst case, that is, for critical instants. Observe that this condition may also not occur since it depends on the initial phases of the tasks, but this fact does not alter the outcome of the following proof. In fact, if a system is schedulable in critical instants, it will remain schedulable for every combination of task phases.

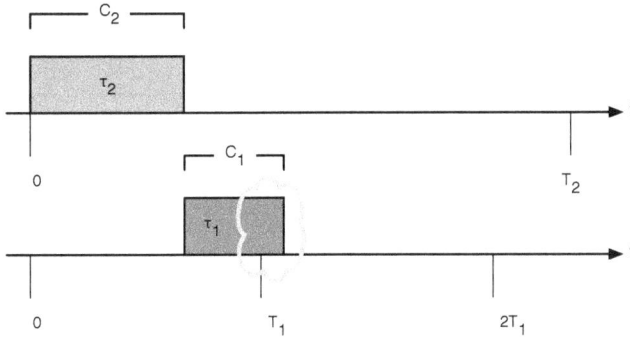

Figure 11.3 Tasks τ_1 and τ_2 not scheduled under RM.

We are now ready to prove the optimality of Rate Monotonic (abbreviated in the following as RM), and we shall do it assuming that all the initial task jobs are released simultaneously at time 0. The optimality of RM is proved by showing that, if a task set is schedulable by an arbitrary (but fixed) priority assignment, then it is also schedulable by RM. This result also implies that if RM cannot schedule a certain task set, no other fixed-priority assignment algorithm can schedule it.

We shall consider first the simpler case in which exactly two tasks are involved, and we shall prove that if the set of two tasks τ_1 and τ_2 is schedulable by any arbitrary, but fixed, priority assignment, then it is schedulable by RM as well.

Let us consider two tasks, τ_1 and τ_2, with $T_1 < T_2$. If their priorities are not assigned according to RM, then τ_2 will have a priority higher than τ_1. At a critical instant, their situation is that shown in Figure 11.3.

The schedule is feasible if (and only if) the following inequality is satisfied:

$$C_1 + C_2 \leq T_1 \tag{11.1}$$

In fact, if the sum of the computation time of τ_1 and τ_2 is greater than the period of τ_1, it is not possible that τ_1 can finish its computation within its deadline.

If priorities are assigned according to RM, then task τ_1 will have a priority higher than τ_2. If we let F be the number of periods of τ_1 *entirely* contained within T_2, that is,

$$F = \left\lfloor \frac{T_2}{T_1} \right\rfloor \tag{11.2}$$

then, in order to determine the feasibility conditions, we must consider two cases (which cover all possible situations):

1. The execution time C_1 is "short enough" so that all the instances of τ_1 within the critical zone of τ_2 are completed before the next release of τ_2.
2. The execution of the last instance of τ_1 that starts within the critical zone of τ_2 overlaps the next release of τ_2.

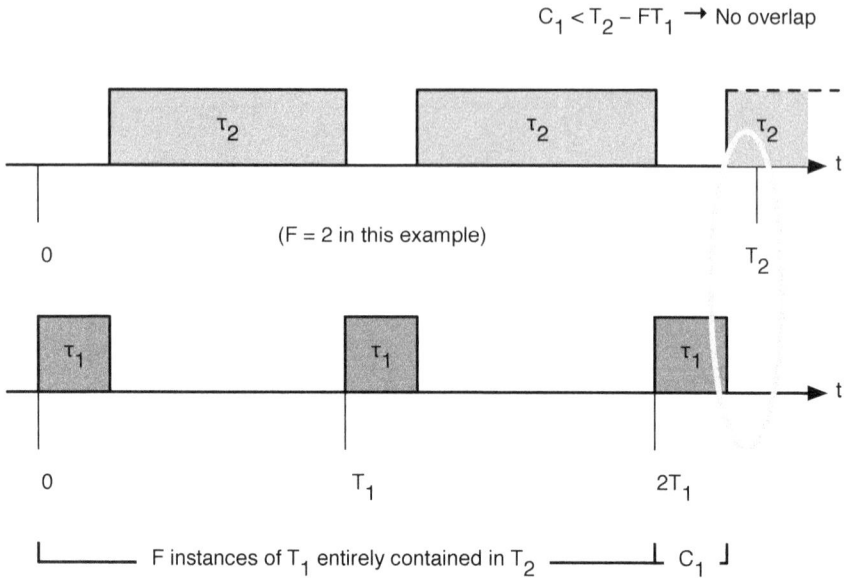

Figure 11.4 Situation in which all the instances of τ_1 are completed before the next release of τ_2.

Let us consider case 1 first, corresponding to Figure 11.4.
The first case occurs when

$$C_1 < T_2 - FT_1 \tag{11.3}$$

From Figure 11.4 , we can see that the task set is schedulable if and only if

$$(F+1)C_1 + C_2 \leq T_2 \tag{11.4}$$

Now consider case 2, corresponding to Figure 11.5. The second case occurs when

$$C_1 \geq T_2 - FT_1 \tag{11.5}$$

From Figure 11.5, we can see that the task set is schedulable if and only if

$$FC_1 + C_2 \leq FT_1 \tag{11.6}$$

In summary, given a set of two tasks, τ_1 and τ_2, with $T_1 < T_2$ we have the following two conditions:

1. When priorities are assigned according to RM, the set is schedulable if and only if
   - $(F+1)C_1 + C_2 \leq T_2$, when $C_1 < T_2 - FT_1$.
   - $FC_1 + C_2 \leq FT_1$, when $C1 \geq T_2 - FT_1$.
2. When priorities are assigned otherwise, the set is schedulable if and only if $C_1 + C_2 \leq T_1$

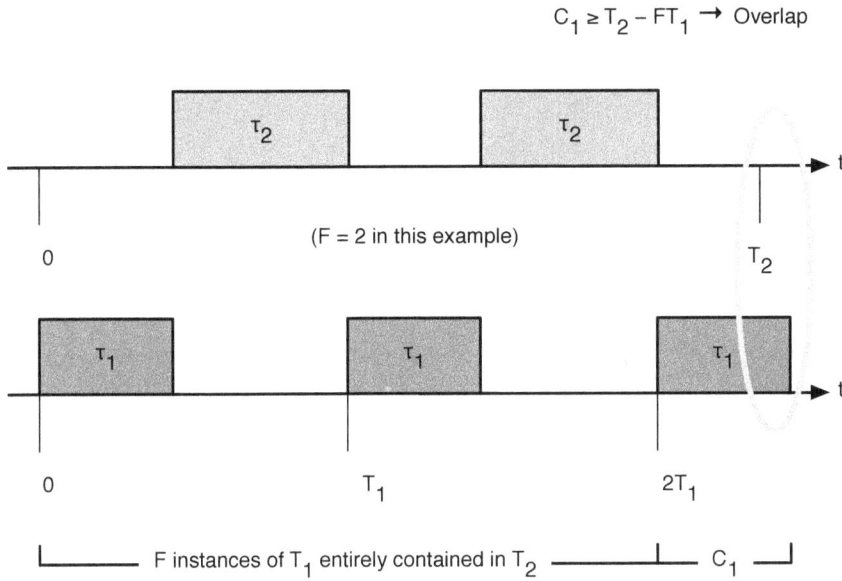

$C_1 \geq T_2 - FT_1 \rightarrow$ Overlap

(F = 2 in this example)

F instances of T_1 entirely contained in T_2 ⎯⎯⎯⎯ C_1

Figure 11.5 Situation in which the last instance of τ_1 that starts within the critical zone of τ_2 overlaps the next release of τ_2.

To prove the optimality of RM with two tasks, we must show that the following two implications hold:

1. If $C_1 < T_2 - FT_1$, then $C_1 + C_2 \leq T_1 \Rightarrow (F+1)C_1 + C_2 \leq T_2$.
2. If $C_1 \geq T_2 - FT_1$, then $C_1 + C_2 \leq T_1 \Rightarrow FC_1 + C_2 \leq FT_1$.

Consider the first implication: if we multiply both members of $C_1 + C_2 \leq T_1$ by F and then add C_1, we obtain

$$(F+1)C_1 + FC_2 \leq FT_1 + C_1 \tag{11.7}$$

We know that $F \geq 1$ (otherwise, it would not be $T_1 < T_2$), and hence,

$$FC_2 \geq C_2 \tag{11.8}$$

Moreover, from the hypothesis we have

$$FT_1 + C_1 < T_2 \tag{11.9}$$

As a consequence, we have

$$(F+1)C_1 + C_2 \leq (F+1)C_1 + FC_2 \leq FT_1 + C_1 \leq T_2 \tag{11.10}$$

which proves the first implication.

Consider now the second implication: if we multiply both members of $C_1 + C_2 \leq T_1$ by F, we obtain

$$FC_1 + FC_2 \leq FT_1 \tag{11.11}$$

We know that $F \geq 1$ (otherwise, it would not be $T_1 < T_2$), and hence

$$FC_2 \geq C_2 \tag{11.12}$$

As a consequence, we have

$$FC_1 + C_2 \leq FC_1 + FC_2 \leq FT_1 \tag{11.13}$$

which concludes the proof of the optimality of RM when considering two tasks.

The optimality of RM is then extended to an arbitrary set of tasks thanks to the following theorem [63]:

Theorem 11.2. *If the task set $\tau_1, \ldots, \tau_n$ (n tasks) is schedulable by any arbitrary, but fixed, priority assignment, then it is schedulable by RM as well.*

The proof is a direct consequence of the previous considerations: let τ_i and τ_j be two tasks of adjacent priorities, τ_i being the higher-priority one, and suppose that $T_i > T_j$. Having adjacent priorities, both τ_i and τ_j are affected in the same way by the interferences coming from the higher-priority tasks (and not at all by the lower-priority ones). Hence, we can apply the result just obtained and state that if we interchange the priorities of τ_i and τ_j, the set is still schedulable. Finally, we notice that the RM priority assignment can be obtained from any other priority assignment by a sequence of pairwise priority reorderings as above, thus ending the proof.

The above problem has far-reaching implications because it gives us a simple way for assigning priorities to real-time tasks knowing that that choice is the best ever possible. At this point we may wonder if it is possible to do better by relaxing the fixed-priority assumption. From the discussion at the beginning of this chapter, the reader may have concluded that dynamic priority should be abandoned when dealing with real-time systems. This is true for the priority assignment algorithms that are commonly used in general purpose operating systems since there is no guarantee that a given job will terminate within a fixed amount of time. There are, however, other algorithms for assigning priority to tasks that do not only ensure a timely termination of the job execution but perform better than fixed-priority scheduling. The next section will introduce the *Earliest Deadline First* dynamic priority assignment policy, which takes into account the absolute deadline of every task in the priority assignment.

11.3   THE EARLIEST DEADLINE FIRST SCHEDULER

The Earliest Deadline First (abbreviated as EDF) algorithm selects tasks according to their absolute deadlines. That is, at each instant, tasks with earlier deadlines will receive higher priorities. Recall that the absolute deadline $d_{i,j}$ of the j-th instance (job) of the task τ_i is formally

$$d_{i,j} = \phi_i + jT_i + D_i \tag{11.14}$$

where ϕ_i is the phase of task τ_i, that is, the release time of its first instance (for which $j = 0$), and T_i and D_i are the period and relative deadlines of task τ_i, respectively. The priority of each task is assigned dynamically, because it depends on the current deadlines of the active task instances. The reader may be concerned about the practical implementation of such dynamic priority assignment: does it require that the scheduler must continuously monitor the current situation in order to arrange task priorities when needed? Luckily, the answer is no: in fact, task priorities may be updated only when a new task instance is released (task instances are released at every task period). Afterward, when time passes, the relative order due to the proximity in time of the next deadline remains unchanged among active tasks, and therefore, priorities are not changed.

As for RM, EDF is an intuitive choice as it makes sense to increase the priority of more "urgent" tasks, that is, for which deadline is approaching. We already stated that intuition is not a mathematical proof, therefore we need a formal way of proving that EDF is the optimal scheduling algorithm, that is, if any task set is schedulable by *any* scheduling algorithm, then it is also schedulable by EDF. This fact can be proved under the following assumption:

- Tasks are scheduled preemptively;
- There is only one processor.

The formal proof will be provided in the next chapter, where it will be shown that any set of tasks whose processor utilization does not exceed the processor capability is schedulable under EDF. The *processor utilization* for a set of tasks $\tau_1, \ldots, \tau_n$ is formally defined as

$$\sum_{i=1}^{n} \frac{C_i}{T_i} \tag{11.15}$$

where each term $\frac{C_i}{T_i}$ represents the fraction of processor time devoted to task τ_i. Clearly, it is not possible to schedule on a single processor a set of tasks for which the above sum is larger than one (in other words, processor utilization cannot be larger than 100%). Otherwise, the set of tasks will be *in any case* schedulable under EDF.

11.4 SUMMARY

This chapter has introduced the basics of task based scheduling, providing two "optimal" scheduling procedures: RM for fixed task priority assignment, and EDF for dynamic task priority assignment. Using a fixed-priority assignment has several advantages over EDF, among which are the following:

- Fixed-priority assignment is easier to implement than EDF, as the scheduling attribute (priority) is static.

- EDF requires a more complex run-time system, which will typically have a higher overhead.
- During overload situations, the behavior of fixed-priority assignment is easier to predict (the lower-priority processes will miss their deadlines first).
- EDF is less predictable and can experience a *domino* effect in which a large number of tasks unnecessarily miss their deadline.

On the other side, EDF is always able to exploit the full processor capacity, whereas fixed-priority assignment, and therefore RM, in the worst case does not.

EDF implementations are not common in commercial real-time kernels because the operating system would need to keep into account a set of parameters that is not considered in general-purpose operating systems. Moreover, EDF refers to a task model (periodic tasks with given deadline) that is more specific than the usual model of process. There is, however, a set of real-time open-source kernels that support EDF scheduling, and a new scheduling mode has been recently proposed for Linux [33]. Both have developed under the FP7 European project ACTORS [1].

Here, each task is characterized by a *budget* and a *period*, which is equal to its relative deadline. At any time, the system schedules the ready tasks having the earliest deadlines. During execution, the budget is decreased at every clock tick, and when a task's budget reaches zero (i.e., the task executed for a time interval equal to its budget), the task is stopped until the beginning of the next period, the deadline of the other tasks changed accordingly, and the task with the shortest deadline chosen for execution.

Up to now, however, the usage of EDF scheduling is not common in embedded systems, and a fixed task priority under RM policy is normally used.

As a final remark, observe that all the presented analysis relies on the assumption that the considered tasks do not interact each other, neither are they suspended, for example, due to an I/O operation. This is a somewhat unrealistic assumption (whole chapters of this book are devoted to interprocess communication and I/O), and such effects must be taken into consideration in real-world systems. This will be the main argument of Chapters 14 and 15, which will discuss the impact in the schedulability analysis of the use of system resources and I/O operations.

11.5 EXERCISES

EXERCISE 1

Task τ_i	Period T_i (ms)	Execution time C_i (ms)
τ_1	60	2
τ_2	20	3
τ_3	30	3
τ_4	20	4

- Assign appropriate priorities to the tasks listed in the table above according to the Rate Monotonic (RM) scheduling algorithm and draw the corresponding scheduling diagram, assuming that all tasks are ready for execution at $t = 0$.

- Identify the critical instant(s) in the scheduling diagram.
- Under the same assumptions, draw the scheduling diagram for the same task set, when scheduled by the Earliest Deadline First (EDF) algorithm.

Let τ_i have a priority higher than the priority of τ_j if $i < j$ to avoid ambiguities in the diagrams, when the scheduling algorithm would otherwise assign the same priority to them. Be sure to understand the differences between these scheduling diagrams and the one of the cycle executive designed for Exercise 2, Chapter 10.

EXERCISE 2

Draw the RM and EDF scheduling diagrams for the following set of tasks, in the interval $0 \le t \le 56\,\text{ms}$. Determine whether these two algorithms are able to successfully schedule the task set in the given time interval. As in the previous exercise, let τ_i have a priority higher than the priority of τ_j if $i < j$, when the scheduling algorithm would otherwise assign the same priority to them.

Task τ_i	Period T_i (ms)	Execution time C_i (ms)
τ_1	40	8
τ_2	20	3
τ_3	30	12
τ_4	20	5

12 Schedulability Analysis Based on Utilization

The previous chapter introduced the basic concepts in process scheduling and analyzed the two classes of scheduling algorithms: *fixed priority* and *variable priority*. When considering fixed-priority scheduling, it has been shown that Rate Monotonic (RM) Scheduling is *optimal*, that is, if a task set is schedulable under any-fixed priority schema, then it will be under RM. For variable-priority assignment, the optimality of Earliest Deadline First (EDF) has been enunciated and will be proved in this chapter.

Despite the elegance and importance of these two results, their practical impact for the moment is rather limited. In fact, what we are interested in practice is to know whether a given task assignment is schedulable, before knowing what scheduling algorithm to use. This is the topic of this chapter and the next one. In particular, a sufficient condition for schedulability will be presented here, which, when satisfied, ensures that the given set of tasks is definitely schedulable. Only at this point do the results of the previous chapter turn out to be useful in practice because they give us an indication of the right scheduling algorithm to use.

We shall discover in this chapter that the schedulability check will be very simple, being based on an upper limit in the processor utilization. This simplicity is, however, paid for by the fact that this condition is only a sufficient one. As a consequence, if the utilization check fails, we cannot state that the given set of tasks is not schedulable. A more accurate but also more complex check will be provided in Chapter 13.

12.1 PROCESSOR UTILIZATION

Our goal is to define a schedulability test for either RMS or EDF based on very simple calculations over the tasksÂ' period and execution time. In the following, we will assume that the basic process model is being used and, in particular, we shall consider *single-processor* systems.

Given a set of N periodic tasks $\Gamma = \{\tau_1, \ldots, \tau_N\}$, the *processor utilization factor* U is the fraction of processor time spent in the execution of the task set, that is,

$$U = \sum_{i=1}^{N} \frac{C_i}{T_i} \tag{12.1}$$

where $\frac{C_i}{T_i}$ is the fraction of processor time spent executing task τ_i. The processor utilization factor is therefore a measure of the computational load imposed on the processor by a given task set and can be increased by increasing the execution times C_i of the tasks. For a given scheduling algorithm A, there exists a maximum value of

DOI: 10.1201/9781003593416-12

U below which the task set Γ is schedulable, but for which any increase in the computational load C_i of any of the tasks in the task set will make it no longer schedulable. This limit will depend on the task set Γ and on the scheduling algorithm A.

A task set Γ is said to *fully utilize* the processor with a given scheduling algorithm A if it is schedulable by A, but any increase in the computational load C_i of any of its tasks will make it no longer schedulable. The corresponding *upper bound* of the utilization factor is denoted as $U_{ub}(\Gamma, A)$.

If we consider now *all the possible task sets* Γ, it is interesting (and useful) to ask how large the utilization factor can be in order to guarantee the schedulability of any task set Γ by a given scheduling algorithm A. In order to do this, we must determine the minimum value of $U_{ub}(\Gamma, A)$ over all task sets Γ that fully utilize the processor with the scheduling algorithm A. This new value, called *least upper bound* and denoted as $U_{lub}(A)$, will only depend on the scheduling algorithm A and is defined as

$$U_{lub}(A) = \min_{\Gamma} U_{ub}(\Gamma, A) \qquad (12.2)$$

where Γ represents the set of all task sets that fully utilize the processor. A pictorial representation of the meaning of $U_{lub}(A)$ is given in Figure 12.1. The least upper

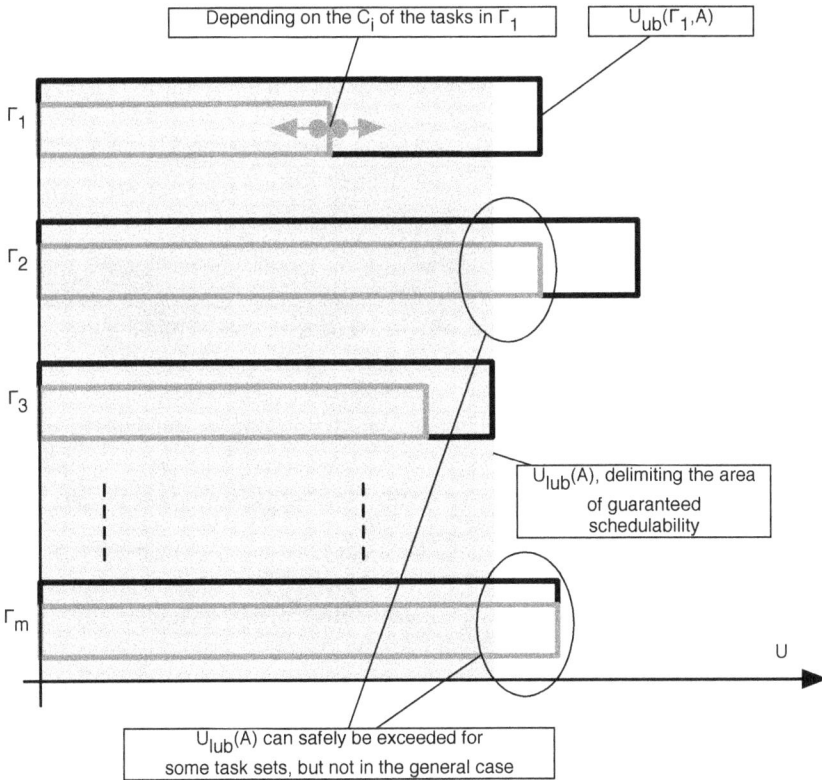

Figure 12.1   Upper Bounds and Least Upper Bound for scheduling algorithm A.

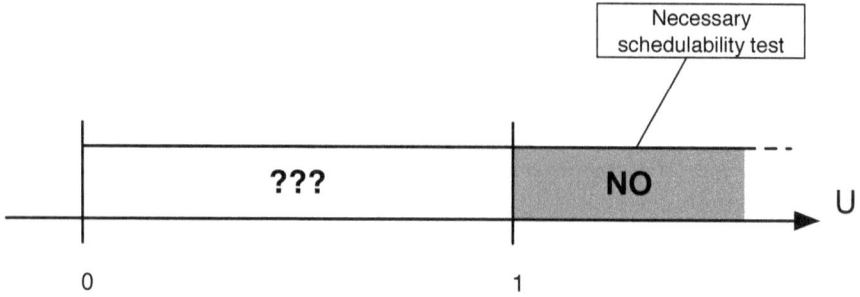

Figure 12.2   Necessary schedulability condition.

bound $U_{lub}(A)$ corresponds to the shaded part of the figure. For every possible task set Γ_i, the maximum utilization depends on both A and Γ. The actual utilization for task set Γ_i will depend on the computational load of the tasks but will never exceed $U_{ub}(\Gamma_i, A)$. Since $U_{lub}(A)$ is the minimum upper bound over all possible task sets, any task set whose utilization factor is below $U_{lub}(A)$ will be schedulable by A. On the other hand, it may happen that $U_{lub}(A)$ can sometimes be exceeded, but not in general case.

Regardless of the adopted scheduling algorithm, there is an upper limit in processor utilization that can never be exceeded, as defined in the following theorem:

Theorem 12.1. *If the processor utilization factor U of a task set Γ is greater than one (that is, if $U > 1$), then the task set is not schedulable, regardless of the scheduling algorithm.*

Even if the theorem can be proved formally, the result is quite intuitive and, stated in words, it says that it is impossible to allocate to the tasks a fraction of CPU time greater than the total Â"quantityÂ" of CPU time available. This, therefore, represents a necessary condition: if the total utilization is above one for a single processor system, then we definitely know that the system is not schedulable, but we cannot say anything in the case where the total utilization is below one, as shown in Figure 12.2.

12.2   SUFFICIENT SCHEDULABILITY TEST FOR RATE MONOTONIC

We will now show how to compute the least upper bound U_{lub} of the processor utilization for RM. From this we will derive a sufficient schedulability test for RM so that, if a given task set Γ satisfies it, its schedulability will be guaranteed by RM. This is a practical result and can be used, for example, in a dynamic real-time system that may accept in run time requests for new tasks to be executed. Based on the expected processor usage of the new task, the system may accept or reject the request: if accepted, there is the guarantee that the real-time requirements of the system are not infringed. Of course, since the test will not be exact, its failure will give us no information about schedulability, and therefore, it may happen that a task will be refused even if it may be safely run. This is the price paid for the simplicity of the utilization-based test. In the next chapter, a more accurate and complex schedulability test will be presented.

In the following, the utilization limit for RM will be formally derived for two tasks. The general result for n tasks will then be enunciated. Readers not interested in the proof details may safely skip to the end of the section where the final result is presented.

12.2.1 U_{LUB} **FOR TWO TASKS**

Let us consider a set of two periodic tasks τ_1 and τ_2, with periods $T_1 < T_2$. According to the RM priority assignment, τ_1 will be the task with the highest priority. We will first compute the upper bound U_{ub} of their utilization factor by setting the task computation times to fully utilize the processor. Then, to obtain U_{lub}, we will minimize U_{ub} over all the other task parameters.

As before, let F be the number of periods of τ_1 entirely contained within T_2:

$$F = \left\lfloor \frac{T_2}{T_1} \right\rfloor \tag{12.3}$$

Without loss of generality, we will adjust C_2 to fully utilize the processor. Again, we must consider two cases:

- The execution time C_1 is Â"short enoughÂ" so that all the instances of τ_1 within the critical zone of τ_2 are completed before the next release of τ_2.
- The execution of the last instance of τ_1 that starts within the critical zone of τ_2 overlaps the next release of τ_2.

Let us consider the first case, shown in Figure 12.3. The largest possible value of C_2 is:

$$C_2 = T_2 - (F+1)C_1 \tag{12.4}$$

If we compute U for this value of C_2, we will obtain U_{ub}. In fact, in this case, the processor is fully utilized, and every increment in either C_1 or C_2 would make the task set no more schedulable.

By definition of U we have

$$
\begin{aligned}
U_{ub} &= \frac{C_1}{T_1} + \frac{C_2}{T_2} &= \frac{C_1}{T_1} + \frac{T_2 - (F+1)C_1}{T_2} \\
&= 1 + \frac{C_1}{T_1} - \frac{(F+1)C_1}{T_2} \\
&= 1 + \frac{C_1}{T_2}[\frac{T_2}{T_1} - (F+1)] \tag{12.5}
\end{aligned}
$$

Since $F = \left\lfloor \frac{T_2}{T_1} \right\rfloor$,

$$FT_1 \leq T_2 < (F+1)T_1 \tag{12.6}$$

$C_1 < T_2 - FT_1 \;\rightarrow\;$ No overlap

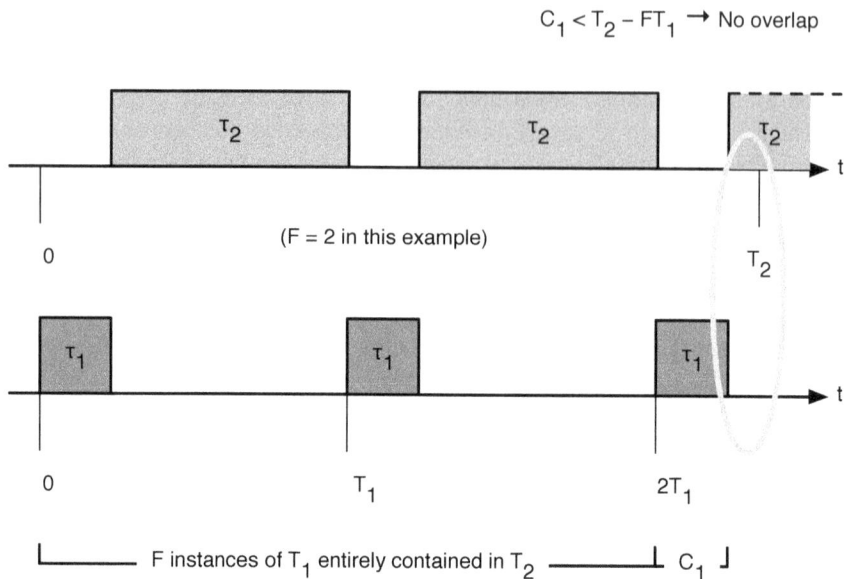

(F = 2 in this example)

F instances of T_1 entirely contained in T_2 ——— C_1

Figure 12.3 No overlap between instances of τ_1 and the next release time of τ_2.

and the quantity between square brackets will be strictly negative. Therefore, U_{ub} is monotonically decreasing with respect to C_1.

Consider now the second case, in which the execution of the last instance of τ_1 that starts within the critical zone of τ_2 overlaps the next release of τ_2. This case is shown in Figure 12.4. The largest possible value of C_2 in this case is

$$C_2 = FT_1 - FC_1 \tag{12.7}$$

Again, if we compute U for this value of C_2, we will obtain U_{ub}. By definition of U we have

$$
\begin{aligned}
U_{ub} &= \frac{C_1}{T_1} + \frac{FT_1 - FC_1}{T_2} \\
&= F\frac{T_1}{T_2} + \frac{C_1}{T_1} - F\frac{C_1}{T_2} \\
&= F\frac{T_1}{T_2} + \frac{C_1}{T_2}[\frac{T_2}{T_1} - F]
\end{aligned} \tag{12.8}
$$

Since $F = \left\lfloor \frac{T_2}{T_1} \right\rfloor$, then $F \le \frac{T_2}{T_1}$, and the quantity between square brackets will be either positive or zero. Therefore, U_{ub} is monotonically nondecreasing with respect to C_1. Considering the minimum possible value of U_{ub} we have, for each of the two cases above

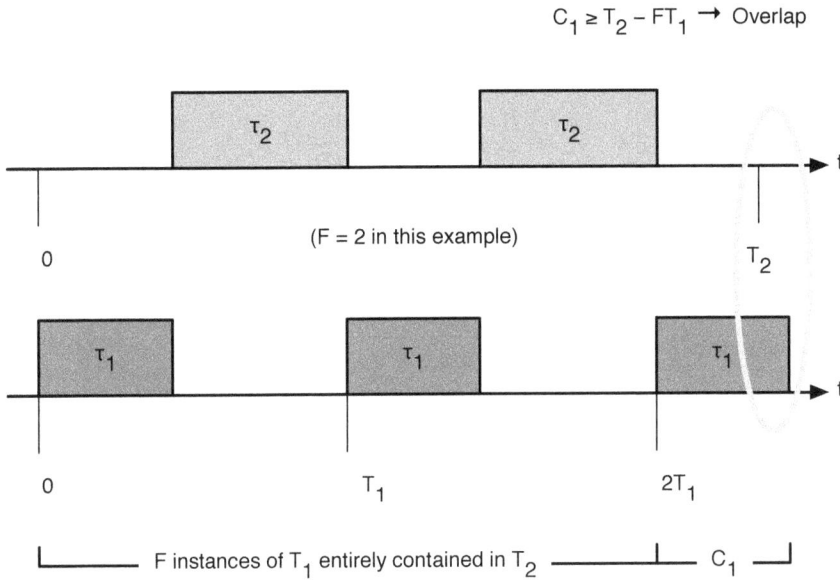

Figure 12.4 Overlap between instances of τ_1 and the next release time of τ_2.

- In the first case, since U_{ub} is monotonically decreasing with respect to C_1, its value will be at its minimum when C_1 assumes its maximum allowed value.
- In the second case, since U_{ub} is monotonically nondecreasing with respect to C_1, its value will be at its minimum when C_1 assumes its minimum allowed value.

Observe now that as $C_1 < T_2 - FT_1$ by hypothesis in the first case (see Figure 12.3), and $C_1 \geq T_2 - FT_1$ in the second case (see Figure 12.4), U_{ub} is at its minimum at the *boundary* between the two cases, that is, when

$$C_1 = T_2 - FT_1 \tag{12.9}$$

At this point, we can take either one of the expressions we derived for U_{ub} and substitute $C_1 = T_2 - FT_1$ into it. In fact, both refer to the same situation from the scheduling point of view, and hence, they must both give the same result.

It should be noted that the resulting expression for U_{ub} will still depend on the task periods T_1 and T_2 through F, and hence, we will have to minimize it with respect to F in order to find the *least* upper bound U_{lub}. By substituting $C_1 = T_2 - FT_1$ into

(12.8), we get

$$
\begin{aligned}
U &= F\frac{T_1}{T_2} + \frac{T_2 - FT_1}{T_2}(\frac{T_2}{T_1} - F) \\
&= F\frac{T_1}{T_2} + (1 - F\frac{T_1}{T_2})(\frac{T_2}{T_1} - F) \\
&= F\frac{T_1}{T_2} + \frac{T_1}{T_2}(\frac{T_2}{T_1} - F)(\frac{T_2}{T_1} - F) \\
&= \frac{T_1}{T_2}[F + (\frac{T_2}{T_1} - F)(\frac{T_2}{T_1} - F)]
\end{aligned}
\tag{12.10}
$$

Let us define now G as

$$
G = \frac{T_2}{T_1} - F
\tag{12.11}
$$

Since, by definition, $F = \left\lfloor \frac{T_2}{T_1} \right\rfloor$, $0 \le G < 1$. It will be $G = 0$ when T_2 is an integer multiple of T_1. By back substitution, we obtain

$$
\begin{aligned}
U &= \frac{T_1}{T_2}(F + G^2) \\
&= \frac{F + G^2}{T_2/T_1} \\
&= \frac{F + G^2}{(T_2/T_1 - F) + F} \\
&= \frac{F + G^2}{F + G} \\
&= \frac{(F + G) - (G - G^2)}{F + G} \\
&= 1 - \frac{G(1 - G)}{F + G}
\end{aligned}
\tag{12.12}
$$

Since $0 \le G < 1$, then $0 < (1 - G) \le 1$ and $0 \le G(1 - G) \le 1$. As a consequence, U is monotonically nondecreasing with respect to F and will be minimum when F is minimum, that is, when $F = 1$. Therefore, we can substitute $F = 1$ in the previous equation to obtain

$$
\begin{aligned}
U &= 1 - \frac{G(1 - G)}{1 + G} \\
&= \frac{(1 + G) - G(1 - G)}{1 + G} \\
&= \frac{1 + G^2}{1 + G}
\end{aligned}
\tag{12.13}
$$

We arrived at expressing the full utilization as a function of a single and continuous variable G, and therefore, we can find its minimum using its derivative

$$
\begin{aligned}
\frac{dU}{dG} &= \frac{2G(1+G) - (1+G^2)}{(1+G)^2} \\
&= \frac{G^2 + 2G - 1}{(1+G)^2}
\end{aligned}
\tag{12.14}
$$

$\frac{dU}{dG}$ will be zero when $G^2 + 2G - 1 = 0$, that is, when

$$
G = -1 \pm \sqrt{2}
\tag{12.15}
$$

Of these solutions, only $G = -1 + \sqrt{2}$ is acceptable because the other one is negative. Finally, the least upper bound of U is given by

$$
\begin{aligned}
U_{lub} &= U|_{G=\sqrt{2}-1} \\
&= \frac{1 + (\sqrt{2} - 1)^2}{1 + (\sqrt{2} - 1)} \\
&= \frac{4 - 2\sqrt{2}}{\sqrt{2}} \\
&= 2(\sqrt{2} - 1)
\end{aligned}
\tag{12.16}
$$

In summary, considering two tasks and RM scheduling, we have the following value for the least upper utilization:

$$
U_{lub} = 2(\sqrt{2} - 1)
\tag{12.17}
$$

12.2.2 U_{LUB} FOR N TASKS

The result just obtained can be extended to an arbitrary set of N tasks. The original proof by Liu and Layland [63] was not completely convincing; it was later refined by Devillers and Goossens [24].

Theorem 12.2. *For a set of N periodic tasks scheduled by the Rate Monotonic algorithm, the least upper bound of the processor utilization factor U_{lub} is*

$$
U_{lub} = N(2^{1/N} - 1)
\tag{12.18}
$$

This theorem gives us a sufficient schedulability test for the RM algorithm: a set of N periodic tasks will be schedulable by the RM algorithm if

$$
\sum_{i=1}^{N} \frac{C_i}{T_i} \leq N(2^{1/N} - 1)
\tag{12.19}
$$

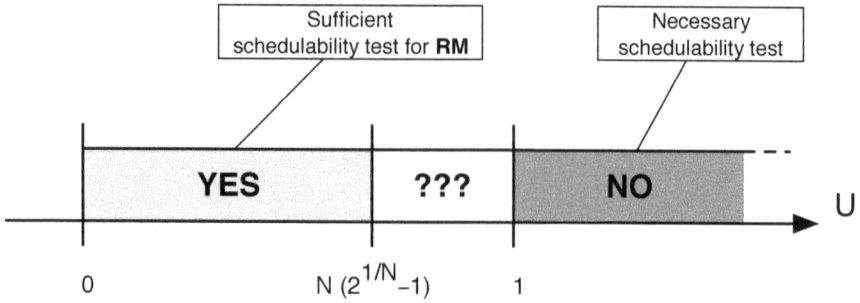

Figure 12.5 Schedulability conditions for Rate Monotonic.

Table 12.1

A task set definitely schedulable by RM

Task τ_i	Period T_i	Computation Time C_i	Priority	Utilization
τ_1	50	20	Low	0.400
τ_2	40	4	Medium	0.100
τ_3	16	2	High	0.125

We can summarize this result as shown in Figure 12.5. With respect to Figure 12.2, the area of uncertain utilization has been restricted. Only for utilization values falling into the white area, are we not yet able to state schedulability.

The next three examples will illustrate in practice the above concepts. Here we shall assume that both T_i and C_i are measured with the same, arbitrary time unit.

Consider first the task set of Table 12.1. In this task assignment, the combined processor utilization factor is $U = 0.625$. For three tasks, from (12.19) we have $U_{lub} = 3(2^{1/3} - 1) \approx 0.779$ and, since $U < U_{lub}$, we conclude from the sufficient schedulability test that the task set is schedulable by RM.

Consider now the set of tasks described in Table 12.2. The priority assignment does not change in respect of the previous example because periods T_i are

---

Table 12.2

A task set for which the sufficient RM scheduling condition does not hold

Task τ_i	Period T_i	Computation Time C_i	Priority	Utilization
τ_1	50	10	Low	0.200
τ_2	30	6	Medium	0.200
τ_3	20	10	High	0.500

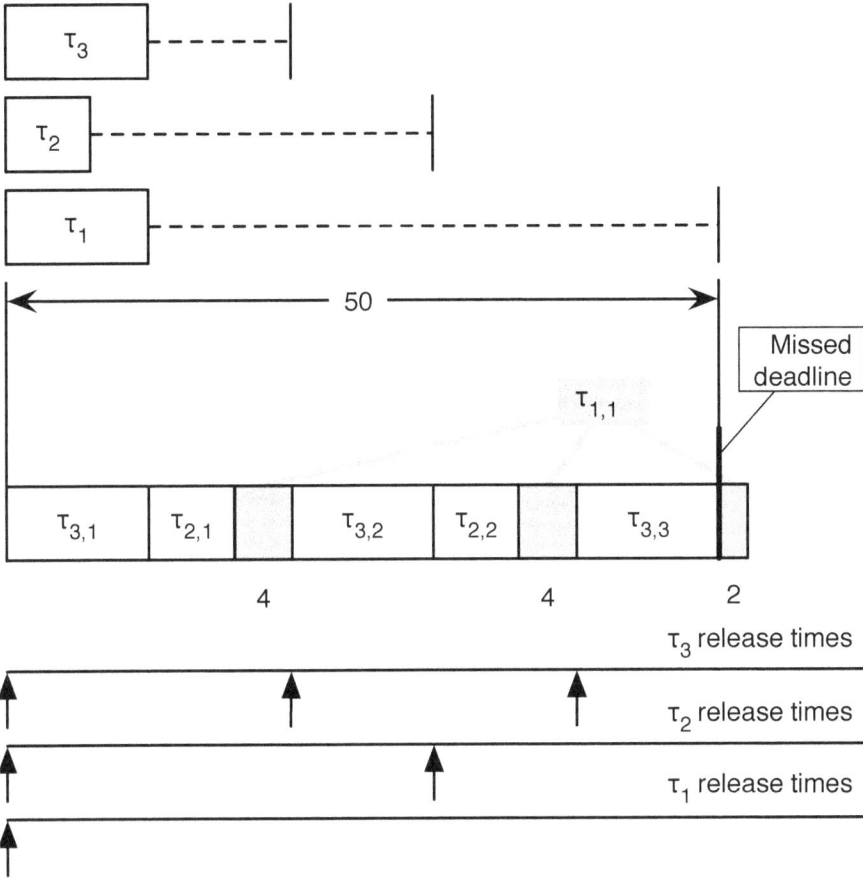

Figure 12.6 RM scheduling for a set of tasks with $U = 0.900$.

still ordered as before. The combined processor utilization factor now becomes $U = 0.900$ and, since $U > U_{lub}$, the sufficient schedulability test does not tell us anything useful in this case.

A snapshot of the scheduling sequence is given in Figure 12.6, where all the tasks are released at time 0. In fact, we know that if all tasks fulfill their deadlines when they are released at their critical instant, that is, simultaneously, then the RM schedule is feasible. However, it is easy to show that task τ_1 misses its deadline, and hence the task set is not schedulable.

Let us now consider yet another set of processes, listed in Table 12.3. As before, the priority assignment does not change with respect to the previous example because periods T_i are still ordered in the same way. The combined processor utilization factor is now the maximum value allowed by the necessary schedulability condition, that

Table 12.3

Another task set for which the sufficient RM scheduling condition does not hold.

Task τ_i	Period T_i	Computation Time C_i	Priority	Utilization
τ_1	40	14	Low	0.350
τ_2	20	5	Medium	0.250
τ_3	10	4	High	0.400

is, $U = 1$. We already know that for a larger utilization, the task set would definitely be not schedulable. However, since $U > U_{lub}$, the sufficient schedulability test does not tell us anything useful, even in this case.

We can check the actual behavior of the scheduler as shown in Figure 12.7, and we discover that all deadlines are met in this case, even if the utilization is larger than in the previous example.

In summary, the utilization check for RM consists in computing the current value of the utilization U and comparing it with $U_{lub} = N(2^{1/N} - 1)$. U_{lub} is monotonically decreasing with respect to the number N of tasks and, for large values of N, it asymptotically approaches $\ln 2 \approx 0.693$, as shown in Figure 12.8. From this observation which is simpler, but more pessimistic, a sufficient condition can be stated for any N: any task set with a combined utilization factor of less than $\ln 2$ will always be schedulable by the RM algorithm.

12.3   SCHEDULABILITY TEST FOR EDF

In the previous section we have derived a utilization limit for RM that turns out to be less than 1. Informally speaking, this is a "penalty" to be paid for the fact that tasks have a fixed priority and therefore it is not possible to make runtime adjustments when a task needs to be processed more "urgently" than another one whose fixed priority is larger. EDF scheduling does exactly this and decides *run time*, which is the task that needs to be served first.

Therefore, we may expect that the utilization limit below which the task set is definitely schedulable will be greater in EDF scheduling compared to RM. Here we shall discover that this limit is exactly 1, that is, every task set is either nonschedulable by any scheduling algorithm because its utilization is greater than 1, or it can be safely scheduled under EDF. This fact is described by the following theorem, from Liu and Layland [63]:

Theorem 12.3. *A set of N periodic tasks is schedulable with the Earliest Deadline First algorithm if and only if*

$$\sum_{i=1}^{N} \frac{C_i}{T_i} \leq 1 \tag{12.20}$$

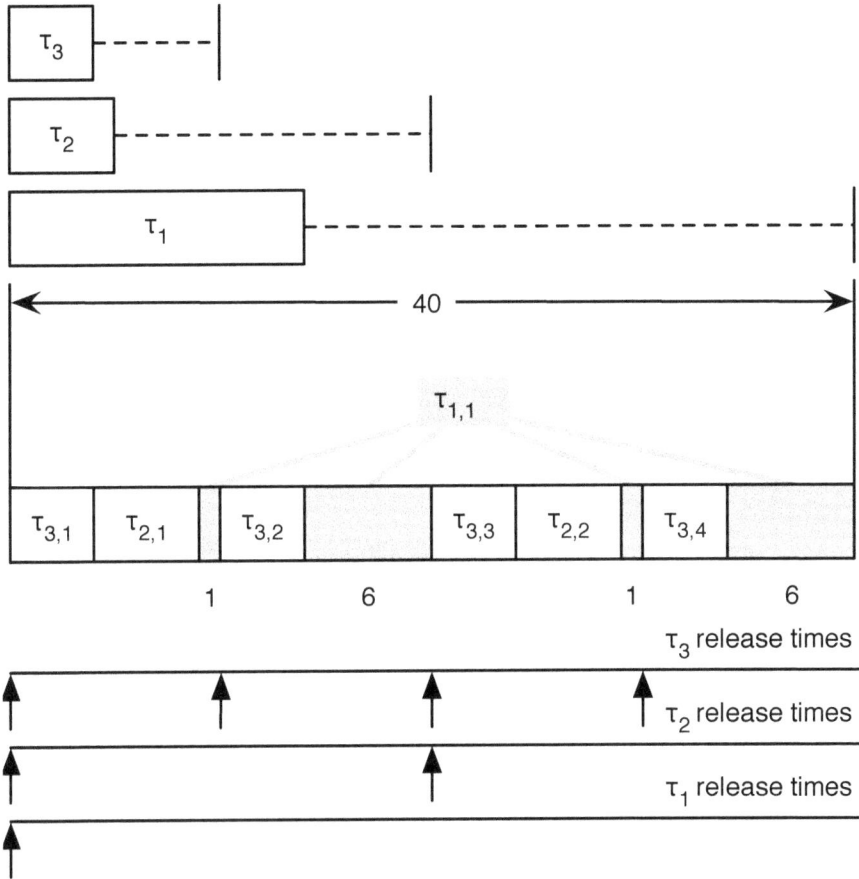

Figure 12.7 RM scheduling for a set of tasks with $U = 1$.

The proof of the *only if* part (necessity) is an immediate consequence of the necessary schedulability condition. To prove the *if* part (sufficiency), we use a *reductio ad absurdum*, that is, we assume that the condition $U \leq 1$ is satisfied and yet the task set is *not schedulable*. Then, we show that, starting from these hypotheses, we come to a contradiction.

Consider any task set that is not schedulable: this means that there will be at least one overflow. Let t_2 be the instant at which the first overflow occurs. Now, go backward in time and choose a suitable t_1 so that $[t_1, t_2]$ is the longest interval of *continuous* utilization before the overflow so that only task instances $\tau_{i,j}$ with an absolute deadline $d_{i,j} \leq t_2$ are executed within it. Observe that t_2 is the deadline for the task for which the overflow occurs: tasks with greater absolute deadline will have a lower priority and therefore cannot be executed in $[t_1, t_2]$. By definition, t_1 will be the release time of some task instance. This situation is shown in Figure 12.9,

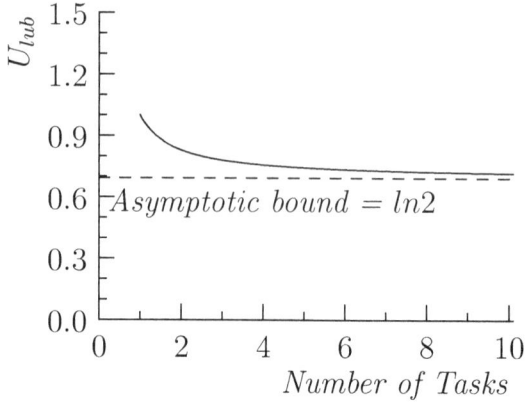

Figure 12.8 U_{lub} value versus the number of tasks in the system.

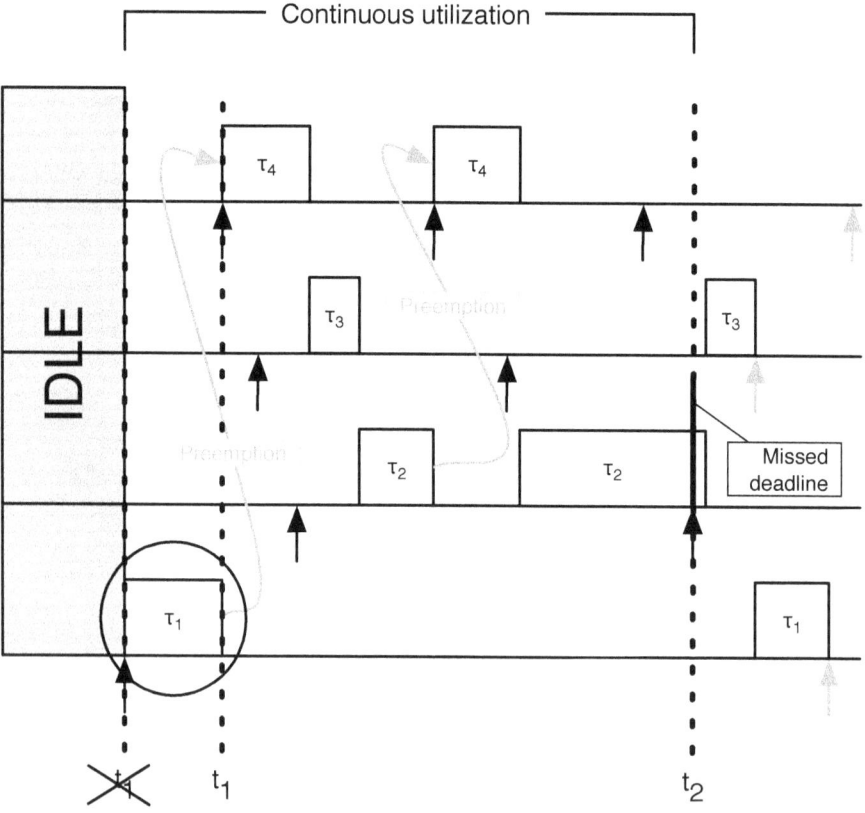

Figure 12.9   A sample task set where an overflow occurs.

highlighting such an interval $[t_1, t_2]$. In the figure, it is also shown that the processor may have been busy for a longer period due to task τ_1 (in the figure, task τ_1 has a lower priority than τ_4). The deadline of τ_1 is however *outside* the $[t_1, t_2]$ interval and, therefore, is not considered in $[t_1, t_2]$.

Let $C_p(t_1, t_2)$ be the total computation time demand in the time interval $[t_1, t_2]$. It can be computed as

$$C_p(t_1, t_2) = \sum_{i \mid r_{i,j} \geq t_1 \land d_{i,j} \leq t_2} C_i \tag{12.21}$$

The maximum number of instances of each τ_i to be considered in the foregoing formula is equal to the number of periods of τ_i *entirely* contained within the time interval $[t_1, t_2]$, that is,

$$\left\lfloor \frac{t_2 - t_1}{T_i} \right\rfloor \tag{12.22}$$

Observe, in fact, that if we add another instance of τ_i, either the first instance would have a release time before t_1, or the last one would have a deadline after t_2, and therefore, it would not be considered here. For N tasks, we can define $C_p(t_1, t_2)$ more explicitly as

$$C_p(t_1, t_2) = \sum_{i=1}^{N} \left\lfloor \frac{t_2 - t_1}{T_i} \right\rfloor C_i \tag{12.23}$$

From the definition of $\lfloor \ \rfloor$ and U we have

$$
\begin{aligned}
C_p(t_1, t_2) &= \sum_{i=1}^{N} \left\lfloor \frac{t_2 - t_1}{T_i} \right\rfloor C_i \\
&\leq \sum_{i=1}^{N} \frac{t_2 - t_1}{T_i} C_i \quad \text{(by definition of } \lfloor \ \rfloor \text{)} \\
&= (t_2 - t_1) \sum_{i=1}^{N} \frac{C_i}{T_i} \\
&= (t_2 - t_1) U \quad \text{(by definition of } U) \tag{12.24}
\end{aligned}
$$

In summary, we have:

$$C_p(t_1, t_2) \leq (t_2 - t_1) U \tag{12.25}$$

On the other hand, *since there is an overflow* at t_2, then $C_p(t_1, t_2)$ (the total computation time demand) must exceed $t_2 - t_1$, which is the time interval in which that demand takes place, that is,

$$C_p(t_1, t_2) > (t_2 - t_1) \tag{12.26}$$

By combining the two inequations just derived, we obtain

$$(t_2 - t_1) U \geq C_p(t_1, t_2) > (t_2 - t_1) \tag{12.27}$$

That is, dividing both sides by $t_2 - t_1$,

$$U > 1 \tag{12.28}$$

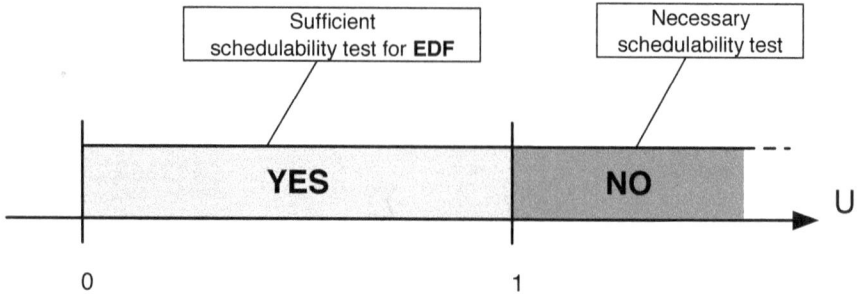

Figure 12.10   Utilization-based schedulability check for EDF.

This is absurd because the conclusion contradicts one of the hypotheses, namely, $U \leq 1$. The contradiction comes from having supposed the task set to be not schedulable. Hence, we can conclude that the condition

$$\sum_{i=1}^{N} \frac{C_i}{T_i} \leq 1 \qquad (12.29)$$

is *both* necessary and sufficient to guarantee the schedulability of a task set with the EDF algorithm.

The EDF algorithm is optimum in the sense that, if any task set is schedulable by any scheduling algorithm, under the hypotheses just set out, then it is also schedulable by EDF. In fact,

- If a task set Γ is schedulable by an arbitrary algorithm A, then it must satisfy the necessary schedulability condition, that is, it must be $U \leq 1$.
- Since Γ has $U \leq 1$, then it is schedulable with the EDF algorithm, because it satisfies the sufficient schedulability test just proved.

The schedulability condition for EDF is graphically expressed in Figure 12.10. With respect to the sufficient schedulability test for the RM algorithm, the corresponding test for the EDF algorithm is conceptually simpler, and there is no "grey area" of uncertainty.

12.4   SUMMARY

This chapter has presented two simple schedulability tests based on the processor utilization for RM and EDF scheduling algorithms. The major advantage of such tests is simplicity: it is possible to execute at run-time a schedulability acceptance test whenever a new task is dynamically added in a running system. As already stated before, fixed priority is the scheduling algorithm supported by most current operating system, and RM can be implemented in this case by

1. Setting the priority of the tasks as fixed;
2. Assigning a priority to each task that is inversely proportional to its period.

As an example of periodic task assignment, consider an embedded system that supervises the execution of a number of control loops. Every loop cycle is triggered, for example, by the availability of a new (set of) input samples whose acquisition has been triggered by an external clock determining the period for that control cycle. Whenever new data are available, the system must acquire and process them to produce one or more control signals for the controlled plant. The deadline for this task is typically the occurrence of the next sampling clock, to ensure that the system never overruns. More than one control cycle may be hosted in the same embedded system, and new controls may be turned on or off run time. In this case, the RM schedulability test allows us to safely assign control tasks, provided we have a reliable estimation of the computational load of each task. Two main practical factor must, however, be considered:

1. The model assumes that no interaction occurs between tasks: this assumption may be true or false depending on the nature of the control being performed. If the system consists of a number of *independent* controls, such an assumption is satisfied. More sophisticated controls, however, may require some degree of information sharing and therefore may introduce logical dependencies among separate tasks.
2. The real computational load may be difficult to estimate since it depends not only on the processor load required to carry out the required control computation (whose execution time may be nevertheless affected by external parameters such as the cache performance) but also on I/O operations that may interfere with I/O due to the other tasks in the system. For this reason, a conservative estimate of task utilization must be chosen, covering the worst case in execution. The next chapters will analyze this aspect in more detail.

Another fact to be considered is that the results presented in this chapter cannot be extended to multiprocessor systems. Considering, in fact, a system with N tasks and M processors, the necessary schedulability condition becomes, as expected

$$\sum_{i=1}^{N} \frac{C_i}{T_i} \leq M \tag{12.30}$$

In this case, however, RM and EDF scheduling, provably optimum for single-processor systems, are not necessarily optimum. On the other side, multicore computers are becoming more and more widespread even in the embedded systems market. A good compromise between the computational power offered by multicore systems and the required predictability in real-time applications is to *statically assign* task sets to cores so that schedulability checks can be safely performed. Provided the execution of the single cores can be considered independent from the other activities of the system, real-time requirements can be satisfied even for such systems. This assumption is often met in practice in modern multicore systems, especially when the amount of data exchanged with I/O devices and memory is not large.

12.5   EXERCISES

EXERCISE 1

Calculate the processor utilization factor U of the task sets of Exercises 1 and 2, Chapter 11.

EXERCISE 2

Consider the following set of tasks, calculate its processor utilization factor U, and answer the questions.

Task τ_i	Period T_i (ms)	Execution time C_i (ms)
τ_1	60	20
τ_2	10	2
τ_3	30	3

- Is the task set schedulable with the RM algorithm? If needed, draw a scheduling diagram to obtain conclusive results.
- Is the same task set schedulable with the EDF algorithm?

EXERCISE 3

Consider the following set of tasks and answer the questions.

Task τ_i	Period T_i (ms)	Execution time C_i (ms)
τ_1	60	20
τ_2	10	5
τ_3	30	3

- Calculate its processor utilization factor U.
- Use U to determine whether or not the RM algorithm can schedule the task set successfully. If needed, draw a scheduling diagram to obtain conclusive results.
- Would the EDF algorithm be able to schedule the same task set?

13 Schedulability Analysis Based on Response Time Analysis

The previous chapter introduced an approach to schedulability analysis based on a single quantity, the utilization factor U, which is very easy to compute even for large task sets. The drawback of this simple approach is a limit in the accuracy of the analysis, which provides only necessary or sufficient conditions for fixed-priority schedulability. Moreover, utilization-based analysis cannot be extended to more general process models, for example, when the relative deadline D_i of task τ_i is lower than its period T_i. In this chapter, a more sophisticated approach to schedulability analysis will be presented, which will allow coping with the above limitations. This new method for analysis will then be used to analyze the impact in the system of *sporadic tasks*, that is, tasks that are not periodic but for which it is possible to state a minimum interarrival time. Such tasks typically model the reaction of the system to external events, and their schedulability analysis is therefore important in determining overall real-time performance.

13.1 RESPONSE TIME ANALYSIS

Response Time Analysis (RTA) [8, 9] is an exact (necessary and sufficient) schedulability test for any fixed-priority assignment scheme on single-processor systems. It allows prediction of the worst-case response time of each task, which depends on the interference due to the execution of higher-priority tasks. The worst-case response times are then compared with the corresponding task deadlines to assess whether all tasks meet their deadline or not.

The task organization that will be considered here still defines a fixed-priority, preemptive scheduler under the basic process model, but the condition that the relative deadline corresponds to the task's period is now relaxed into condition $D_i \leq T_i$.

During execution, the preemption mechanism Â"grabsÂ" the processor from a task whenever a higher-priority task is released. For this reason, all tasks (except the highest-priority one) suffer a certain amount of interference from higher-priority tasks during their execution. Therefore, the worst-case response time R_i of task τ_i is computed as the sum of its computation time C_i and the worst-case interference I_i it experiences, that is,

$$R_i = C_i + I_i \tag{13.1}$$

Observe that the interference must be considered *over any possible interval* $[t, t+R_i]$, that is, for any t, to determine the worst case. We already know, however, that the

DOI: 10.1201/9781003593416-13

worst case occurs when all the higher-priority tasks are released *at the same time* as task τ_i. In this case, t becomes a critical instant and, without loss of generality, it can be assumed that all tasks are released simultaneously at the critical instant $t = 0$.

The contribution of each higher-priority task to the overall worst-case interference will now be analyzed individually by considering the interference due to any single task τ_j of higher priority than τ_i. Within the interval $[0, R_i]$, τ_j will be released one (at $t = 0$) or more times. The exact number of releases can be computed by means of a ceiling function, as

$$\left\lceil \frac{R_i}{T_j} \right\rceil \tag{13.2}$$

Since each release of τ_j will impose on τ_i an interference of C_j, the worst-case interference imposed on τ_i by τ_j is

$$\left\lceil \frac{R_i}{T_j} \right\rceil C_j \tag{13.3}$$

This because if task τ_j is released at any time $t < R_i$, than its execution must have finished *before* R_i, as τ_j has a larger priority, and therefore, that instance of τ_j must have terminated before τ_i can resume.

Let $hp(i)$ denote the set of task indexes with a priority higher than τ_i. These are the tasks from which τ_i will suffer interference. Hence, the total interference endured by τ_i is

$$I_i = \sum_{j \in hp(i)} \left\lceil \frac{R_i}{T_j} \right\rceil C_j \tag{13.4}$$

Recalling that $R_i = C_i + I_i$, we get the following recursive relation for the worst-case response time R_i of τ_i:

$$R_i = C_i + \sum_{j \in hp(i)} \left\lceil \frac{R_i}{T_j} \right\rceil C_j \tag{13.5}$$

No simple solution exists for this equation since R_i appears on both sides, and is inside $\lceil . \rceil$ on the right side. The equation may have more than one solution: the smallest solution is the actual worst-case response time.

The simplest way of solving the equation is to form a *recurrence relationship* of the form

$$w_i^{(k+1)} = C_i + \sum_{j \in hp(i)} \left\lceil \frac{w_i^{(k)}}{T_j} \right\rceil C_j \tag{13.6}$$

where $w_i^{(k)}$ is the k-th estimate of R_i and the $(k+1)$-th estimate from the k-th in the above relationship. The initial approximation $w_i^{(0)}$ is chosen by letting $w_i^{(0)} = C_i$ (the smallest possible value of R_i).

The succession $w_i^{(0)}$, $w_i^{(1)}$, ..., $w_i^{(k)}$, ... is monotonically nondecreasing. This can be proved by induction, that is by proving that

1. $w_i^{(0)} \le w_i^{(1)}$ (Base Case)
2. If $w_i^{(k-1)} \le w_i^{(k)}$, then $w_i^{(k)} \le w_i^{(k+1)}$ for $k > 1$ (Inductive Step)

The base case derives directly from the expression of $w_i^{(1)}$:

$$w_i^{(1)} = C_i + \sum_{j \in hp(i)} \left\lceil \frac{w_i^{(0)}}{T_j} \right\rceil C_j \geq w_i^{(0)} = C_i \qquad (13.7)$$

because every term in the summation is not negative.

To prove the inductive step, we shall prove that $w_i^{(k+1)} - w_i^{(k)} \geq 0$. From (13.6),

$$w_i^{(k+1)} - w_i^{(k)} = \sum_{j \in hp(i)} \left(\left\lceil \frac{w_i^{(k)}}{T_j} \right\rceil - \left\lceil \frac{w_i^{(k-1)}}{T_j} \right\rceil \right) C_j \geq 0 \qquad (13.8)$$

In fact, since, by hypothesis, $w_i^{(k)} \geq w_i^{(k-1)}$, each term of the summation is either 0 or a positive integer multiple of C_j. Therefore, the succession $w_i^{(0)}$, $w_i^{(1)}$, ..., $w_i^{(k)}$, ... is monotonically nondecreasing.

Two cases are possible for the succession $w_i^{(0)}$, $w_i^{(1)}$, ..., $w_i^{(k)}$, ...:

- If the equation has no solutions, the succession does not converge, and it will be $w_i^{(k)} > D_i$ for some k. In this case, τ_i clearly does not meet its deadline.
- Otherwise, the succession converges to R_i, and it will be $w_i^{(k)} = w_i^{(k-1)} = R_i$ for some k. In this case, τ_i meets its deadline if and only if $R_i \leq D_i$.

It is possible to assign a physical meaning to the current estimate $w_i^{(k)}$. If we consider a point of release of task τ_i, from that point and until that task instance completes, the processor will be busy and will execute only tasks with the priority of τ_i or higher. $w_i^{(k)}$ can be seen as a time window that is moving down the busy period. Consider the initial assignment $w_i^{(0)} = C_i$: in the transformation from $w_i^{(0)}$ to $w_i^{(1)}$, the results of the ceiling operations will be (at least) 1. If this is indeed the case, then

$$w_i^{(1)} = C_i + \sum_{j \in hp(i)} C_j \qquad (13.9)$$

Since at $t = 0$ it is assumed that all higher-priority tasks have been released, this quantity represents the length of the busy period *unless* some of the higher-priority tasks are *released again* in the meantime. If this is the case, the window will need to be pushed out further by computing a new approximation of R_i. As a result, the window always expands, and more and more computation time falls into the window. If this expansion continues indefinitely, then the busy period is unbounded, and there is no solution. Otherwise, at a certain point, the window will not suffer any additional hit from a higher-priority task. In this case, the window length is the true length of the busy period and represents the worst-case response time R_i.

We can now summarize the complete RTA procedure as follows:

1. The worst-case response time R_i is individually calculated for each task $\tau_i \in \Gamma$.

Table 13.1

A sample task set

Task τ_i	Period T_i	Computation Time C_i	Priority
τ_1	8	3	High
τ_2	14	4	Medium
τ_3	22	5	Low

2. If, at any point, either a diverging succession is encountered or $R_i > D_i$ for some i, then Γ is not schedulable because τ_i misses its deadline.
3. Otherwise, Γ is schedulable, and the worst-case response time is known for all tasks.

It is worth noting that this method no longer assumes that the relative deadline D_i is equal to the task period T_i but handles the more general case $D_i \leq T_i$. Moreover, the method works with any fixed-priority ordering, and not just with the RM assignment, as long as $hp(i)$ is defined appropriately for all i and we use a preemptive scheduler.

To illustrate the computation of RTA consider the task set listed in Table 13.1, with $D_i = T_i$.

The priority assignment is Rate Monotonic and the CPU utilization factor U is

$$U = \sum_{i=1}^{3} \frac{C_i}{T_i} = \frac{3}{8} + \frac{4}{14} + \frac{5}{22} \simeq 0.89 \tag{13.10}$$

The necessary schedulability test ($U \leq 1$) does not deny schedulability, but the sufficient test for RM is of no help in this case because $U > 1/3(2^{1/3} - 1) \simeq 0.78$.

The highest-priority task τ_1 does not endure interference from any other task. Hence, it will have a response time equal to its computation time, that is, $R_1 = C_1$. In fact, considering (13.6), $hp(1) = \emptyset$ and, given $w_1^{(0)} = C_1$, we trivially have $w_1^{(1)} = C_1$. In this case, $C_1 = 3$, hence $R_1 = 3$ as well. Since $R_1 = 3$ and $D_1 = 8$, then $R_1 \leq D_1$ and τ_1 meets its deadline.

For τ_2, $hp(2) = \{1\}$ and $w_2^{(0)} = C_2 = 4$. The next approximations of R_2 are

$$w_2^{(1)} = 4 + \left\lceil \frac{4}{8} \right\rceil 3 = 7$$

$$w_2^{(2)} = 4 + \left\lceil \frac{7}{8} \right\rceil 3 = 7 \tag{13.11}$$

Since $w_2^{(2)} = w_2^{(1)} = 7$, then the succession converges, and $R_2 = 7$. In other words, widening the time window from 4 to 7 time units did not introduce any additional interference. Task τ_2 meets its deadline, too, because $R_2 = 7$, $D_2 = 14$, and thus $R_2 \leq D_2$.

For τ_3, $hp(3) = \{1,2\}$. It gives rise to the following calculations:

$$
\begin{aligned}
w_3^{(0)} &= 5 \\
w_3^{(1)} &= 5 + \left\lceil \frac{5}{8} \right\rceil 3 + \left\lceil \frac{5}{14} \right\rceil 4 = 12 \\
w_3^{(2)} &= 5 + \left\lceil \frac{12}{8} \right\rceil 3 + \left\lceil \frac{12}{14} \right\rceil 4 = 15 \\
w_3^{(3)} &= 5 + \left\lceil \frac{15}{8} \right\rceil 3 + \left\lceil \frac{15}{14} \right\rceil 4 = 19 \\
w_3^{(4)} &= 5 + \left\lceil \frac{19}{8} \right\rceil 3 + \left\lceil \frac{19}{14} \right\rceil 4 = 22 \\
w_3^{(5)} &= 5 + \left\lceil \frac{22}{8} \right\rceil 3 + \left\lceil \frac{22}{14} \right\rceil 4 = 22
\end{aligned}
\tag{13.12}
$$

$R_3 = 22$ and $D_3 = 22$, and thus $R_3 \le D_3$ and τ_3 (just) meets its deadline.

Figure 13.1 shows the scheduling of the three tasks: τ_1 and τ_2 are released 3 and 2 times, respectively, within the period of τ_3, which, in this example, corresponds also to the worst response time for τ_3. The worst-case response time for all the three tasks is summarized in Table 13.2.

In this case RTA guarantees that all tasks meet their deadline.

13.2 COMPUTING THE WORST-CASE EXECUTION TIME

We have just seen how the worst-case response time can be derived by the knowledge of the task (fixed) priority assignment and from the worst-case execution time, that is, the time only due to the task in the worst case, without considering interference by higher-priority tasks.

The worst-case execution time can be obtained by two distinct methods often used in combination: *measurement* and *analysis*. Measurement is often easy to perform, but it may be difficult to be sure that the worst case has actually been observed. There are, in fact, several factors that may lead the production system to behave in a different way from the test system used for measurement.

For example, the task execution time can depend on its input data, or unexpected interference may arise, for example, due to the interrupts generated by some I/O device not considered in testing.

On the other side, analysis may produce a tight estimate of the worst-case execution time, but is more difficult to perform. It requires, in fact, an effective model of the processor (including pipelines, caches, memory, etc.) and sophisticated code analysis techniques. Moreover, as for testing, external factors not considered in analysis may arise in the real-world system. Most analysis techniques involve several distinct activities:

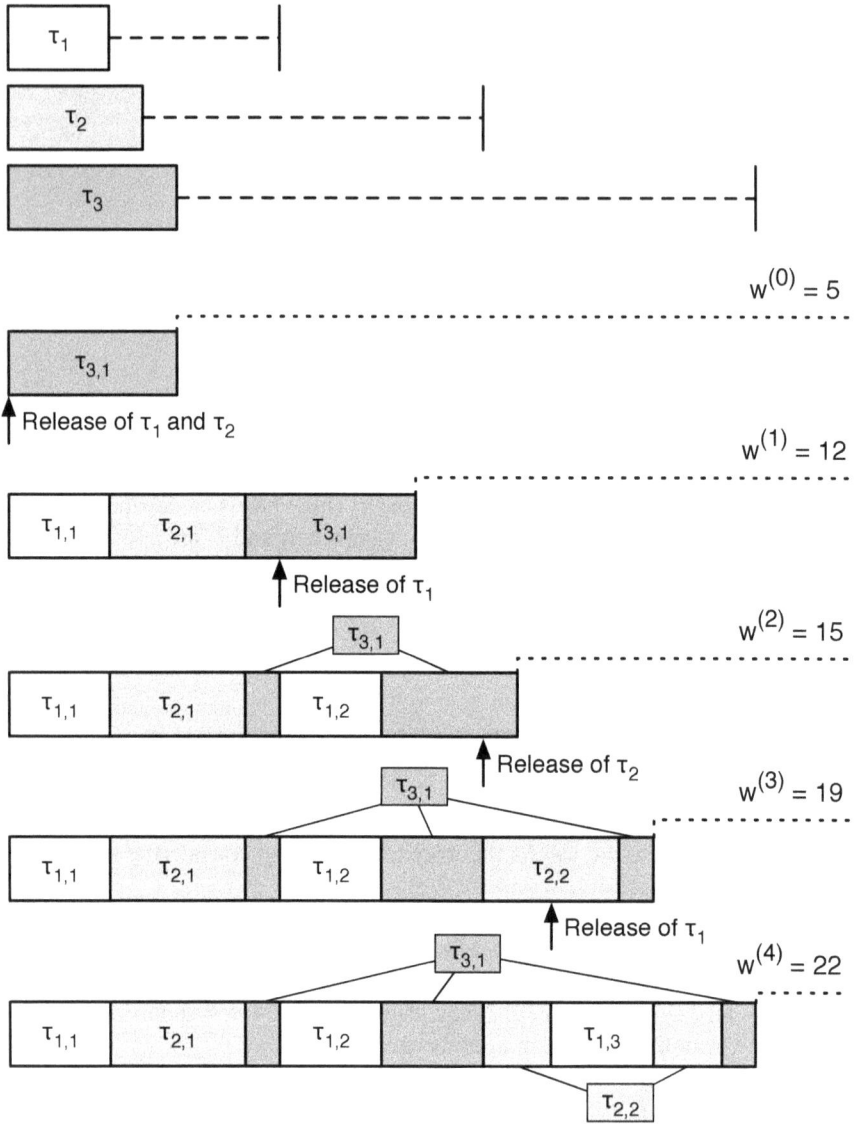

Figure 13.1 Scheduling sequence of the tasks of Table 13.1 and RTA analysis for task τ_3.

1. Decompose the code of the task into a directed graph of basic blocks. Each basic block is a Â"straightÂ" segment of code (without tests, loops, and other conditional statements).
2. Consider each basic block and, by means of the processor model, determine its worst-case execution time.

Table 13.2
Worst-case response time for the sample task set

Task τ_i	Period T_i	Computation Time C_i	Priority	Worst-Case Resp. Time R_i
τ_1	8	3	High	3
τ_2	14	4	Medium	7
τ_3	22	5	Low	22

3. Collapse the graph by means of the available semantic information about the program and, possibly, additional annotations provided by either the programmer or the compiler.

For example, the graph of an "if P then S1 else S2" statement can be collapsed into a single block whose worst-case execution time is equal to the maximum of the worst-case execution times of blocks S1 and S2. If that statement is enclosed in a loop to be executed 10 times, a straight estimation of the worst-case execution time would be derived by multiplying the worst-case execution time of the if block by 10. However, if it were possible to deduce, say, by means of more sophisticated analysis techniques, that the predicate P can be true on at the most three occasions, it would be possible to compute a more accurate overall worst-case execution time for the whole loop by multiplying the execution time of the block by 3.

Often, some restrictions must be placed on the structure of the code so that it can be safely analyzed to derive accurate values of its worst execution time. The biggest challenge in estimating worst execution time derives, however, from the influence of several hardware components commonly found in modern processors, whose intent is to increase computation throughput. These devices aim to reduce the *average* execution time, and normally perform very well in this respect. As a consequence, ignoring them makes the analysis very pessimistic. On the other hand, their impact on the worst-case execution time can be hard to predict and may sometimes be worse than in a basic system not using such mechanisms. In fact, in order to improve average performance, it may turn out more convenient to "sacrifice" performance in situations that seldom occur, in favor of others that happen more frequently. Average performance optimization methods, commonly used in modern processors, are *Caches, Translation Lookaside Buffers* , and *Branch Prediction*.

The cache consists of a fast memory, often mounted on the same processor chip, limited in size, which is intended to contain data most frequently accessed by the processor. Whenever a memory location is accessed, two possible situations arise:

1. The data item is found (cache *hit*) in the cache, and therefore, there is no need to access the external memory. Normally, the access is carried out in a single clock cycle.

2. The data item is not present in the cache (cache *miss*). In this case, the data item must be accessed in the external memory. At the same time, a set of contiguous data are read from the memory and stored in a portion of the cache (called cache *line*) so that the next time close data are accessed, they are found in the cache. The time required to handle a cache miss can be orders of magnitude larger than in a cache hit.

The performance of the cache heavily depends on the locality of memory access, that is, on the likelihood that accesses in memory are performed at contiguous addresses. This is indeed a common situation: program instructions are fetched in sequence unless code branches are performed, program variables are located on the stack at close addresses, and arrays are often accessed sequentially. It is, however, not possible to exclude situations in which a sequence of cache misses occurs, thus obtaining a worst-case execution time that is much larger than the average one. Therefore, even if it is possible to satisfactorily assess the performance of a cache from the statistical point of view, the behavior of a cache with respect to a specific data access is often hard to predict. Moreover, cache behavior depends in part on events external to the task under analysis, such as the allocation of cache lines due to memory accesses performed by other tasks.

Translation Lookaside Buffers (TLB) is a mechanism for speeding memory access in virtual memory systems. We have seen in Chapter 2 that virtual memory address translation implies a double memory access: a first access to the page table to get the address translation, and then to the selected memory page. As for cache access, very often memory access is local, and therefore, if a memory page has been recently accessed, then it is likely that it will be accessed again in a short time. For this reason, it is convenient to store in a small and fast associative memory the page translation information for a subset of the memory pages recently accessed to decrease overall memory access time. Observe that, in virtual memory systems, address translations is always performed when accessing memory regardless of the availability of the requested datum in the cache. If the page translation information is found in the TLB (TLB hit), there is no need to read the page table entry to get the physical page number. As for caches, even if the average performance is greatly improved, it is usually hard to predict whether a particular memory access will give rise to a TLB hit or to a miss. In the latter case, the memory access time may grow by an order of magnitude. Moreover, as for caches, TLB behavior depends in part on events external to the task under analysis. In particular, when a context switch occurs due, for example, to task preemption, the TLB is flushed because the virtual address translation is changed.

Branch prediction is a technique used to reduce the so-called "branch hazards" in pipelined processors. All modern processors adopt a pipelined organization, that is, the execution of the machine instructions is split into stages, and every stage is carried out by a separate pipeline component. The first stage consists in fetching the instruction code from the memory and, as soon as the instruction has been fetched, it is passed on to another element of the pipeline for decoding it, while the first pipeline component can start fetching in parallel the following instruction. After a startup time corresponding to the execution of the first instruction, all the pipeline components can work in parallel, obtaining a speed in overall throughput of a factor of N, where N is the number of components in the processor pipeline. Several facts inhibit, however, the ideal pipeline organization, and one of the main reason for this is due to the branch instructions. After a branch instruction is fetched from memory, the next one is fetched while the former starts being processed. However, if the branch condition turns out to be true (something that may be discovered some stages later), the following instructions that are being processed in the pipeline must be discarded because they refer to a wrong path in program execution. If it were possible to know in advance, as soon as a branch instruction is fetched, at what address the next instruction will be fetched (i.e., whether or not the branch will occur), there would be no need to flush the pipeline whenever the branch condition has been detected to be true. Modern processors use sophisticated techniques of branch prediction based on the past execution flow of the program, thus significantly reducing the average execution time. Prediction is, however, based on statistical assumptions, and therefore, prediction errors may occur, leading to the flush of the pipeline with an adverse impact on the task execution time.

The reader should be convinced at this point that, using analysis alone, it is in practice not possible to derive the worst-case execution time for modern processors. However, given that most real-time systems will be subject to considerable testing anyway, for example, for safety reasons, a combined approach that combines testing and measurement for basic blocks and path analysis for complete components can often be appropriate.

13.3   APERIODIC AND SPORADIC TASKS

Up to now we have considered only periodic tasks, where every task consists of an infinite sequence of identical activities called instances, or jobs, that are regularly released, or activated, at a constant rate.

An *aperiodic task* consists of an infinite sequence of identical jobs. However, unlike periodic tasks, their release does not take place at a regular rate. Typical examples of aperiodic asks are

- User interaction. Events generated by user interaction (key pressed, mouse clicked) and which require some sort of system response.
- Event reaction. External events, such as alarms, may be generated at unpredictable times whenever some condition either in the system or in the controlled plant occurs.

An aperiodic task for which it is possible to determine a minimum inter-arrival time interval is called a *sporadic task*. Sporadic tasks can model many situations that occur in practice. For example, a minimum interarrival time can be safely assumed for events generated by user interaction, because of the reaction time of the human brain, and, more in general, system events can be filtered in advance to ensure that, after the occurrence of a given event, no new instance will be issued until after a given dead time.

One simple way of expanding the basic process model to include sporadic tasks is to interpret the period T_i as the *minimum* interarrival time interval. Of course, much more sophisticated methods of handling sporadic tasks do exist, but their description is beyond the scope of an introductory textbook like this one. Interested readers are referred References [18, 19, 64] for a more comprehensive treatment of this topic.

For example, a sporadic task τ_i with $T_i = 20$ ms is guaranteed not to arrive more frequently than once in any 20 ms interval. Actually, it may arrive much less frequently, but a suitable schedulability analysis test will ensure (if passed) that the *maximum rate* can be sustained.

For these tasks, assuming $D_i = T_i$, that is, a relative deadline equal to the minimum interarrival time, is unreasonable because they usually encapsulate error handlers or respond to alarms. The fault model of the system may state that the error routine will be invoked rarely but, when it is, it has a very short deadline. For many periodic tasks it is useful to define a deadline shorter than the period.

The RTA method just described is adequate for use with the extended process model just introduced, that is, when $D_i \leq T_i$. Observe that the method works with any fixed-priority ordering, and not just with the RM assignment, as long as the set $hp(i)$ of tasks with priority larger than task τ_i is defined appropriately for all i and we use a preemptive scheduler. This fact is especially important to make the technique applicable also for sporadic tasks. In fact, even if RM was shown to be an optimal fixed-priority assignment scheme when $D_i = T_i$, this is no longer true for $Di \leq T_i$.

The following theorem (Leung and Whitehead, 1982) [62] introduces another fixed-priority assignment no more based on the period of the task but on their relative deadlines.

Theorem 13.1. *The deadline monotonic priority order (DMPO), in which each task has a fixed priority inversely proportional to its deadline, is optimum for a preemptive scheduler under the basic process model extended to let $D_i \leq T_i$.*

The optimality of DMPO means that, if any task set Γ can be scheduled using a preemptive, fixed-priority scheduling algorithm A, then the same task set can also be scheduled using the DPMO. As before, such a priority assignment sounds to be good choice since it makes sense to give precedence to more "urgent" tasks. The formal proof of optimality will involve transforming the priorities of Γ (as assigned by A), until the priority ordering is Deadline Monotonic (DM). We will show that each transformation step will preserve schedulability.

Let τ_i and τ_j be two tasks in Γ, with adjacent priorities, that are Â"in the wrong orderÂ" for DMPO under the priority assignment schema A. That is, let $P_i > P_j$ and $D_i > D_j$ under A, where P_i (P_j) denotes the priority of τ_i (τ_j). We shall define now a new priority assignment scheme A' to be identical to A, except that the priorities of τ_i and τ_j are swapped and we prove that Γ is still schedulable under A'.

We observe first that all tasks with a priority higher than P_i (the maximum priority of the tasks being swapped) will be unaffected by the swap. All tasks with priorities lower than P_j (the minimum priority of the tasks being swapped) will be unaffected by the swap, too, because the amount of interference they experience from τ_i and τ_j is the same before and after the swap.

Task τ_j has a higher priority after the swap, and since it was schedulable, by hypothesis, under A, it will suffer after the swap either the same or less interference (due to the priority increase). Hence, it must be schedulable under A', too. The most difficult step is to show now that task τ_i, which was schedulable under A and has had its priority lowered, is still schedulable under A'.

We observe first that once the tasks have been switched, the new worst-case response time of τ_i becomes equal to the old response time of τ_j, that is, $R'_i = R_j$. Under the previous priority assignment A, we had

- $R_j \leq D_j$ (schedulability)
- $D_j < D_i$ (hypothesis)
- $D_i \leq T_i$ (hypothesis)

Hence, under A, τ_i only interferes once during the execution of τ_j because $R_j < T_i$, that is, during the worst-case execution time of τ_j, no new releases of τ_i can occur. Under both priority orderings, $C_i + C_j$ amount of computation time is completed with the same amount of interference from higher-priority processes.

Under A', since τ_j was released only once during R_j and $R'_i = R_j$, it interferes only once during the execution of τ_i. Therefore, we have

- $R'_i = R_j$ (just proved)
- $R_j \leq D_j$ (schedulability under A)
- $D_j < D_i$ (hypothesis)

Hence, $R'_i < D_i$, and it can be concluded that τ_i is *still schedulable* after the switch.

In conclusion, the DM priority assignment can be obtained from any other priority assignment by a sequence of pairwise priority reorderings as above. Each such reordering step preserves schedulability.

The following example illustrates a successful application of DMPO for a task set where RM priority assignment fails. Consider the task set listed in Table 13.3, where the RM and DM priority assignments differ for some tasks.

Table 13.3
RM and DM priority assignment

Task Parameters				Priority	
Task	T_i	Deadline D_i	C_i	RM	DM
τ_1	19	6	3	Low	High
τ_2	14	7	4	Medium	Medium
τ_3	11	11	3	High	Low
τ_4	20	19	2	Very Low	Very Low

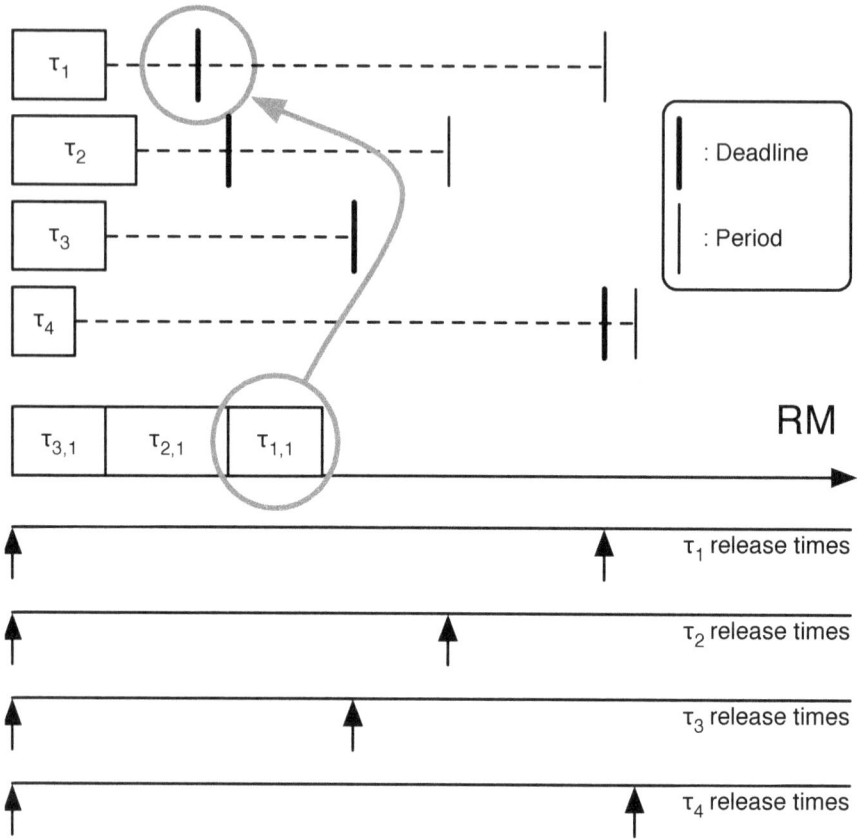

Figure 13.2 RM scheduling fails for the tasks of Table 13.3.

The behaviors of RM and DM for this task set will be now examined and compared.

From Figures 13.2 and 13.3 we can see that RM is unable to schedule the task set, whereas DM succeeds. We can derive the same result performing RTA analysis for the RM schedule.

For the RM schedule, we have, for τ_3, $hp(3) = \emptyset$. Hence, $R_3 = C_3 = 3$, and τ_3 (trivially) meets its deadline.

For τ_2, $hp(2) = \{3\}$ and:

$$
\begin{aligned}
w_2^{(0)} &= 4 \\
w_2^{(1)} &= 4 + \left\lceil \frac{4}{11} \right\rceil 3 = 7 \\
w_2^{(2)} &= 4 + \left\lceil \frac{7}{11} \right\rceil 3 = 7 = R_2
\end{aligned}
\tag{13.13}
$$

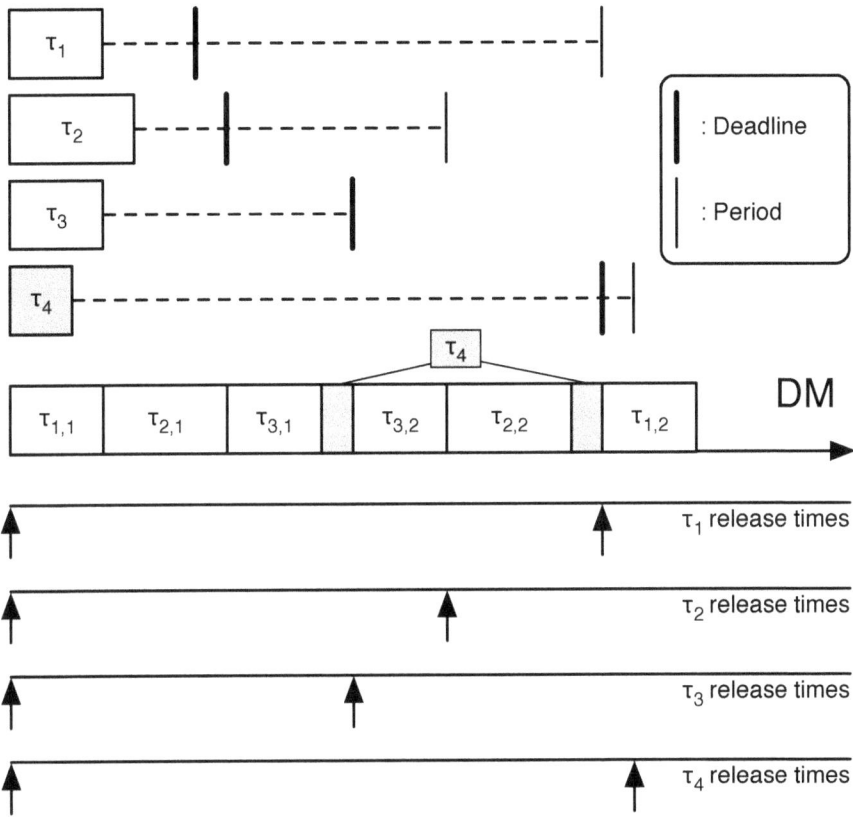

Figure 13.3 DM scheduling succeeds for the tasks of Table 13.3.

Since $R_2 = 7$ and $D_2 = 7$, τ_2 meets its deadline.

For τ_1, $hp(1) = \{3, 2\}$ and

$$
\begin{aligned}
w_1^{(0)} &= 3 \\
w_1^{(1)} &= 3 + \left\lceil \frac{3}{11} \right\rceil 3 + \left\lceil \frac{3}{14} \right\rceil 4 = 10 \\
w_1^{(2)} &= 3 + \left\lceil \frac{10}{11} \right\rceil 3 + \left\lceil \frac{10}{14} \right\rceil 4 = 10 = R_1
\end{aligned}
$$

$$(13.14)$$

Since $R_1 = 10$ and $D_1 = 6$, τ_1 misses its deadline: RM is unable to schedule this task set.

Consider now RTA for the DM schedule: for τ_1, $hp(1) = \emptyset$. Hence, $R_1 = C_1 = 3$, and τ_1 (trivially) meets its deadline.

For τ_2, $hp(2) = \{1\}$ and

$$
\begin{aligned}
w_2^{(0)} &= 4 \\
w_2^{(1)} &= 4 + \left\lceil \frac{4}{19} \right\rceil 3 = 7 \\
w_2^{(2)} &= 4 + \left\lceil \frac{7}{19} \right\rceil 3 = 7 = R_2
\end{aligned}
\tag{13.15}
$$

Since $R_2 = 7$ and $D_2 = 7$, τ_2 just meets its deadline.

For τ_3, $hp(3) = \{1, 2\}$ and

$$
\begin{aligned}
w_3^{(0)} &= 3 \\
w_3^{(1)} &= 3 + \left\lceil \frac{3}{19} \right\rceil 3 + \left\lceil \frac{3}{14} \right\rceil 4 = 10 \\
w_3^{(2)} &= 3 + \left\lceil \frac{10}{19} \right\rceil 3 + \left\lceil \frac{10}{14} \right\rceil 4 = 10 = R_3
\end{aligned}
\tag{13.16}
$$

Since $R_3 = 10$ and $D_3 = 11$, τ_3 meets its deadline, too.

For τ_4, $hp(4) = \{1, 2, 3\}$ and

$$
\begin{aligned}
w_4^{(0)} &= 2 \\
w_4^{(1)} &= 2 + \left\lceil \frac{2}{19} \right\rceil 3 + \left\lceil \frac{2}{14} \right\rceil 4 + \left\lceil \frac{2}{11} \right\rceil 3 = 12 \\
w_4^{(2)} &= 2 + \left\lceil \frac{12}{19} \right\rceil 3 + \left\lceil \frac{12}{14} \right\rceil 4 + \left\lceil \frac{12}{11} \right\rceil 3 = 15 \\
w_4^{(3)} &= 2 + \left\lceil \frac{15}{19} \right\rceil 3 + \left\lceil \frac{15}{14} \right\rceil 4 + \left\lceil \frac{15}{11} \right\rceil 3 = 19 \\
w_4^{(4)} &= 2 + \left\lceil \frac{19}{19} \right\rceil 3 + \left\lceil \frac{19}{14} \right\rceil 4 + \left\lceil \frac{19}{11} \right\rceil 3 = 19 = R_4
\end{aligned}
\tag{13.17}
$$

Finally, τ_4 meets its deadline also because $R_4 = 19$ and $D_4 = 19$. This terminates the RTA analysis, proving the schedulability of Γ for the DM priority assignment.

13.4 SUMMARY

This chapter introduced RTA, a check for schedulability that allows a finer resolution in respect of the other utilization-based checks. It is worth noting that, even when RTA fails, the task set may be schedulable because RTA assumes that all the tasks are released at the same critical instant. It is, however, always convenient and safer to consider the worst case (critical instant) in scheduling dynamics because it would be very hard to make sure that critical instants never occur due to the variability in the task execution time on computers, for example, to cache misses and pipeline hazards. Even if not so straight as the utilization-based check, RTA represents a practical schedulability check because, even in case of nonconvergence, it can be stopped as long as the currently computed response time for any task exceeds the task deadline.

The RTA method can be also be applied to Earliest Deadline First (EDF), but is considerably more complex than for the fixed-priority case and will not be considered in this book due its very limited applicability in practical application.

RTA is based on an estimation of the worst-case execution time for each considered task, and we have seen that the exact derivation of this parameter is not easy, especially for general-purpose processors, which adopt techniques for improving average execution speed, and for which the worst-case execution time can be orders of magnitude larger that the execution time in the large majority of the executions. In this case, basing schedulability analysis on the worst case may sacrifice most of the potentiality of the processor, with the risk of having a very low total utilization and, therefore, of wasting computer resources. On the other side, considering lower times may produce occasional, albeit rare, deadline misses, so a trade-off between deadline miss probability and efficient processor utilization is normally chosen. The applications where absolutely no deadline miss is acceptable are, in fact, not so common. For example, if the embedded system is used within a feedback loop, the effect of occasional deadline misses can be considered as a disturb (or noise) in either the controlled process, the detectors, or the actuators, and can be handled by the system, provided enough stability margin in achieved control.

Finally, RTA also allows dealing with the more general case in which the relative deadline D_i for task τ_i is lower than the task period T_i. This is the case for sporadic jobs that model a set of system activities such as event and alarm handling. As a final remark, observe that, with the inclusion of sporadic tasks, we are moving toward a more realistic representation of real-time systems. The major abstraction so far is due to the task model, which assumes that tasks do not depend on each other, an assumption often not realistic. The next two chapters will cover this aspect, taking into account the effect due to the use of resources shared among tasks and the consequent effect in synchronization.

13.5   EXERCISES

EXERCISE 1

Apply the RTA method to the task sets of Exercises 1 and 2, Chapter 11. Compare RTA results with the corresponding scheduling diagrams.

EXERCISE 2

Apply the RTA method to the task sets of Exercises 2 and 3, Chapter 12. Compare RTA results with the corresponding utilization-based schedulability tests and scheduling diagrams.

14 Task Interactions and Blocking

The basic process model, first introduced in Chapter 10, was used in Chapters 11 through 13 as an underlying set of hypotheses to prove several interesting properties of real-time scheduling algorithms and, most importantly, all the schedulability analysis results we discussed so far. Unfortunately, as it has already been remarked at the end of Chapter 13, some aspects of the basic process model are not fully realistic, and make those results hard to apply to real-world problems.

The hypothesis that tasks are completely *independent* from each other regarding execution is particularly troublesome because it sharply goes against the basics of all the interprocess communication methods introduced in Chapters 5 and 6. In one form or another, they all require tasks to interact and coordinate, or synchronize, their execution. In other words, tasks will sometimes be forced to *block* and wait until some other task performs an action in the future.

For example, tasks may either have to wait at a critical region's boundary to keep a shared data structure consistent, or wait for a message from another task before continuing. In all cases, their execution will clearly no longer be independent from what the other tasks are doing at the moment.

We shall see that adding task interaction to a real-time system raises some unexpected issues involving task priority ordering—another concept of paramount importance in real-time programming—that must be addressed adequately. In this chapter, the discussion will mainly address task interactions due to *mutual exclusion*, a ubiquitous necessity when dealing with shared data. The next chapter will instead analyze the situation in which a task is forced to wait for other reasons, for instance, an I/O operation.

14.1 THE PRIORITY INVERSION PROBLEM

In a real-time system, task interaction must be designed with great care, above all when the tasks being synchronized have different priorities. Informally speaking, when a high-priority task is waiting for a lower-priority task to complete some required computation, the task priority scheme is, in some sense, being hampered because the high-priority task would take precedence over the lower-priority one in the model.

This happens even in very simple cases, for example, when several tasks access a shared resource by means of a critical region protected by a mutual exclusion semaphore. Once a lower-priority task enters its critical region, the semaphore mechanism will block any higher-priority task wanting to enter its own critical region protected by the same semaphore and force it to wait until the former exits.

DOI: 10.1201/9781003593416-14

This phenomenon is called *priority inversion* and, if not adequately addressed, can have adverse effects on the schedulability of the system, to the point of making the response time of some tasks completely unpredictable because the priority inversion region may last for an *unbounded* amount of time. Accordingly, such as a situation is usually called *unbounded priority inversion.*

It can easily be seen that priority inversion can (and should) be reduced to a minimum by appropriate software design techniques aimed at avoiding, for instance, redundant task interactions when the system can be designed in a different way. At the same time, however, it is also clear that the problem cannot completely solved in this way except in very simple cases. It can then be addressed in two different ways:

1. Modify the mutual exclusion mechanism by means of an appropriate technique, to be discussed in this chapter. The modification shall guarantee that the blocking time endured by each individual task in the system has a known and finite upper bound. This worst-case blocking time can then be used as an additional ingredient to improve the schedulability analysis methods discussed so far.
2. Depart radically from what was discussed in Chapter 5 and devise a way for tasks to exchange information through a shared memory in a meaningful way without resorting to mutual exclusion or, more in general, without ever forcing them to wait [2, 3, 4, 38, 39, 56].

Before going further with the discussion, presenting a very simple example of unbounded priority inversion is useful to better understand what the priority inversion problem really means from a practical standpoint. The same example will also be used in the following to gain a better understanding of how the different methods address the priority inversion issue.

Let us consider the execution of three real-time tasks τ_H (high priority), τ_M (middle priority), and τ_L (low priority), executed under the control of a fixed-priority scheduler on a single-processor system. We will also assume, as shown in Figure 14.1, that τ_H and τ_L share some information, stored in a certain shared memory area M.

Being written by proficient concurrent programmers (who carefully perused Chapter 5) both τ_H and τ_L make access to M only within a suitable critical region, protected by a mutual exclusion semaphore m. On the contrary, τ_M has nothing to do with τ_H and τ_L, that is, it does not interact with them in any way. The only relationship among τ_M and the other two tasks is that "by chance" it was assigned a priority that happens to be between the priority of τ_H and τ_L.

In the previous statement, the meaning of the term "by chance" is that the peculiar priority relationship may very well be unknown to the programmers who wrote τ_H, τ_M, and τ_L. For example, we saw that, according to the Rate Monotonic (RM) priority assignment, the priority of a task depends only on its period. When different software modules—likely written by distinct groups of programmers and made of several tasks each—are put together to build the complete application, it may be very hard to predict how the priority of a certain task will be located, with respect to the others.

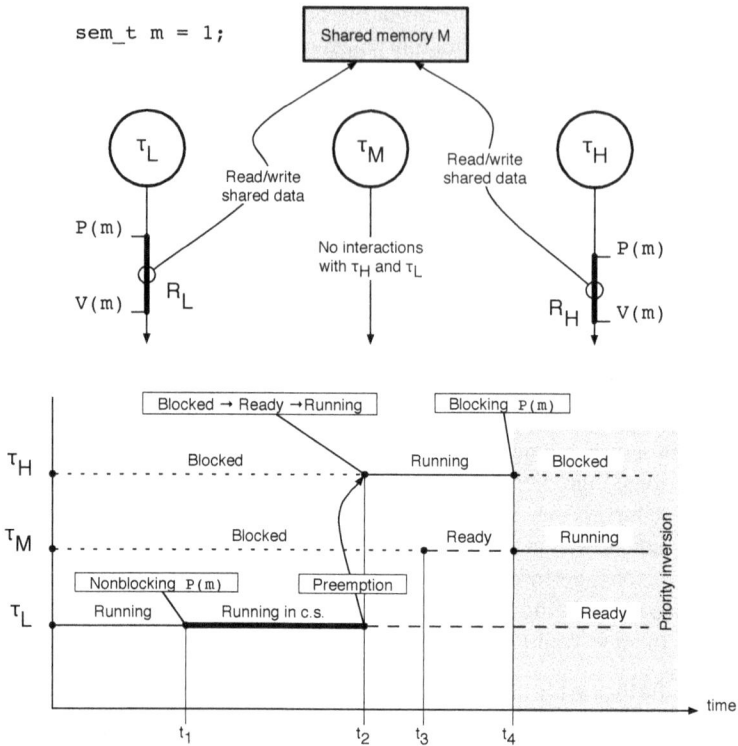

Figure 14.1   A simple example of unbounded priority inversion involving three tasks.

The following sequence of events may happen:

- Initially, neither τ_H nor τ_M are ready for execution. They may be, for instance, periodic tasks waiting for their next execution instance or they may be waiting for the completion of an I/O operation.
- On the other hand, τ_L is ready; the fixed-priority scheduler moves it into the Running state and executes it.
- During its execution, at t_1 in the figure, τ_L enters into its critical region R_L, protected by semaphore m. The semaphore primitive P(m) at the critical region's boundary is nonblocking because no other tasks are accessing the shared memory M at the moment. Therefore, τ_L is allowed to proceed immediately and keeps running within the critical region.
- If τ_H becomes ready while τ_L still is in the critical region, the fixed-priority scheduler stops executing τ_L, puts it back into the Ready state, moves τ_H into the Running state and executes it. This action has been called *preemption* in Chapter 11 and takes place at t_2 in the figure.
- At t_3, task τ_M becomes ready for execution and moves into the Ready state, too, but this has no effect on the execution of τ_H, because the priority of τ_M is lower than the priority of τ_H.

- As τ_H proceeds with its execution, it may try to enter its critical region. In the figure, this happens at t_4. At this point, τ_H is blocked by the semaphore primitive P(m) because the value of semaphore m is now zero. This behavior is correct since τ_L is within its own critical region R_L, and the semaphore mechanism is just enforcing mutual exclusion between R_L and R_H, the critical region τ_H wants to enter.
- Since τ_H is no longer able to run, the scheduler chooses another task to execute. As τ_M and τ_L are both Ready but the priority of τ_M is greater than the priority of τ_L, the former is brought into the Running state and executed.

Therefore, starting from t_4 in Figure 14.1, a *priority inversion* region begins: τ_H (the highest priority task in the system) is blocked by τ_L (a lower priority task) and the system executed τ_M (yet another lower priority task). In the figure, this region is highlighted by a gray background.

Although the existence of this priority inversion region cannot be questioned because it entirely depends on how the mutual exclusion mechanism for the shared memory area M has been *designed* to work, an interesting question is *for how long* it will last.

Contrary to expectations, the answer is that the duration of the priority inversion region does not depend at all on the two tasks directly involved in it, that is, τ_H and τ_L. In fact, τ_H is in the Blocked state and, by definition, it cannot perform any further action until τ_L exits from R_L and executes V(m) to unblock it. On its part, τ_L is Ready, but it is not being executed because the fixed-priority scheduler does not give it any processor time. Hence, it has no chance of proceeding through R_L and eventually execute V(m).

The length of the priority inversion region depends instead on how much time τ_M keeps running. Unfortunately, as discussed above, τ_M has nothing to do with τ_H and τ_L. The programmers who wrote τ_H and τ_L may even be unaware that τ_M exists. The existence of multiple middle-priority tasks $\tau_{M1}, \ldots, \tau_{Mn}$ instead of a single one makes the situation even worse. In a rather extreme case, those tasks could take turns entering the Ready state and being executed so that, even if none of them keeps running for a long time individually, taken as a group there is always at least one task τ_{Mk} in the Ready state at any given instant. In that scenario, τ_H will be blocked for an *unbounded* amount of time by $\tau_{M1}, \ldots, \tau_{Mn}$ even if they all have a lower priority than τ_H itself.

In summary, we are willing to accept that a certain amount of blocking of τ_H by some lower-priority tasks cannot be removed. By intuition, when τ_H wants to enter its critical region R_H in the example just discussed, it must be prepared to wait up to the maximum time needed by τ_L to execute within critical region R_L. This is a direct consequence of the mutual exclusion mechanism, which is necessary to access the shared resources in a safe way. However, it is also necessary for the blocking time to have a computable and finite upper bound. Otherwise, the overall schedulability of the whole system, and of τ_H in particular, will be compromised in a rather severe way.

As for many other concurrent programming issues, it must also be remarked that this is not a systematic error. Rather, it is a *time-dependent* issue that may go undetected when the system is bench tested.

14.2   THE PRIORITY INHERITANCE PROTOCOL

Going back to the example shown in Figure 14.1, it is easy to notice that the root cause of the unbounded priority inversion was the preemption of τ_L by τ_H while τ_L was executing within R_L. If the context switch from τ_L to τ_H had been somewhat delayed until after the execution of V(m) by τ_L—that is, until the end of R_L—the issue would not have occurred.

On a single-processor system, a very simple (albeit drastic) solution to the unbounded priority inversion problem is to forbid preemption completely during the execution of all critical regions. This may be obtained by disabling the operating system scheduler or, even more drastically, turning interrupts off within critical regions. In other words, with this approach any task that successfully enters a critical region implicitly gains the highest possible priority in the system so that no other task can preempt it. The task goes back to its regular priority when it exits from the critical region.

One clear advantage of this method is its extreme simplicity. It is also easy to convince oneself that it really works. Informally speaking, if we prevent any tasks from unexpectedly losing the processor while they are holding any mutual exclusion semaphores, they will not block any higher-priority tasks for this reason should they try to get the same semaphores. At the same time, however, the technique introduces a new kind of blocking, of a different nature.

That is, any higher-priority task τ_M that becomes ready while a low-priority task τ_L is within a critical region will not get executed—and we therefore consider it to be blocked by τ_L—until τ_L exits from the critical region. This happens even if τ_M does not interact with τ_L at all. The problem has been solved anyway because the amount of blocking endured by τ_M is indeed bounded. The upper bound is the maximum amount of time τ_L may actually spend running within its critical region. Nevertheless, we are now blocking some tasks, like τ_M, which were not blocked before.

For this reason, this way of proceeding is only appropriate for very short critical regions, because it causes much unnecessary blocking. A more sophisticated approach is needed in the general case, although introducing additional kinds of blocking into the system in order to set an upper bound on the blocking time is a trade-off common to all the solutions to the unbounded priority inversion problem that we will present in this chapter. We shall see that the approach just discussed is merely a strongly simplified version of the priority ceiling emulation protocol, to be described in Section 14.3.

In any case, the underlying idea is useful: the unbounded priority inversion problem can be solved by means of a better cooperation between the *synchronization* mechanism used for mutual exclusion and the processor *scheduler*. This cooperation can be implemented, for instance, by allowing the mutual exclusion mechanism to *temporarily change* task priorities. This is exactly the way the *priority inheritance* algorithm, or protocol, works.

The priority inheritance protocol has been proposed by Sha, Rajkumar, and Lehoczky [83], and offers a straightforward solution to the problem of unbounded priority inversion. The general idea is to dynamically increase the priority of a task as soon as it is blocking some higher-priority tasks. In particular, if a task τ_L is blocking a set of n higher-priority tasks τ_{H1}, ..., τ_{Hn} at a given instant, it will temporarily inherit the highest priority among them. This prevents any middle-priority task from preempting τ_L and unduly make the blocking experienced by τ_{H1}, ..., τ_{Hn} any longer than necessary.

In order to define the priority inheritance protocol in a more rigorous way and look at its most important properties, it is necessary to set forth some additional hypotheses and assumptions about the system being considered. In particular,

- It is first of all necessary to distinguish between the *initial*, or *baseline*, priority given to a task by the scheduling algorithm and its *current*, or *active*, priority. The baseline priority is used as the initial, default value of the active priority but, as we just saw, the latter can be higher if the task being considered is blocking some higher-priority tasks.
- The tasks are under the control of a fixed-priority scheduler and run within a single-processor system. The scheduler works according to active priorities.
- If there are two or more highest-priority tasks ready for execution, the scheduler picks them in First-Come First-Served (FCFS) order.
- Semaphore wait queues are ordered by active priority as well. In other words, when a task executes a V(s) on a semaphore s and there is at least one task waiting on s, the highest-priority waiting task will be put into the Ready state.
- Semaphore waits due to mutual exclusion are the only source of blocking in the system. Other causes of blocking such as, for example, I/O operations, must be taken into account separately, as discussed in Chapter 15.

The priority inheritance protocol itself consists of the following set of rules:

1. When a task τ_H attempts to enter a critical region that is "busy"—that is, its controlling semaphore has already been taken by another task τ_L—it blocks, but it also *transmits* its active priority to the task τ_L that is blocking it if the active priority of τ_L is lower than τ_H's.
2. As a consequence, τ_L will execute the rest of its critical region with a priority at least equal to the priority it just inherited. In general, a task inherits the highest active priority among all tasks it is blocking.
3. When a task τ_L exits from a critical region and it is no longer blocking any other task, its active priority returns back to the baseline priority.
4. Otherwise, if τ_L is still blocking some tasks—this happens when critical regions are nested into each other—it inherits the highest active priority among them.

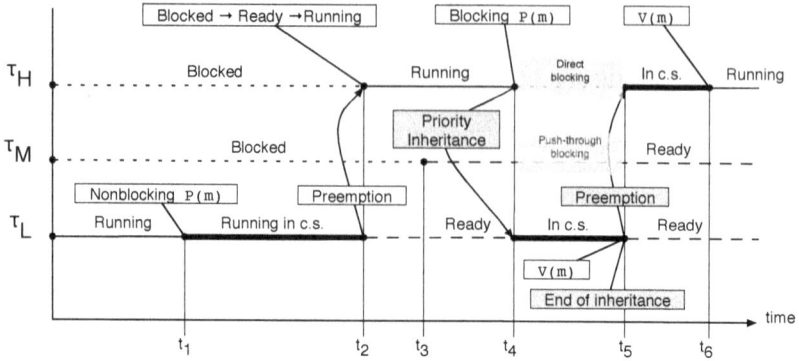

Figure 14.2 A simple application of the priority inheritance protocol involving three tasks.

Although this is not a formal proof at all, it can be useful to apply the priority inheritance protocol to the example shown in Figure 14.1 to see that it really works, at least in a very simple case. The result is shown in Figure 14.2. The most important events that are different with respect to the previous figure are highlighted with a gray background:

- From t_1 to t_4 the system behaves as before. The priority inheritance protocol has not been called into action yet, because no tasks are blocking any other, and all tasks have got their initial, or baseline, priority.
- At t_4, τ_H is blocked by the semaphore primitive P(m) because the value of semaphore m is zero. At this point, τ_H is blocked by τ_L because τ_L is within a critical region controlled by the same semaphore m. Therefore, the priority inheritance protocol makes τ_L inherit the priority of τ_H.
- Regardless of the presence of one (or more) middle-priority tasks like τ_M, at t_4 the scheduler resumes the execution of τ_L because its active priority is now the same as τ_H's priority.
- At t_5, τ_L eventually finishes its work within the critical region R_L and releases the mutual exclusion semaphore with a V(m). This has two distinct effects: the first one pertains to task synchronization, and the second one concerns the priority inheritance protocol:
  1. Task τ_H acquires the mutual exclusion semaphore and returns to the Ready state;
  2. Task τ_L returns to its baseline priority because it is no longer blocking any other task, namely, it is no longer blocking τ_H.

  Consequently, the scheduler immediately preempts the processor from τ_L and resumes the execution of τ_H.
- Task τ_H executes within its critical region from t_5 until t_6. Then it exits from the critical region, releasing the mutual exclusion semaphore with V(m), and keeps running past t_6.

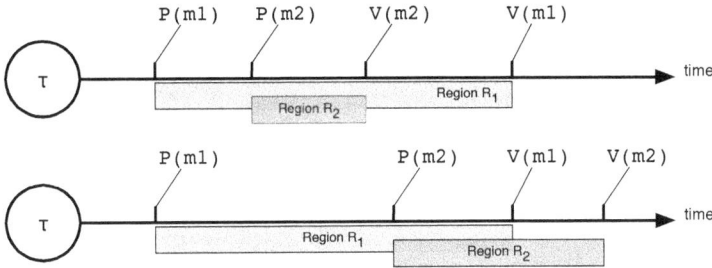

Figure 14.3   In a task τ, critical regions can be properly (above) or improperly (below) nested.

Even from this simple example, it is clear that the introduction of the priority inheritance protocol makes the concept of blocking more complex than it was before. Looking again at Figure 14.2, there are now *two* distinct kinds of blocking rather than one, both occurring between t_4 and t_5:

1. *Direct blocking* occurs when a high-priority task tries to acquire a resource held by a lower-priority task. In this case, τ_H is blocked by τ_L. Direct blocking was already present in the system and is necessary for mutual exclusion, to ensure the *consistency* of the shared resources.
2. Instead, *push-through blocking* is a consequence of the priority inheritance protocol. It occurs when an intermediate-priority task (τ_M in this example) cannot run because a lower-priority task (τ_L in our case) has temporarily inherited a higher priority. This kind of blocking may affect a task even if it does not use any shared resource, just as it happens to τ_M in the example. Nevertheless, it is necessary to avoid *unbounded* priority inversion.

In the following, we will present the main properties of the priority inheritance protocol. The final goal will be to prove that the maximum blocking time that each task may experience is *bounded* in all cases. The same properties will also be useful to define several algorithms that *calculate* the worst-case blocking time for each task in order to analyze the schedulability of a periodic task set. As the other schedulability analysis algorithms discussed in Chapters 12 and 13, these algorithms entail a trade-off between the tightness of the bound they compute and their complexity.

For simplicity, in the following discussion the fact that critical regions can be *nested* into each other will be neglected. Under the assumption that critical regions are properly nested, the set of critical regions belonging to the same task is partially ordered by region inclusion. Proper nesting means that, as shown in Figure 14.3, if two (or more) critical regions R_1 and R_2 overlap in a given task, then either the boundaries of R_2 are entirely contained in R_1 or the other way around.

For each task, it is possible to restrict the attention to the set of maximal blocking critical regions, that is, regions that can be a source of blocking for some other task but are not included within any other critical region with the same property. It can be shown that most results discussed in the following are still valid even if only maximal critical regions are taken into account, unless otherwise specified. The interested reader should refer to Reference [83] for more information about this topic.

The first lemma we shall discuss establishes under which conditions a high-priority task τ_H can be blocked by a lower-priority task.

Lemma 14.1. *Under the priority inheritance protocol, a task τ_H can be blocked by a lower-priority task τ_L only if τ_L is executing within a critical region Z that satisfies either one of the following two conditions when τ_H is released:*

> *1The critical region Z is guarded by the same semaphore as a critical region of τ_H. In this case, τ_L can block τ_H directly as soon as τ_H tries to enter that critical region.*
> *2The critical region Z can lead τ_L to inherit a priority higher than or equal to the priority of τ_H. In this case, τ_L can block τ_H by means of push-through blocking.*

Proof. The lemma can be proved by observing that if τ_L is not within a critical region when τ_H is released, it will be preempted immediately by τ_H itself. Moreover, it cannot block τ_H in the future, because it does not hold any mutual exclusion semaphore that τ_H may try to acquire, and the priority inheritance protocol will not boost its priority.

On the other hand, if τ_L is within a critical region when τ_H is released but neither of the two conditions is true, there is no way for τ_L to either block τ_H directly or get a priority high enough to block τ_H by means of push-through blocking.

When the hypotheses of Lemma 14.1 are satisfied, then task τ_L can block τ_H. The same concept can also be expressed in two slightly different, but equivalent, ways:

1. When the focus of the discussion is on critical regions, it can also be said that that the *critical region Z*, being executed by τ_L when τ_H is released, can block τ_H. Another, equivalent way of expressing this concept is to say that Z is a *blocking critical region* for τ_H.
2. Since critical regions are always protected by mutual exclusion semaphores in our framework, if critical region Z is protected by semaphore S_Z, it can also be said that the *semaphore S_Z can block τ_H.*

Now that the conditions under which tasks can block each other are clear, another interesting question is *how many times* a high-priority task τ_H can be blocked by lower-priority tasks during the execution of one of its instances, and *for how long*. One possible answer is given by the following lemma:

Lemma 14.2. *Under the priority inheritance protocol, a task τ_H can be blocked by a lower-priority task τ_L for* at most *the duration of one critical region that belongs to τ_L and is a blocking critical region for τ_H regardless of the number of semaphores τ_H and τ_L share.*

Proof. According to Lemma 14.1, for τ_L to block τ_H, τ_L must be executing a critical region that is a blocking critical region for τ_H. Blocking can by either direct or push-through.

When τ_L eventually exits from that critical region, its active priority will certainly go back to a value less than the priority of τ_H. From this point on, τ_H can preempt τ_L, and it cannot be blocked by τ_L again. Even if τ_H will be blocked again on another critical region, the blocking task will inherit the priority of τ_H itself and thus prevent τ_L from being executed.

The only exception happens when τ_H relinquishes the processor for other reasons, thus offering τ_L a chance to resume execution and acquire another mutual exclusion semaphore. However, this is contrary to the assumption that semaphore waits due to mutual exclusion are the only source of blocking in the system.

In the general case, we consider n lower-priority tasks $\tau_{L1}, \ldots, \tau_{Ln}$ instead of just one. The previous lemma can be extended to cover this case and conclude that the worst-case blocking time experienced by τ_H is bounded even in that scenario.

Lemma 14.3. *Under the priority inheritance protocol, a task τ_H for which there are n lower-priority tasks $\tau_{L1}, \ldots, \tau_{Ln}$ can be blocked for at most the duration of* one *critical region that can block τ_H for each τ_{Li}, regardless of the number of semaphores used by τ_H.*

Proof. Lemma 14.2 states that *each* lower-priority task τ_{Li} can block τ_H for at most the duration of one of its critical sections. The critical section must be one of those that can block τ_H according to Lemma 14.1.

In the worst case, the same scenario may happen for *all* the n lower-priority tasks, and hence, τ_H can be blocked at most n times, regardless of how many semaphores τ_H uses.

More important information we get from this lemma is that, provided all tasks only spend a finite amount of time executing within their critical regions in all possible circumstances, then the maximum blocking time is *bounded*. This additional hypothesis is reasonable because, by intuition, if we allowed a task to enter a critical region and execute within it for an unbounded amount of time without ever leaving, the mutual exclusion framework would no longer work correctly anyway since no other tasks would be allowed to get into any critical region controlled by the same semaphore in the meantime.

It has already been discussed that push-through blocking is an additional form of blocking, introduced by the priority inheritance protocol to keep the worst-case blocking time bounded. The following lemma gives a better characterization of this kind of blocking and identifies which semaphores can be responsible for it.

Lemma 14.4. *A semaphore S can induce push-through blocking onto task τ_H only if it is accessed* both *by a task that has a priority lower than the priority of τ_H, and by a task that either* has *or can inherit a priority higher than the priority of τ_H.*

Proof. The lemma can be proved by showing that, if the conditions set forth by the lemma do *not* hold, then push-through blocking cannot occur.

If S is not accessed by any task τ_L with a priority lower than the priority of τ_H, then, by definition, push-through blocking cannot occur.

Let us then suppose that S is indeed accessed by a task τ_L, with a priority lower than the priority of τ_H. If S is *not* accessed by any task that has or can inherit a priority higher than the priority of τ_H, then the priority inheritance mechanism will never give to τ_L an active priority higher than τ_H. In this case, τ_H can always preempt τ_L and, again, push-through blocking cannot take place.

If both conditions hold, push-trough blocking of τ_H by τ_L may occur, and the lemma follows.

It is also worth noting that, in the statement of Lemma 14.4, it is crucial to perceive the difference between saying that a task *has* a certain priority or that it *can inherit* that priority:

- When we say that a task *has* a certain priority, we are referring to its baseline priority.
- On the other hand, a task *can inherit* a certain priority higher than its baseline priority through the priority inheritance mechanism.

Lemma 14.3 states that the number of times a certain task τ_H can be blocked is bounded by n, that is, how many *tasks* have a priority lower than its own but can block it. The following lemma provides a different bound, based on how many *semaphores* can block τ_H. As before, the definition of "can block" must be understood according to what is stated in Lemma 14.1.

The two bounds are not equivalent because, in general, there is no one-to-one correspondence between tasks and critical regions, as well as between critical regions and semaphores. For example, τ_H may pass through more than one critical region but, if they are guarded by the same semaphore, it will be blocked at most once. Similarly, a single task τ_L may have more than one blocking critical region for τ_H, but it will nevertheless block τ_H at most once.

Lemma 14.5. *If task τ_H can endure blocking from m distinct semaphores $S_1, \ldots, S_m$, then τ_H can be blocked at most for the duration of m critical regions, once for each of the m semaphores.*

Proof. Lemma 14.1 establishes that a certain lower-priority task—let us call it τ_L— can block τ_H only if it is currently executing within a critical region that satisfies the hypotheses presented in the lemma itself and is therefore a blocking critical region for τ_H.

Since each critical region is protected by a mutual exclusion semaphore, τ_L must necessarily have acquired that mutual exclusion semaphore upon entering the critical region and no other tasks can concurrently be within another critical region associated with the same semaphore. Hence, only *one* of the lower-priority tasks, τ_L

in our case, can be within a blocking critical region protected by any given semaphore S_i.

Due to Lemma 14.2, as soon as τ_L leaves the blocking critical region, it can be preempted by τ_H and can no longer block it. Therefore, for each semaphore S_i, at the most *one* critical region can induce a blocking on τ_H. Repeating the argument for the m semaphores proves the lemma.

At this point, by combining Lemmas 14.3 and 14.5, we obtain the following important theorem, due to Sha, Rajkumar, and Lehoczky [83].

Theorem 14.1. *Under the priority inheritance protocol, a task τ_H can be blocked for at most the worst-case execution time, or* duration, *of* $\min(n, m)$ *critical regions in the system, where*

- *n is the number of lower-priority tasks that can block τ_H, and*
- *m is the number of semaphores that can block τ_H.*

It should be stressed that a critical region or a semaphore can block a certain task τ_H *even if* the critical region does not belong to τ_H, or τ_H does not use the semaphore.

An important phenomenon that concerns the priority inheritance protocol and makes its analysis more difficult is that priority inheritance must be *transitive*. A transitive priority inheritance occurs when a high-priority task τ_H is directly blocked by an intermediate-priority task τ_M, which in turn is directly blocked by a low-priority task τ_L.

In this case, the priority of τ_H must be transitively transmitted not only to τ_M but to τ_L, too. Otherwise, the presence of any other intermediate-priority task could still give rise to an unbounded priority inversion by preempting τ_L at the wrong time. The scenario is illustrated in Figure 14.4. There, if there is no transitive priority inheritance, τ_H's blocking may be unbounded because τ_B can preempt τ_L. Therefore, a critical region of a task can block a higher-priority task via transitive priority inheritance, too.

Fortunately, the following lemma makes it clear that transitive priority inheritance can occur only in a single, well-defined circumstance.

Lemma 14.6. *Transitive priority inheritance can occur* only *in presence of* nested *critical regions.*

Proof. In the proof, we will use the same task nomenclature just introduced to define transitive priority inheritance. Since τ_H is directly blocked by τ_M, then τ_M must hold a semaphore, say S_M. But, by hypothesis, τ_M is also directly blocked by a third task τ_L on a different semaphore held by τ_L, say S_L.

As a consequence, τ_M must have performed a blocking P() on S_L *after* successfully acquiring S_M, that is, *within* the critical region protected by S_M. This corresponds to the definition of properly nested critical regions.

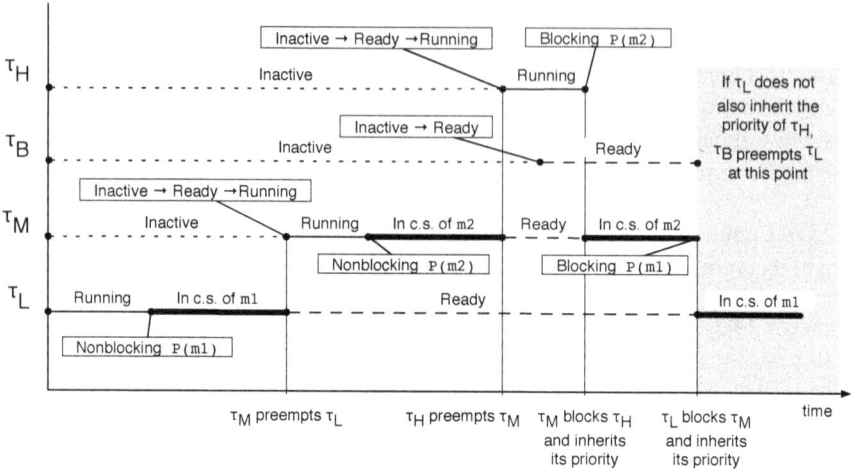

Figure 14.4 An example of transitive priority inheritance involving three tasks. If transitive inheritance did *not* take place, τ_H's blocking would be unbounded due to the preemption of τ_L by τ_B.

If transitive priority inheritance is ruled out with the help of Lemma 14.6, that is, if nested critical regions are forbidden, a stronger version of Lemma 14.4 holds:

Lemma 14.7. *In the absence of nested critical regions, a semaphore S can block a task τ_H only if it is used both by (at least) one task with a priority less than τ_H, and (at least) one task with a priority higher than or equal to τ_H.*

Proof. As long as push-through blocking is concerned, the statement of the lemma is the same as Lemma 14.4, minus the "can inherit" clause. The reasoning is valid because Lemma 14.6 rules out transitive priority inheritance if critical regions are not nested. Moreover, transitive inheritance is the only way for a task to acquire a priority higher than the highest-priority task with which it shares a resource.

The conditions set forth by this lemma also cover direct blocking because semaphore S can directly block a task τ_H only if it is used by another task with a priority less than the priority of τ_H, and by τ_H itself. As a consequence, the lemma is valid for all kinds of blocking (both direct and push-through).

If nested critical regions are forbidden, as before, the following theorem provides an easy way to compute an upper bound on the worst-case blocking time that a task τ_i can possibly experience.

Theorem 14.2. *Let K be the total number of semaphores in the system. If critical regions cannot be nested, the worst-case blocking time experienced by each activation* *of task τ_i under the priority inheritance protocol is bounded by B_i:*

$$B_i = \sum_{k=1}^{K} \text{usage}(k,i)C(k)$$

where

- usage(k,i) *is a function that returns* 1 *if semaphore S_k is used by (at least) one task with a priority less than the priority of τ_i and (at least) one task with a priority higher than or equal to the priority of τ_i. Otherwise,* usage(k,i) *returns* 0.
- $C(k)$ *is the worst-case execution time among all critical regions corresponding to, or guarded by, semaphore S_k.*

Proof. The proof of this theorem descends from the straightforward application of the previous lemmas. The function usage(k,i) captures the conditions under which semaphore S_k can block τ_i by means of either direct or push-through blocking, set forth by Lemma 14.7.

A single semaphore S_k may guard more than one critical region, and it is generally unknown which specific critical region will actually cause the blocking for task τ_i. It is even possible that, on different execution instances of τ_i, the region will change. However, the worst-case blocking time is still bounded by the worst-case execution time among *all* critical regions guarded by S_k, that is, $C(k)$.

Eventually, the contributions to the worst-case blocking time coming from the K semaphores are added together because, as stated by Lemma 14.5, τ_i can be blocked at most once for each semaphore.

It should be noted that the algorithm discussed in Theorem 14.2 is *not optimal* for the priority inheritance protocol, and the bound it computes is not so tight because

- It assumes that if a certain semaphore *can block* a task, it *will actually block* it.
- For each semaphore, the blocking time suffered by τ_i is assumed to be equal to the execution time of the longest critical region guarded by that semaphore even if that particular critical region is not a blocking critical region for τ_i.

However, it is an acceptable compromise between the tightness of the bound it calculates and its computational complexity. Better algorithms exist and are able to provide a tighter bound of the worst-case blocking time, but their complexity is much higher.

14.3   THE PRIORITY CEILING PROTOCOL

Even if the priority inheritance protocol just described enforces an upper bound on the number and the duration of blocks a high-priority task τ_H can encounter, it has several shortcomings:

- In the worst case, if τ_H tries to acquire n mutual exclusion semaphores that have been locked by n lower-priority tasks, it will be blocked for the duration of n critical regions. This is called *chained blocking*.
- The priority inheritance protocol does not prevent *deadlock* from occurring. Deadlock must be avoided by some other means, for example, by imposing a total order on semaphore accesses, as discussed in Chapter 4.

All of these issues are addressed by the *priority ceiling protocols*, also proposed by Sha, Rajkumar, and Lehoczky [83]. In this chapter we will discuss the original priority ceiling protocol and its *immediate* variant; both have the following properties:

1. A high-priority task can be blocked *at most once* during its execution by lower-priority tasks.
2. They prevent *transitive blocking* even if critical regions are nested.
3. They prevent *deadlock*.
4. *Mutual exclusive* access to resources is ensured by the protocols themselves.

The basic idea of the priority ceiling protocol is to *extend* the priority inheritance protocol with an additional rule for granting lock requests to a *free* semaphore. The overall goal of the protocol is to ensure that, if task τ_L holds a semaphore and it could lead to the blocking of a higher-priority task τ_H, then no other semaphores that could also block τ_H are to be acquired by any task other than τ_L itself.

A side effect of this approach is that a task can be blocked, and hence delayed, not only by attempting to lock a busy semaphore but also when granting a lock to a free semaphore could lead to multiple blocking on higher-priority tasks.

In other words, as we already did before, we are trading off some useful properties for an additional form of blocking that did not exist before. This new kind of blocking that the priority ceiling protocol introduces in addition to direct and push-through blocking, is called *ceiling blocking*.

The underlying hypotheses of the original priority ceiling protocol are the same as those of the priority inheritance protocol. In addition, it is assumed that each semaphore has a static *ceiling* value associated with it. The ceiling of a semaphore can easily be calculated by looking at the application code and is defined as the *maximum initial priority* of all tasks that use it.

As in the priority inheritance protocol, each task has a current (or active) priority that is greater than or equal to its initial (or baseline) priority, depending on whether it is blocking some higher-priority tasks or not. The priority inheritance rule is exactly the same in both cases.

A task can immediately acquire a semaphore only if its active priority is *higher than the ceiling* of any currently locked semaphore, *excluding* any semaphore that the task has already acquired in the past and not released yet. Otherwise, it will be blocked. It should be noted that this last rule can block the access to busy as well as free semaphores.

The first property of the priority ceiling protocol to be discussed puts an upper bound on the priority a task may get when it is preempted within a critical region.

Lemma 14.8. *If a task τ_L is preempted within a critical region Z_L by another task τ_M, and then τ_M enters a critical region Z_M, then, under the priority ceiling protocol, τ_L cannot inherit a priority higher than or equal to the priority of τ_M until τ_M completes.*

Proof. The easiest way to prove this lemma is by contradiction. If, contrary to our thesis, τ_L inherits a priority higher than or equal to the priority of τ_M, then it must block a task τ_H. The priority of τ_H must necessarily be higher than or equal to the priority of τ_M. If we call P_H and P_M the priorities of τ_H and τ_M, respectively, it must be $P_H \geq P_M$.

On the other hand, since τ_M was allowed to enter Z_M without blocking, its priority must be strictly higher than the maximum ceiling of the semaphores currently locked by any task except τ_M itself. Even more so, if we call C^* the maximum ceiling of the semaphores currently locked by tasks with a priority *lower* than the priority of τ_M, thus including τ_L, it must be $P_M > C^*$.

The value of C^* cannot undergo any further changes until τ_M completes because no tasks with a priority lower than its own will be able to run and acquire additional semaphores as long as τ_M is Ready or Running. The same is true also if τ_M is preempted by any higher-priority tasks. Moreover, those higher-priority tasks will never transmit their priority to a lower-priority one because, being $P_M > C^*$, they do not share any semaphore with it. Last, if τ_M ever blocks on a semaphore, it will transmit its priority to the blocking task, which will then get a priority at least equal to the priority of τ_M.

By combining the two previous inequalities by transitivity, we obtain $P_H > C^*$. From this, we can conclude that τ_H *cannot be blocked* by τ_L because the priority of τ_H is higher than the ceiling of all semaphores locked by τ_L. This is a contradiction and the lemma follows.

Then, we can prove that no transitive blocking can ever occur when the priority ceiling protocol is in use because

Lemma 14.9. *The priority ceiling protocol prevents transitive blocking.*

Proof. Again, by contradiction, let us suppose that a transitive blocking occurs. By definition, there exist three tasks τ_H, τ_M, and τ_L with decreasing priorities such that τ_L blocks τ_M and τ_M blocks τ_H.

Then, also by definition, τ_L must inherit the priority of τ_H by transitive priority inheritance. However, this *contradicts* Lemma 14.8. In fact, its hypotheses are fulfilled, and it states that τ_L cannot inherit a priority higher than or equal to the priority of τ_M until τ_M completes.

Another important property of the priority ceiling protocol concerns deadlock prevention and is summarized in the following theorem.

Theorem 14.3. *The priority ceiling protocol prevents deadlock.*

Proof. We assume that a task cannot deadlock "by itself," that is, by trying to acquire again a mutual exclusion semaphore it already acquired in the past. In terms of program code, this would imply the execution of two consecutive P() on the same semaphore, with no V() in between.

Then, a deadlock can only be formed by a *cycle* of $n \geq 2$ tasks $\{\tau_1, \ldots, \tau_n\}$ waiting for each other according to the *circular wait* condition discussed in Chapter 4. Each of these tasks must be within one of its critical regions; otherwise, deadlock cannot occur because the *hold & wait* condition is not satisfied.

By Lemma 14.9, it must be $n = 2$; otherwise, transitive blocking would occur, and hence, we consider only the cycle $\{\tau_1, \tau_2\}$. For a circular wait to occur, one of the tasks was preempted by the other while it was within a critical region because they are being executed by one single processor. Without loss of generality, we suppose that τ_2 was firstly preempted by τ_1 while it was within a critical region. Then, τ_1 entered its own critical region.

In that case, by Lemma 14.8, τ_2 cannot inherit a priority higher than or equal to the priority of τ_1. On the other hand, since τ_1 is blocked by τ_2, then τ_2 must inherit the priority of τ_1, but this is a contradiction and the theorem follows.

This theorem is also very useful from the practical standpoint. It means that, under the priority ceiling protocol, programmers can put into their code an arbitrary number of critical regions, possibly (properly) nested into each other. As long as each task does not deadlock with itself, there will be *no deadlock* at all in the system.

The next goal is, as before, to compute an upper bound on the worst-case blocking time that a task τ_i can possibly experience. First of all, it is necessary to ascertain how many times a task can be blocked by others.

Theorem 14.4. *Under the priority ceiling protocol, a task τ_H can be blocked for, at most, the duration of* one *critical region.*

Proof. Suppose that τ_H is blocked by *two* lower-priority tasks τ_L and τ_M, where $P_L \leq P_H$ and $P_M \leq P_H$. Both τ_L and τ_M must be in a critical region.

Let τ_L enter its critical region first, and let C_L^* be the highest-priority ceiling among all semaphores locked by τ_L at this point. In this scenario, since τ_M was allowed to enter its critical region, it must be $P_M > C_L^*$; otherwise, τ_M would be blocked.

Moreover, since we assumed that τ_H can be blocked by τ_L, it must necessarily be $P_H \leq C_L^*$. However, this implies that $P_M > P_H$, leading to a contradiction.

The next step is to identify the critical regions of interest for blocking, that is, which critical regions can block a certain task.

Lemma 14.10. *Under the priority ceiling protocol, a critical region Z, belonging to task τ_L and guarded by semaphore S, can block another task τ_H only if $P_L < P_H$, and the priority ceiling of S, C_S^* is greater than or equal to P_H, that is, $C_S^* \geq P_H$.*

Proof. If it were $P_L \geq P_H$, then τ_H could not preempt τ_L, and hence, could not be blocked by S. Hence, it must be $P_L < P_H$ for τ_H to be blocked by S.

Let us assume that $C_S^* < P_H$, that is, the second part of the hypothesis, is not satisfied, but τ_H is indeed blocked by S. Then, its priority must be less than or equal to the maximum ceiling C^* among all semaphores acquired by tasks other than itself, that is, $P_H \leq C^*$. But it is $P_H > C_S^*$, and hence, $C_S^* < C^*$ and *another semaphore*, not S, must be the source of the blocking in this case.

It is then possible to conclude that τ_H can be blocked by τ_L, by means of semaphore S, only if $P_L < P_H$ and $C_S^* \geq P_H$.

By combining the previous two results, the following theorem provides an upper bound on the worst-case blocking time.

Theorem 14.5. *Let K be the total number of semaphores in the system. The worst-case blocking time experienced by each activation of task τ_i under the priority ceiling protocol is bounded by B_i:*

$$B_i = \max_{k=1}^{K} \{\text{usage}(k,i)C(k)\}$$

where

- *usage(k,i) is a function that returns 1 if semaphore S_k is used by (at least) one task with a priority less than τ_i and (at least) one task with a priority higher than or equal to τ_i. Otherwise, it returns 0.*
- *$C(k)$ is the worst-case execution time among all critical regions guarded by semaphore S_k.*

Proof. The proof of this theorem descends from the straightforward application of

1. Theorem 14.4, which limits the blocking time to the duration of *one* critical region, the longest critical region among those that *can block* τ_i.
2. Lemma 14.10, which identifies *which* critical regions must be considered for the analysis. In particular, the definition of usage(k,i) is just a slightly different way to state the necessary conditions set forth by the lemma to determine whether a certain semaphore S_K can or cannot block τ_i.

The complete formula is then built with the same method already discussed in the proof of Theorem 14.2.

The only difference between the formulas given in Theorem 14.2 (for priority inheritance) and Theorem 14.5 (for priority ceiling), namely, the presence of a summation instead of a maximum can be easily understood by comparing Lemma 14.5 and Theorem 14.4.

Lemma 14.5 states that, for priority inheritance, a task τ_H can be blocked at most once *for each semaphore* that satisfies the blocking conditions of Lemma 14.7, and hence, the presence of the summation for priority inheritance. On the other hand, Theorem 14.4 states that, when using priority ceiling, τ_H can be blocked at most once, *period*, regardless of how many semaphores can potentially block it. In this case, to be conservative, we take the maximum among the worst-case execution times of all critical regions controlled by all semaphores that can potentially block τ_H.

An interesting variant of the priority ceiling protocol, called *immediate* priority ceiling or priority ceiling *emulation* protocol, takes a more straightforward approach. It raises the priority of a task to the priority ceiling associated with a resource *as soon as* the task acquires it, rather than only when the task is blocking a higher-priority task. Hence, it is defined as follows:

1. Each task has an initial, or baseline, priority assigned.
2. Each semaphore has a static ceiling defined, that is, the maximum priority of all tasks that may use it.
3. At each instant a task has a dynamic, active priority, that is, the maximum of its static, initial priority and the ceiling values of any semaphore it has acquired.

It can be proved that, as a consequence of the last rule, a task will only suffer a block at the very beginning of its execution. The worst-case behavior of the immediate priority ceiling protocol is the same as the original protocol, but

- The immediate priority ceiling is *easier to implement*, as blocking relationships must not be monitored. Also, for this reason, it has been specified in the POSIX standard [52], along with priority inheritance, for mutual exclusion semaphores.
- It leads to *less context switches* as blocking is prior to the first execution.
- On average, it requires *more priority movements*, as this happens with *all semaphore operations* rather than only if a task is actually blocking another.

14.4 SCHEDULABILITY ANALYSIS AND EXAMPLES

In the previous section, an upper bound for the worst-case blocking time B_i that a task τ_i can suffer has been obtained. The bound depends on the way the priority inversion problem has been addressed. The worst-case response time R_i, already introduced in

Chapter 13, can be redefined to take B_i into account as follows:

$$R_i = C_i + B_i + I_i \tag{14.1}$$

In this way, the worst-case response time R_i of task τ_i is expressed as the sum of three components:

1. the worst-case execution time C_i,
2. the worst-case interference I_i, and
3. the worst-case blocking time B_i.

The corresponding recurrence relationship introduced for Response Time Analysis (RTA) becomes

$$w_i^{(k+1)} = C_i + B_i + \sum_{j \in hp(i)} \left\lceil \frac{w_i^{(k)}}{T_j} \right\rceil C_j \tag{14.2}$$

It can be proved that the new recurrence relationship still has the same properties as the original one. In particular, if it converges, it still provides the worst-case response time R_i for an appropriate choice of w_i^0. As before, either 0 or C_i are good starting points.

The main difference is that the new formulation is *pessimistic*, instead of necessary and sufficient, because the bound B_i on the worst-case blocking time is not tight, and hence, it may be impossible for a task to ever actually incur in a blocking time equal to B_i.

Let us now consider an example. We will consider a simple set of tasks and determine the effect of the priority inheritance and the immediate priority ceiling protocols on their worst-case response time, assuming that their periods are large (> 100 time units). In particular, the system includes

- A high-priority task τ_H, released at $t = 4$ time units, with a computation time of $C_H = 3$ time units. It spends the last 2 time units within a critical region guarded by a semaphore, S.
- An intermediate-priority task τ_M, released at $t = 2$ time units, with a computation time of 4 time units. It does not have any critical region.
- A low-priority task τ_L, released at $t = 0$ time units, with a computation time of 4 time units. It shares some data with τ_H, hence it spends its middle 2 time units within a critical region guarded by S.

The upper part of Figure 14.5 sketches the internal structure of the three tasks. Each task is represented as a rectangle with the left side aligned with its release time. The rectangle represents the execution of the task if it were alone in the system; the gray area inside the rectangle indicates the location of the critical region the task contains, if any.

The lower part of the figure shows how the system of tasks being considered is scheduled when nothing is done against unbounded priority inversion. When τ_H is

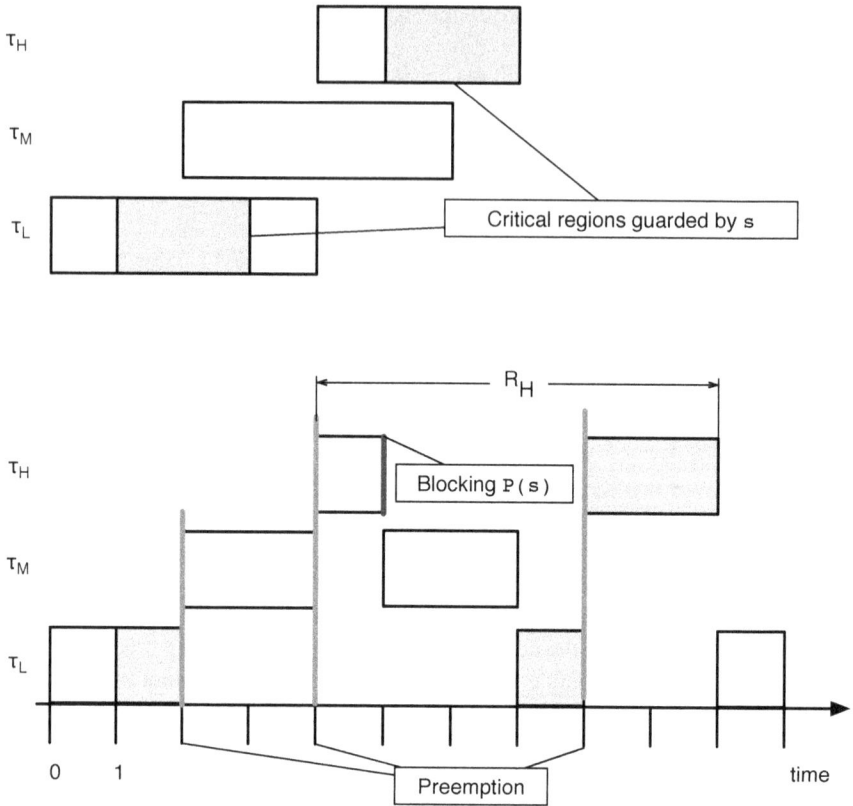

Figure 14.5 An example of task scheduling with unbounded priority inversion.

blocked by its P(s) at $t = 5$ time units, task τ_M is resumed because it is the highest-priority task ready for execution. Hence, τ_L is resumed only after τ_M completes its execution, at $t = 7$ time units.

Thereafter, τ_L leaves its critical region at $t = 8$ and is immediately preempted by τ_H, which is now no longer blocked by τ_L and is ready to execute again. τ_H completes its execution at $t = 10$ time units. Last, τ_L is resumed and completes at $t = 11$ time units.

Overall, the task response times in this case are: $R_H = 10 - 4 = 6$, $R_M = 7 - 2 = 5$, $R_L = 11 - 0 = 11$ time units. Looking at Figure 14.5, it is easy to notice that R_H, the response time of τ_H, includes part of the execution time of τ_M; even if τ_M's priority is lower, there is no interaction at all between τ_H and τ_M, and executing τ_M instead of τ_H is not of help to any other higher-priority tasks.

The behavior of the system with the priority inheritance protocol is shown in Figure 14.6. It is the same as before until $t = 5$ time units, when τ_H is blocked by the execution of P(s). In this case, τ_H is still blocked as before, but it also transmits

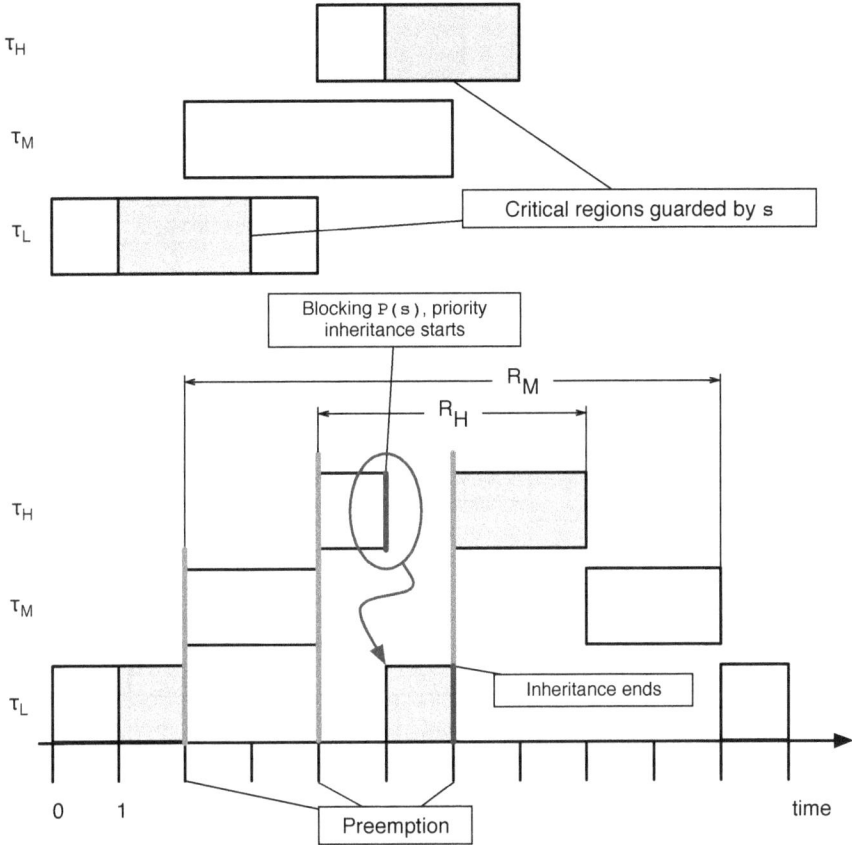

Figure 14.6 An example of task scheduling with priority inheritance.

its priority to τ_L. In other words τ_L, the blocking task, inherits the priority of τ_H, the blocked one.

Hence, τ_L is executed (instead of τ_M) and leaves its critical region at $t = 6$ time units. At that moment, the priority inheritance ends and τ_L returns to its original priority. Due to this, it is immediately preempted by τ_H. At $t = 8$ time units, τ_H completes and τ_M is resumed. Finally, τ_M and τ_L complete their execution at $t = 10$ and $t = 11$ time units, respectively.

With the priority inheritance protocol, the task response times are $R_H = 8 - 4 = 4$, $R_M = 10 - 2 = 8$, $R_L = 11 - 0 = 11$ time units. From Figure 14.6, it can be seen that R_H no longer comprises any part of τ_M. It does include, instead, part of the critical region of τ_L, but this is unavoidable as long as τ_H and τ_L share some data with a mutual exclusion mechanism. In addition, the worst-case contribution it makes to R_H is bounded by the maximum time τ_L spends within its critical region.

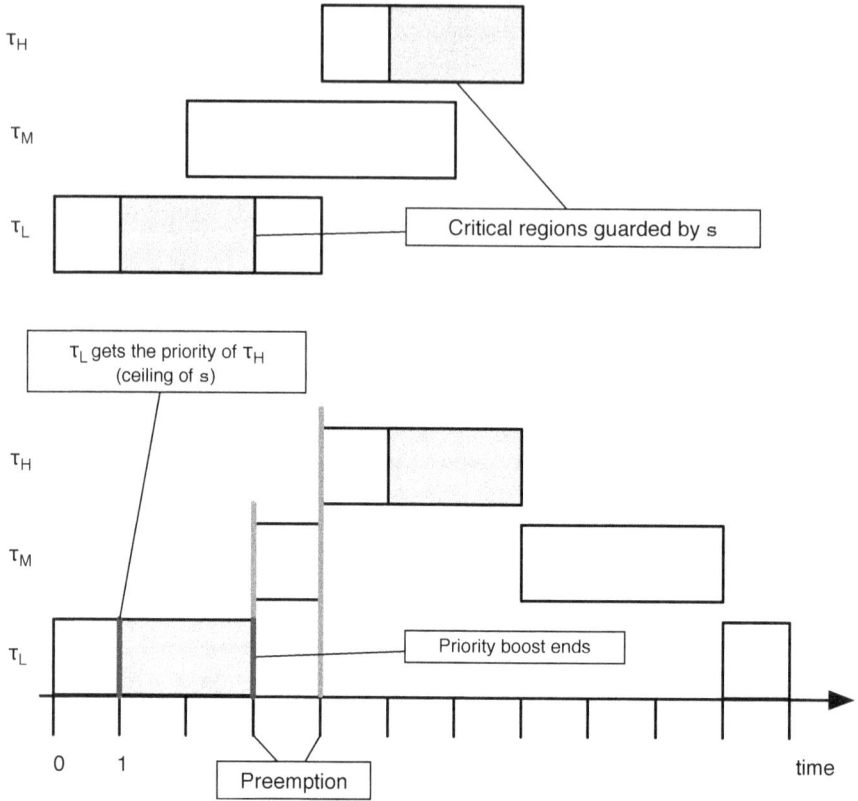

Figure 14.7 An example of task scheduling with immediate priority ceiling.

It can also be noted that R_M, the response time of τ_M, now includes part of the critical region of τ_L, too. This is a very simple case of push-through blocking and is a side effect of the priority inheritance protocol. In any case, the contribution to R_M is still bounded by the maximum time τ_L spends within its critical region.

As shown in Figure 14.7, with the immediate priority ceiling protocol, τ_L acquires a priority equal to the priority of τ_H, the ceiling of s, as soon as it enters the critical region guarded by s itself at $t = 1$ time unit.

As a consequence, even if τ_M is released at $t = 2$, it does *not* preempt τ_L up to $t = 3$ when τ_L leaves the critical region and returns to its original priority. Another preemption, of τ_M in favor of τ_H, occurs at $t = 4$ time units, as soon as τ_H is released.

Then, τ_M is resumed, after the completion of τ_H, at $t = 7$ time units, and completes at $t = 10$ time units. Finally, τ_L completes at $t = 11$ time units. With the immediate priority ceiling protocol, the task response times are $R_H = 7 - 4 = 3$, $R_M = 10 - 2 = 8$, $R_L = 11 - 0 = 11$ time units.

Applying RTA to the example is particularly easy. The task periods are large with respect to their response time, and hence, each interfering task must be taken into

Table 14.1

Task response times for Figures 14.5–14.7

Task	Actual Response Time			Worst Case Response Time
	Nothing	P. Inheritance	Imm. P. Ceiling	
τ_H	6	4	3	5
τ_M	5	8	8	9
τ_L	11	11	11	11

account only *once*, and all the successions converge immediately. Moreover, since there is only *one* critical section, the worst-case blocking times are the same for both protocols.

The worst-case execution time among all critical regions guarded by s is 2 time units. Therefore,

- $B_H = 2$ because s can block τ_H;
- $B_M = 2$ because s can block τ_M;
- $B_L = 0$ because s cannot block τ_L.

Substituting back into (14.2), we get

- For τ_H, $\text{hp}(H) = \emptyset$ and $B_H = 2$, and hence: $R_H = C_H + B_H = 5$ time units.
- For τ_M: $\text{hp}(M) = \{H\}$ and $B_M = 2$, and hence: $R_M = C_M + B_M + C_H = 9$ time units.
- For τ_L: $\text{hp}(L) = \{H,M\}$ and $B_L = 0$, and hence: $R_L = C_L + C_H + C_M = 11$ time units.

Table 14.1 summarizes the actual response time of the various tasks involved in the example, depending on how the priority inversion problem has been addressed. For comparison, the rightmost column also shows the worst-case response times computed by the RTA method, extended as shown in (14.2). The worst-case response times are the same for both priority inheritance and immediate priority ceiling because their B_i are all the same.

Both the priority inheritance and the immediate priority ceiling protocols bound the maximum blocking time experienced by τ_H. It can be seen that its response time gets lower when any of these protocols are in use. On the other hand, the response time of τ_M increases as a side effect.

The response time of τ_L does not change but this can be expected because it is the lowest priority task. As a consequence, it can only suffer interference from the other tasks in the system but not blocking. For this reason, its response time cannot be affected by any protocol that introduces additional forms of blocking. It can also be seen that RTA provides a satisfactory worst-case response time in all cases except when the priority inversion problem has not been addressed at all.

14.5 SUMMARY

In most concurrent applications of practical interest, tasks must interact with each other to pursue a common goal. In many cases, task interaction also implies *blocking*, that is, tasks must synchronize their actions and therefore wait for each other.

In this chapter we saw that careless task interactions may undermine priority assignments and, eventually, jeopardize the ability of the whole system to be scheduled because they may lead to an *unbounded priority inversion*. This happens even if the interactions are very simple, for instance, when tasks manipulate some shared data by means of critical regions.

One way of solving the problem is to set up a better cooperation between the *scheduling* algorithm and the *synchronization* mechanism, that is, to allow the synchronization mechanism to modify task priorities as needed. This is the underlying idea of the algorithms discussed in this chapter: *priority inheritance* and the two variants of *priority ceiling*.

It is then possible to show that all these algorithms are actually able to force the worst-case blocking time of any task in the system to be upper-bounded. Better yet, it is also possible to *calculate* the upper bound, starting from a limited amount of additional information about task characteristics, which is usually easy to collect.

Once the upper bound is known, the RTA method can also be extended to take consider it and calculate the worst-case response time of each task in the system, taking task interaction into account.

14.6 EXERCISES

EXERCISE 1

Consider a low-priority task τ_L and a high-priority task τ_H that share a mutually exclusive resource by means of a semaphore S. Explain how they may worsen the response time of a mid-priority task τ_M even if they do not interact directly with it, assuming that priority inheritance is in effect.

EXERCISE 2

Discuss how is it possible for two low-priority tasks τ_{L1} and τ_{L2} to affect the response time of a mid-priority task τ_M in the following scenarios, assuming that priority inheritance is in effect.

1. τ_{L1} and τ_{L2} do not use any mutual exclusion semaphore.
2. τ_{L1} and τ_{L2} use the same mutual exclusion semaphore S and no other tasks use it. Neither τ_{L1} nor τ_{L2} uses any other semaphore.
3. τ_{L1} uses a semaphore S together with a high-priority task τ_H; moreover, τ_{L1} and τ_{L2} are the only users of another semaphore S'.

15 Self-Suspension and Schedulability Analysis

The schedulability analysis techniques presented in Chapters 11 through 13 are based on the hypothesis that task instances never block for any reason, unless they have been executed completely and the next instance has not been released yet. In other words, they assume that a task may only be in three possible states:

1. ready for execution, but not executing because a higher-priority task is being executed on its place,
2. executing, or
3. not executing, because its previous instance has been completed and the next one has not been released yet.

Then, in Chapter 14, we analyzed the effects of task interaction on their worst-case response time and schedulability. In particular, we realized that when tasks interact, even in very simple ways, they must sometimes *block* themselves until some other tasks perform a certain action. The purpose of this extension was to make our model more realistic and able to better represent the behavior of real tasks. However, mutual exclusion on shared data access was still considered to be the *only* source of blocking in the system.

This is still not completely representative of what can happen in the real world where tasks also invoke external operations and wait for their completion. For instance, it is common for tasks to start an Input–Output (I/O) operation and wait until it completes or a timeout expires; another example would be to send a network message to a task residing on a different system and then wait for an answer.

More generally, all situations in which a task voluntarily suspends itself for a variable amount of time with a known upper bound—such as those just described—are called *self-suspension* or *self-blocking*. Therefore, in this chapter, the formulas to calculate the worst-case blocking time that may affect a task, given in Chapter 14, will be further extended to incorporate self-suspension. The extension is based on [80], addressing schedulability analysis in the context of real-time synchronization for multiprocessor systems.

15.1 SELF-SUSPENSION AND THE CRITICAL INSTANT THEOREM

At a first sight, it may seem that the effects of self-suspension should be *local* to the task that is experiencing it: after all, the task is neither competing for nor consuming any processor time while it is suspended, and hence, it should not hinder the schedulability of any other task because of this. This is unfortunately not true and, although

DOI: 10.1201/9781003593416-15

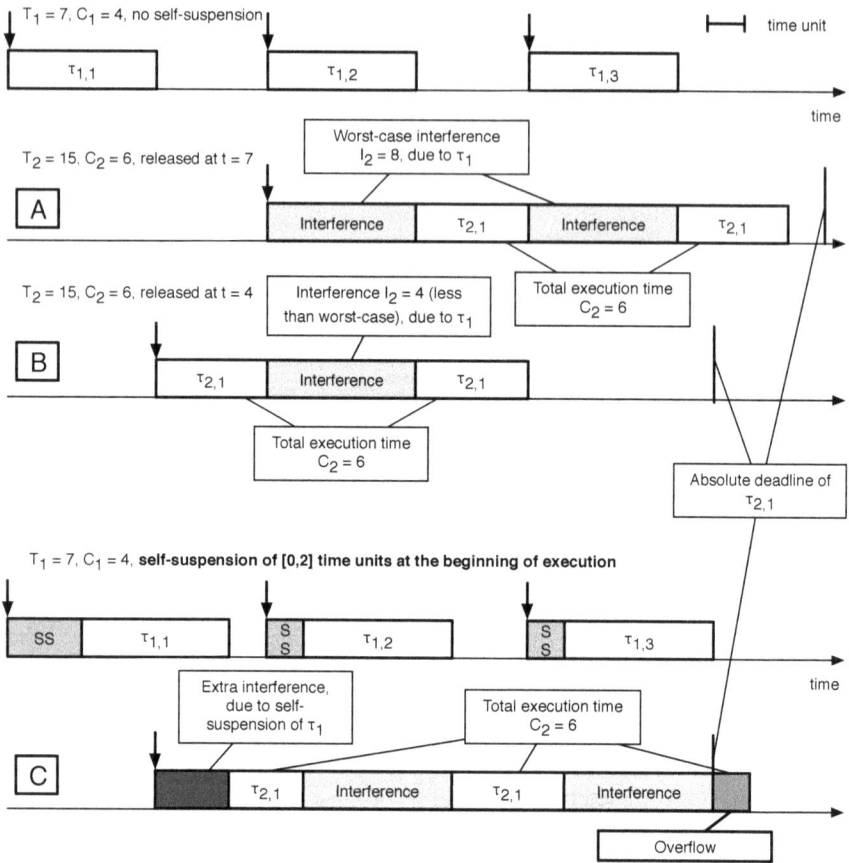

Figure 15.1 When a high-priority task self-suspends itself, the response time of lower-priority tasks may no longer be the worst at a critical instant.

this sounds counterintuitive, it turns out that adding a self-suspension region to a higher-priority task may render a lower-priority task no longer schedulable.

Even the critical instant theorem, on which many other theorems given in Chapter 11 are based, is no longer directly applicable to compute the worst-case interference that a task may be subject to. A counterexample is shown in Figure 15.1.

The upper part of the figure shows how two tasks τ_1 (high priority) and τ_2 (low priority) are scheduled when self-suspension is not allowed. Task τ_1 has a period of $T_1 = 7$ time units and an execution time of $C_1 = 4$ time units. The uppermost time diagram shows how it is executed: as denoted by the downward-pointing arrows, its first three instances are released at $t = 0$, $t = 7$, and $t = 14$ time units, and since they have a priority higher than τ_2, they always run to completion immediately after being released.

Time diagram **A** shows instead how the first instance of task τ_2 is executed when it is released at $t = 7$ time units. This is a critical instant for τ_2, because $\tau_{1,2}$ (the second instance of τ_1) has been released at the same time, too. The execution time of $C_2 = 6$ time units is allotted to $\tau_{2,1}$ in two chunks—from $t = 11$ to $t = 14$, and from $t = 18$ to $t = 21$ time units—because of the interference due to τ_1.

Accordingly, the response time of $\tau_{2,1}$ is 14 time units. If the period of τ_2 is $T_2 = 15$ time units and it is assumed that its relative deadline is equal to the period, that is $D_2 = T_2$, then $\tau_{2,1}$ meets its deadline. The total amount of interference experienced by $\tau_{2,1}$ in this case is $I_2 = 8$ time units, corresponding to two full executions of τ_1 and shown in the figure as light grey rectangles. According to the critical instant theorem, this is the worst possible interference τ_2 can ever experience, and we can also conclude that τ_2 is schedulable.

As shown in time diagram **B**, if $\tau_{2,1}$ is not released at a critical instant, the interference will never be greater than 8 time units. In that time diagram, $\tau_{2,1}$ is released at $t = 4$ time units, and it only experiences 4 time units of interference due to τ_1. Hence, as expected, its response time is 10 time units, less than the worst case value calculated before and well within the deadline.

The lower part of Figure 15.1 shows, instead, what happens if each instance of τ_1 is allowed to self-suspend for a variable amount of time, between 0 and 2 time units, at the very beginning of its execution. In particular, the time diagrams have been drawn assuming that $\tau_{1,1}$ self-suspends for 2 time units, whereas $\tau_{1,2}$ and $\tau_{1,3}$ self-suspend for 1 time unit each. In the time diagram, the self-suspension regions are shown as grey rectangles. Task instance $\tau_{2,1}$ is still released as in time diagram **B**, that is, at $t = 4$ time units.

However, the scheduling of $\tau_{2,1}$ shown in time diagram **C** is very different than before. In particular,

- The self-suspension of $\tau_{1,1}$, lasting 2 time units, has the local effect of shifting its execution to the right of the time diagram. However, the shift has a more widespread effect, too, because it induces an "extra" interference on $\tau_{2,1}$ that postpones the beginning of its execution. In Figure 15.1, this extra interference is shown as a dark gray rectangle.
- Since its execution has been postponed, $\tau_{2,1}$ now experiences 8 time units of "regular" interference instead of 4 due to $\tau_{1,2}$ and $\tau_{1,3}$. It should be noted that this amount of interference is still within the worst-case interference computed using the critical instant theorem.
- The eventual consequence of the extra interference is that $\tau_{2,1}$ is unable to conclude its execution on time and overflows its deadline by 1 time unit.

In summary, the comparison between time diagrams **B** and **C** in Figure 15.1 shows that the same task instance ($\tau_{2,1}$ in this case), released at the same time, is or is not schedulable depending on the absence or presence of self-suspension in *another*, higher-priority task (τ_1).

On the other hand, the self-suspension of a task cannot induce any extra interference on higher-priority tasks. In fact, any lower-priority task is immediately preempted as soon as a higher-priority task becomes ready for execution, and cannot interfere with it in any way.

The worst-case extra interference endured by task τ_i due to its own self-suspension, as well as the self-suspension of other tasks, can be modeled as an additional source of blocking that will be denoted B_i^{SS}. As proved in [80], B_i^{SS} can be written as

$$B_i^{SS} = S_i + \sum_{j \in \text{hp}(i)} \min(C_j, S_j) \qquad (15.1)$$

where S_i is the worst-case self-suspension time of task τ_i and, as usual, $\text{hp}(i)$ denotes the set of task indexes with a priority higher than τ_i.

According to this formula, the total blocking time due to self-suspension endured by task τ_i is given by the sum of its own worst-case self-suspension time S_i plus a contribution from each of the higher-priority tasks. The contribution of task τ_j to B_i^{SS} is given by its worst-case self-suspension time S_j, but it cannot exceed its execution time C_j.

In the example, from (15.1) we have

$$B_1^{SS} = S_1 + 0 = 2 \qquad (15.2)$$

because the worst-case self-suspension time of τ_1 is indeed 2 time units and the set $\text{hp}(1)$ is empty.

Moreover, it is also

$$B_2^{SS} = 0 + \min(C_1, S_1) = \min(4, 2) = 2 \qquad (15.3)$$

because τ_2 never self-suspends itself (thus $S_2 = 0$) but may be affected by the self-suspension of τ_1 (in fact, $\text{hp}(2) = \{1\}$) for up to 2 time units.

15.2 SELF-SUSPENSION AND TASK INTERACTION

The self-suspension of a task has an impact on how it interacts with other tasks, and on the properties of the interaction, too. That is, many theorems about the worst-case blocking time proved in Chapter 14, lose part of their validity when tasks are allowed to self-suspend.

Figure 15.2 shows how two periodic, interacting tasks, τ_1 and τ_2, are scheduled with and without allowing self-suspension. The general structure of a task instance is the same for both tasks and is shown in the uppermost part of the figure:

- The total execution time of a task instance is $C_1 = C_2 = 10$ time units and comprises two critical regions.
- At the beginning, the task instance executes for 2 time units.
- Then, it enters its first critical region, shown as a dark grey rectangle in the figure. If we are using a semaphore m for synchronization, the critical region is delimited, as usual, by the primitives P(m) (at the beginning) and V(m) (at the end).

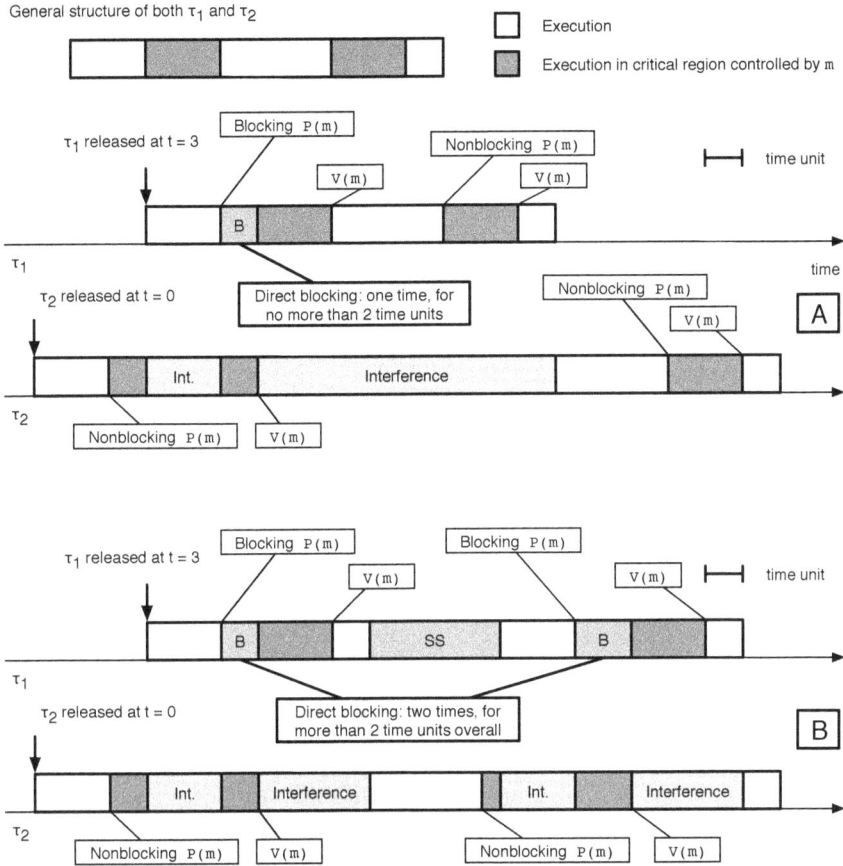

Figure 15.2 When a task self-suspends itself, it may suffer more blocking than predicted by the theorems of Chapter 14.

- The task instance executes within the first critical region for 2 time units.
- After concluding its first critical region, the task instance further executes for 3 time units before entering its second critical region.
- It stays in the second critical region for 2 time units.
- After the end of the second critical region, the task instance executes for 1 more time unit, and then it concludes.

Time diagram **A** of Figure 15.2 shows how the two tasks are scheduled, assuming that τ_1 is released at $t = 3$ time units, τ_2 is released at $t = 0$ time units, and the priority of τ_1 is higher than the priority of τ_2. It is also taken for granted that the period of both tasks is large enough—let us say $T_1 = T_2 = 30$ time units—so that only one instance of each is released for the whole length of the time diagram, and *no self-suspension* is allowed. In particular,

- Task τ_2 is executed from $t = 0$ until $t = 3$ time units. At $t = 2$, it enters its first critical region without blocking because no other task currently holds the mutual exclusion semaphore m.
- At $t = 3$, τ_2 is preempted because τ_1 is released and hence becomes ready for execution. The execution of τ_1 lasts 2 time units until it blocks at the first critical region's boundary with a blocking P(m).
- Task τ_2 resumes execution (possibly with an increased active priority if either priority inheritance or priority ceiling are in use) until it exits from the critical region at $t = 6$ time units.
- At this time, τ_1 resumes—after 1 time unit of direct blocking—and is allowed to enter into its first critical region. From this point on, the current instance of τ_1 runs to run completion without blocking again.
- Thereafter, the system resumes τ_2 and concludes it.

It should be noted that the system behavior, in this case, can still be accurately predicted by the theorems of Chapters 11 through 14, namely:

1. The interference underwent by τ_2 (the lower-priority task) is equal to the worst-case interference predicted by the critical instant theorem and Response Time Analysis (RTA), that is, $C_1 = 10$ time units. In Figure 15.2, interference regions are shown as light gray rectangles. Thus, the response time of τ_2 is 20 time units, 10 time units of execution plus 10 time units of interference.
2. Task τ_1 (the higher-priority task) does not suffer any interference due to τ_2; it undergoes blocking, instead. We know that τ_1 may block *at most once*, due to Lemmas 14.2, Lemma 14.5, or Theorem 14.1 (for priority inheritance), as well as Theorem 14.4 (for priority ceiling).
3. The *worst-case blocking time* can be calculated by means of Theorem 14.2 (for priority inheritance), or Theorem 14.5 (for priority ceiling). The result is the same in both cases and is equal to the maximum execution time of τ_2 within a critical region, that is, 2 time units. In the example, τ_1 only experiences 1 time unit of blocking, less than the worst case. Accordingly, the response time of τ_1 is 11 time units, 10 time units of execution plus 1 time unit of direct blocking.

Time diagram **B** of Figure 15.2 shows instead what happens if task τ_1 self-suspends for 3.5 time units after 5 time units of execution:

- Task τ_2 is now executed while τ_1 is suspended, but it is still subject to the same total amount of interference as before. Therefore, unlike in the example discussed in Section 15.1, the response time of τ_2 stays the same in this case.
- The scheduling of τ_1 is radically different and implies an increase of its response time, well beyond the additional 3.5 time units due to self-suspension. In fact, the response time goes from 11 to 16 time units.

- The reason for this can readily be discerned from the time diagram. When τ_1 is resumed after self-suspension and tries to enter its second critical region, it is *blocked again* because τ_2 entered its second critical region while τ_1 itself was suspended.
- The second block of τ_1 lasts for 1.5 time units, that is, until τ_2 eventually exits from the second critical region and unblocks it.
- The total response time of τ_1 is now given by 10 time units of execution plus 3.5 time units of self-suspension (the gray block marked SS in the figure) plus 2.5 time units of blocking time (the two gray blocks marked B). In particular, 1 time unit of blocking is due to the first block and 1.5 units to the second one.

In summary,

1. It is no longer true that τ_1 may block *at most once*. In fact, τ_1 was blocked twice in the example. As a consequence, it is clear that Lemmas 14.2, Lemma 14.5, and Theorem 14.1 (for priority inheritance), as well as Theorem 14.4 (for priority ceiling) are no longer valid.
2. Similarly, Theorem 14.2 (for priority inheritance) and Theorem 14.5 (for priority ceiling) are no longer adequate to compute the *worst-case blocking time* of τ_1. Even in this simple example, we have just seen that the total blocking time of τ_1 is 2.5 time units, that is, more than the 2 time units of worst-case blocking time predicted by those theorems.

There are several different ways to take self-suspension into account during worst-case blocking time calculation. Perhaps the most intuitive one, presented in References [64, 80], considers task *segments*—that is, the portions of task execution delimited by a self-suspension—to be completely independent from each other for what concerns blocking. Stated in an informal way, it is as if each task went back to the worst possible blocking scenario after each self-suspension.

In the previous example, τ_1 has two segments because it contains one self-suspension. Thus, each of its instances executes for a while (this is the first segment), self-suspends, and then executes until it completes (second segment). More generally, if task τ_i contains Q_i self-suspensions, it has $Q_i + 1$ segments.

According to this approach, let us call B_i^{1S} the worst-case blocking time calculated as specified in Chapter 14, without taking self-suspension into account; that is,

$$B_i^{1S} = \sum_{k=1}^{K} \text{usage}(k,i)C(k) \quad \text{(for priority inheritance)} \qquad (15.4)$$

or

$$B_i^{1S} = \max_{k=1}^{K} \{\text{usage}(k,i)C(k)\} \quad \text{(priority ceiling)} \qquad (15.5)$$

where, as before,

- K is the total number of semaphores in the system.
- $\text{usage}(k, i)$ is a function that returns 1 if semaphore S_k is used by (at least) one task with a priority less than τ_i and (at least) one task with a priority higher than or equal to τ_i; otherwise, it returns 0.
- $C(k)$ is the worst-case execution time among all critical regions guarded by semaphore S_k.

This quantity now becomes the worst-case blocking time endured by *each individual* task segment. Therefore, the total worst-case blocking time of task τ_i due to task interaction, B_i^{TI}, is given by

$$B_i^{TI} = (Q_i + 1)B_i^{1S} \tag{15.6}$$

where Q_i is the number of self-suspensions of task τ_i.

In our example, from either (15.4) or (15.5) we have

$$B_1^{1S} = 2 \tag{15.7}$$

because the worst-case execution time of any task within a critical region controlled by semaphore m is 2 time units and τ_1 can be blocked by τ_2.

We also have

$$B_2^{1S} = 0 \tag{15.8}$$

because τ_2 is the lowest-priority task in the system and cannot be blocked by any other task. In fact, $\text{usage}(k, 2) = 0$ for any k.

According to (15.6), the worst-case blocking times are

$$B_1^{TI} = 2B_1^{1S} = 4 \tag{15.9}$$

for τ_1, and

$$B_2^{TI} = B_2^{1S} = 0 \tag{15.10}$$

for τ_2. This is of course not a formal proof, but it can be seen that B_1^{TI} is indeed a correct upper bound of the actual amount of blocking seen in the example.

An additional advantage of this approach is that it is very simple and requires very little knowledge about task self-suspension itself. It is enough to know *how many* self-suspensions each task contains, information quite easy to collect. However, the disadvantage of using such a limited amount of information is that it makes the method extremely conservative. Thus, B_i^{TI} is not a tight upper bound for the worst-case blocking time and may widely overestimate it in some cases.

More sophisticated and precise methods do exist, such as that described in Reference [58]. However, as we have seen in several other cases, the price to be paid for a tighter upper bound for the worst-case blocking is that much more information is needed. For instance, in the case of [58], we need to know not only how many self-suspensions each task has got, but also their exact location within the task. In other words, we need to know the execution time of each, individual task segment, instead of the task execution time as a whole.

15.3   EXTENSION OF THE RESPONSE TIME ANALYSIS METHOD

In the basic formulation of RTA given in Chapter 13, the worst-case response time R_i of task τ_i was expressed according to (13.1) as

$$R_i = C_i + I_i \tag{15.11}$$

where C_i is its execution time, and I_i is the worst-case interference the task may experience due to the presence of higher-priority tasks.

Then, in Chapter 14, we argued that the same method can be extended to also handle task interactions—as well as the blocking that comes from them—by considering an additional contribution to the worst-case response time as written in (14.1):

$$R_i = C_i + B_i + I_i \tag{15.12}$$

where B_i is the worst-case blocking time of τ_i. It can be proved that (15.12) still holds even when tasks are allowed to self-suspend, if we redefine B_i to include the additional sources of blocking discussed in Sections 15.1 (extra interference) and 15.2 (additional blocking after each self-suspension).

Namely, B_i must now be expressed as:

$$B_i = B_i^{SS} + B_i^{TI} \tag{15.13}$$

In summary, referring to (15.1) and (15.6) for the definition of B_i^{SS} and B_i^{TI}, respectively, and then to (15.4) and (15.5) for the definition of B_i^{IS}, we can write

$$B_i = S_i + \sum_{j \in hp(i)} \min(C_j, S_j) + (Q_i + 1) \sum_{k=1}^{K} \text{usage}(k, i) C(k) \tag{15.14}$$

for priority inheritance, as well as

$$B_i = S_i + \sum_{j \in hp(i)} \min(C_j, S_j) + (Q_i + 1) \max_{k=1}^{K} \{\text{usage}(k, i) C(k)\} \tag{15.15}$$

for priority ceiling.

The recurrence relationship used by RTA and its convergence criteria are still the same as (14.2) even with this extended definition of B_i, that is,

$$w_i^{(k+1)} = C_i + B_i + \sum_{j \in hp(i)} \left\lceil \frac{w_i^{(k)}}{T_j} \right\rceil C_j \tag{15.16}$$

A suitable seed for the recurrence relationship is

$$w_i^0 = C_i + B_i \tag{15.17}$$

Let us now apply the RTA method to the examples presented in Sections 15.1 and 15.2. Table 15.1 summarizes the attributes of the tasks shown in Figure 15.1 as

Table 15.1

Attributes of the tasks shown in Figure 15.1 when τ_1 is allowed to self-suspend

Task	C_i	T_i	S_i	B_i^{SS}	B_i^{TI}
τ_1 (high priority)	4	7	2	2	0
τ_2 (low priority)	6	15	0	2	0

calculated so far. The blocking time due to task interaction, B_i^{TI}, is obviously zero for both tasks in this case, because they do not interact in any way.

For what concerns the total blocking time B_i, from (15.13) we simply have:

$$B_1 \quad = \quad 2+0=2 \tag{15.18}$$

$$B_2 \quad = \quad 2+0=2 \tag{15.19}$$

For τ_1, the high-priority task, we simply have

$$w_1^0 \quad = \quad C_1+B_1=4+2=6 \tag{15.20}$$

and the succession converges immediately. Therefore, it is $R_1 = 6$ time units. On the other hand, for the low-priority task τ_2, it is

$$w_2^0 \quad = \quad C_2+B_2=6+2=8 \tag{15.21}$$

$$w_2^1 \quad = \quad 8+\left\lceil\frac{w_2^{(0)}}{T_1}\right\rceil C_1 = 8+\left\lceil\frac{8}{7}\right\rceil 4 = 16 \tag{15.22}$$

$$w_2^2 \quad = \quad 8+\left\lceil\frac{16}{7}\right\rceil 4 = 20 \tag{15.23}$$

$$w_2^3 \quad = \quad 8+\left\lceil\frac{20}{7}\right\rceil 4 = 20 \tag{15.24}$$

Therefore, it is $R_2 = 20$ time units.

Going to the second example, Table 15.2 summarizes the attributes of the tasks shown in Figure 15.2. The worst-case blocking times due to task interaction B_i^{TI} have already been derived in Section 15.2, while the worst-case blocking times due to self-suspension, B_i^{SS}, can be calculated using (15.1) as:

$$B_1^{SS} \quad = \quad S_1=3.5 \tag{15.25}$$

$$B_2^{SS} \quad = \quad S_2+\min(C_1,S_1)=0+\min(10,3.5)=3.5 \tag{15.26}$$

The total blocking time to be considered for RTA is therefore:

$$B_1 \quad = \quad 3.5+4=7.5 \tag{15.27}$$

$$B_2 \quad = \quad 3.5+0=3.5 \tag{15.28}$$

Table 15.2

Attributes of the tasks shown in Figure 15.2 when τ_1 is allowed to self-suspend

Task	C_i	T_i	S_i	B_i^{SS}	B_i^{TI}
τ_1 (high priority)	10	30	3.5	3.5	4
τ_2 (low priority)	10	30	0	3.5	0

As before, the high-priority task τ_1 does not endure any interference, and hence, we simply have:

$$w_1^0 \quad = \quad C_1 + B_1 = 10 + 7.5 = 17.5 \tag{15.29}$$

and the succession converges immediately, giving $R_1 = 17.5$ time units. For the low-priority task τ_2, we have instead:

$$w_2^0 \quad = \quad C_2 + B_2 = 10 + 3.5 = 13.5 \tag{15.30}$$

$$w_2^1 \quad = \quad 13.5 + \left\lceil \frac{w_2^{(0)}}{T_1} \right\rceil C_1 = 13.5 + \left\lceil \frac{13.5}{30} \right\rceil 10 = 23.5 \tag{15.31}$$

$$w_2^2 \quad = \quad 13.5 + \left\lceil \frac{23.5}{30} \right\rceil 10 = 23.5 \tag{15.32}$$

The worst-case response time of τ_2 is therefore $R_2 = 23.5$ time units.

15.4 SUMMARY

This chapter complements the previous one and further extends the schedulability analysis methods at our disposal to also consider task *self-suspension*. This is a rather common event in a real-world system because it occurs whenever a task voluntarily suspends itself for a certain variable, but upper-bounded, amount of time. A typical example would be a task waiting for an I/O operation to complete.

Surprisingly, we saw that the self-suspension of a task has not only a local effect on the response time of that task—this is quite obvious—but it may also give rise to an *extra interference* affecting lower-priority tasks, and further *increase the blocking times* due to task interaction well beyond what is predicted by the analysis techniques discussed in Chapter 14.

The main goal of the chapter was therefore to look at all these effects, calculate their worst-case contribution to the blocking time of each task, and further extend the RTA method to take them into account. This last extension concludes our journey to make our process model, as well as the analysis techniques associated with it, closer and closer to how tasks really behave in an actual real-time application.

15.5 EXERCISES

EXERCISE 1

Which of the following task actions count as self-suspension?

1. The calculation of a square root on a processor not equipped with a floating-point unit.
2. Reading some data from a disk file.
3. Trying to acquire a mutual exclusion semaphore.
4. Performing a blocking P() on a synchronization semaphore.

EXERCISE 2

A task T waits on a message queue for messages coming from two other tasks V_1 and V_2. Whenever a message arrives, from any source, T processes it and waits again, in an infinite loop. Tasks V_1 and V_2 are both periodic. They send a message to T every $t_1 = 50\,\text{ms}$ and $t_2 = 40\,\text{ms}$, respectively. What is the worst-case self-suspension time of T? Might V_1 and V_2 also self-suspend? When?

Section III

Case Studies

16 General-Purpose IoT/Embedded Controller

In this case study we use a typical embedded controller board to build a *gateway* between an Ethernet-based UDP/IP network, briefly described in Chapter 9, and a Controller Area Network (CAN) [53], a network technology mainly used in the automotive and industrial automation domains. Ethernet and CAN interfaces cannot be directly connected because they make use of completely different physical layer technology and communication protocols.

Generally speaking, a gateway is a component whose main purpose is to solve this exact problem. Namely, a gateway has two or more network interfaces, each connected to a different network, and enables nodes on these networks to communicate, although they lack a direct connection, by forwarding traffic among its interfaces. The internal complexity of a gateway depends on how similar (or how different) the networks it interconnects are and on the sophistication of the forwarding rules it puts into effect.

In this particular case, the gateway encapsulates the CAN frames it receives into UDP datagrams to broadcast them on the Ethernet side and vice versa. In this way, for instance, an Ethernet segment and two gateways can work as a backbone between two CAN buses. In addition, the gateway enables systems that are typically not equipped with a CAN interface but do support Ethernet connectivity, like a personal computer, to send and receive messages on a CAN bus.

To keep the complexity of the case study manageable, the gateway does not perform any filtering on CAN frames and just forwards them without any modifications. For the same reason, it does not aggregate multiple CAN frames into a single UDP diagram, although this would improve throughput. Moreover, when two gateways are coupled together through Ethernet, they do not merge the two CAN segments they interconnect into the same arbitration domain. As a consequence, readers are advised that traditional CAN schedulability analysis [23] is not going to be applicable to such an arrangement. The gateway is also equipped with an optional graphical user interface, to give readers an informal idea of how this kind of interface can be realized on embedded systems with limited processing power and even more limited graphics capabilities.

Overall, this exercise gives us the opportunity to discuss how this kind of system can be designed and implemented. Moreover, we will describe how to avoid one of the most common design pitfalls, that is, the deadlock discussed in Chapter 4.

DOI: 10.1201/9781003593416-16

16.1 HARDWARE AND FIRMWARE OVERVIEW

The evaluation and development board selected for the case study presented in this chapter is the Olimex ESP32-EVB open-source board [72], but other equivalent boards are likely to be adequate as well. As shown in Figure 16.1, the board revolves around the ESP32-WROOM-32E module produced by Espressif [31] and depicted on the right side of the figure.

The main components of the module are a dual-core 32-bit Xtensa LX6 processor [20], along with volatile and non-volatile memory, as well as a variety of peripheral components. The module also supports wireless networking, both Wi-Fi and Bluetooth. These components are shown in the figure for the sake of completeness, but they will not be discussed in detail because they are irrelevant to the case study.

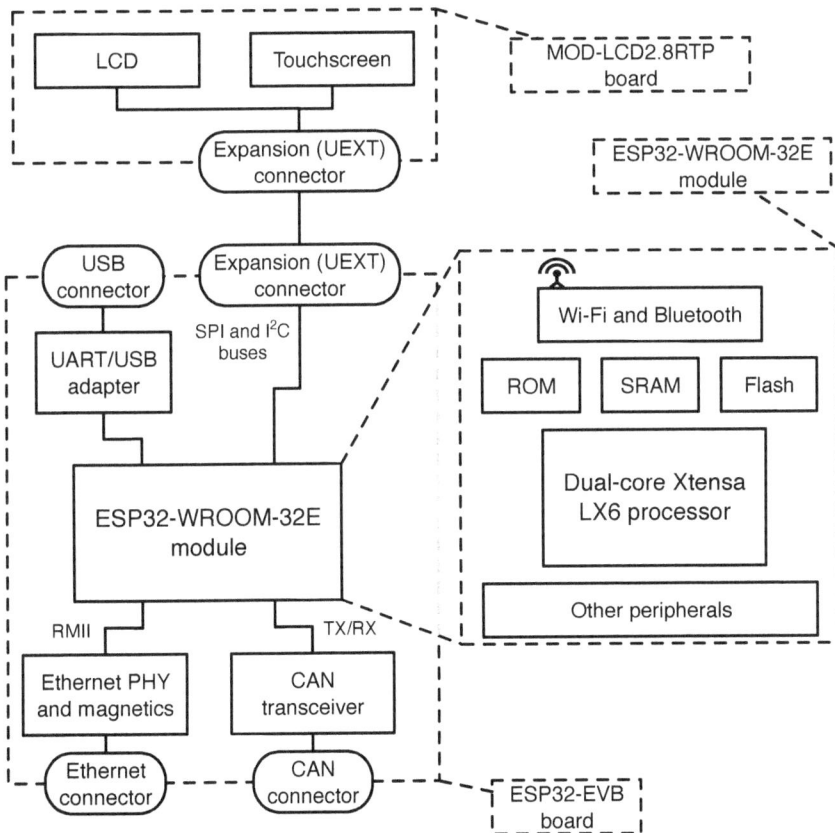

Figure 16.1 Simplified block diagram of the ESP32-based board used in the case study discussed in this chapter.

Besides the processor and wireless networking components already mentioned, the other main components of the ESP32-WROOM-32E module of interest for this case study are:

1. 448 KB of Read-Only Memory (ROM), mainly used to store the first-stage bootloader and other low-level code and data. This memory is not user-programmable and it is mostly hidden from the programmer's view.
2. 520 KB of Static Random-Access Memory (SRAM), which contains non-constant application data. Part of the SRAM, 64 KB in size, can also be used to cache external memory accesses, like the ones involving the Flash memory to be described next.
3. From 4 MB to 16 MB of Flash memory, connected to the main microcontroller chip by means of a dedicated, local Serial Peripheral Interface (SPI) bus distinct from the SPI bus that is made available outside the module. Being SPI a serial bus, the processor cannot read from and write into Flash memory directly. Therefore, specialized hardware and system software mediate Flash memory access, dynamically loading a subset of its contents into the portion of SRAM used as a cache. A simple Memory Management Unit (MMU) maps virtual addresses that belong to the Flash memory address space to physical SRAM addresses as appropriate.
4. A variety of peripherals that connect the module to the outside world. Among them, we mention only the Universal Asynchronous Receiver-Transmitter (UART), the SPI controller, the Inter-Integrated Circuit (I^2C) bus controller, the Ethernet controller, and the CAN controller because they are relevant to the case study at hand.

Zooming out to examine the ESP32-EVB board as a whole, we find the following additional components:

1. An adapter that connects the ESP32-WROOM-32E UART to the on-board Universal Serial Bus (USB) port. Besides enabling the board to get power from the USB host it is connected to, this port is also used to upload firmware into Flash memory, collect log messages emitted by the firmware while it is running, and perform other board control and management functions, for instance, reset it when needed. All these functions are performed by the idf.py tool, which is part of the ESP32 software development toolchain [30].
2. An expansion (UEXT) connector, which allows external access to the SPI and I^2C buses mentioned previously. A typical use of this connector is to link the ESP32-EVB board to a SPI Liquid Crystal Display (LCD) controller and an I^2C touchscreen controller mounted on an auxiliary board. In this way, the firmware can display a Graphical User Interface (GUI) on various LCD boards such as, for instance, the Olimex MOD-LCD2.8RTP board [73] adopted for this case study.
3. A LAN8710A Ethernet Physical Layer (PHY) chip, as well as the related magnetics and connector. The PHY chip is connected to the Ethernet

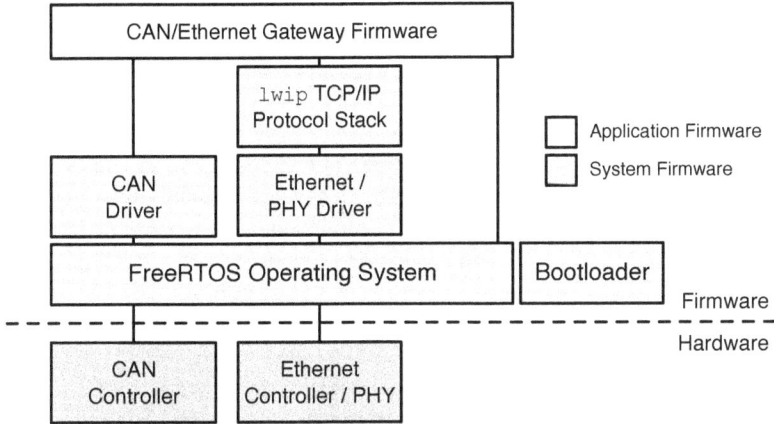

Figure 16.2 Main system and application firmware components of this case study and their relationship with hardware devices.

controller embedded in the ESP32 chip by means of a Reduced Media-Independent Interface (RMII) bus. All together, these components form a fully functional Ethernet interface.

4. A SIT1050T CAN transceiver, which performs a similar function for the ESP32 CAN controller. Namely, besides protecting the ESP32 from voltage spikes on the CAN bus, it transforms the transmit/receive signals from the CAN controller into a differential signal suitable for the bus and vice versa.

The most important firmware components used in this case study are summarized in Figure 16.2, except for the GUI-related components, which will be not be discussed in the book although the firmware does have a GUI. Interested readers can refer to the comments embedded in the source code for more information.

The same figure also shows how these components relate to the Ethernet and CAN controllers. These are the main hardware devices the case study makes use of, besides the aforementioned LCD and touchscreen controllers. In the figure, blocks are colored in different manners depending on their nature: Hardware components are gray, system firmware components are light gray and, finally, user-written application firmware components are white. In particular, we can see:

1. The bootloader, which takes control of the processor when it is powered up or reset. It is responsible for initializing the system and eventually starting the application firmware. The bootloader is composed of two parts, or *stages*, of which the first is stored in ROM and the second in Flash memory. The first-stage bootloader takes care of low-level chip initialization and sets up the SPI bus that connects the microcontroller chip to Flash memory, so that the processor can execute code and read data from it. Then, it transfers control to the second-stage bootloader. In turn, the second-stage bootloader

validates and starts the application firmware, which is also stored in Flash memory. While the board is being programmed, the bootloader is also responsible for receiving the new application image from the UART and storing it into Flash memory.

2. The FreeRTOS real-time operating system, whose API and main features have been presented in Chapter 8. Its main purpose is to coordinate the execution of system and user tasks and let them synchronize and exchange data by means of a variety of interprocess communication primitives. Although standard FreeRTOS only supports single-core processors, the Espressif-provided version has been extended to handle dual-core LX6 CPUs.

3. A CAN driver that interfaces with the ESP32 on-chip hardware CAN controller and provides a hardware-independent API to use it. This driver is provided by the chip manufacturer and its API is described in [30].

4. An Ethernet driver that interfaces with the ESP32 on-chip Ethernet controller and a PHY driver that communicates with the PHY chip through a Media Independent Interface Management (MIIM) bus, also called MDIO or SMI, driven by the Ethernet controller. The MIIM bus is a *control* path, mainly used to configure the PHY, and is distinct from the RMII *data* path discussed previously. The Ethernet driver always stays the same, because the Ethernet controller is an on-chip component. The PHY driver, instead, depends on the PHY actually mounted on the specific board model in use and may change depending on it. In this particular case, both drivers are provided directly by Espressif. A third-party PHY driver will be needed, instead, if the PHY installed on the board is not among the few that Espressif supports.

5. The lwIP [29] protocol stack. In this case study it is responsible for UDP communication and makes use of the Ethernet driver just discussed to send and receive Ethernet frames. Although lwIP has its own, native API for network communication, it also provides the more ordinary socket-based interface that has been discussed in Chapter 9. Using the sockets API also brings the additional advantage of making the application code more portable to a variety of other operating systems and protocol stacks.

Finally, the application firmware that implements the high-level logic of the gateway sits at the top of Figure 16.2. It makes use of the other components discussed so far to perform its duties, as is typical of software systems designed in a modular way. Figure 16.3 shows a slightly simplified diagram of the application firmware as a whole and highlights its relationship with the system firmware just described. The arrows in the figure show how network packets travel from one port of the gateway to the other as they are transferred between firmware components. White blocks represent application firmware components while light-gray blocks correspond to system firmware components. Objects that are part of the per-port shared context are depicted as dark gray blocks.

The gateway architecture is completely symmetrical around the vertical, dashed line of Figure 16.3, except for the presence of an additional component—the *Canudp*

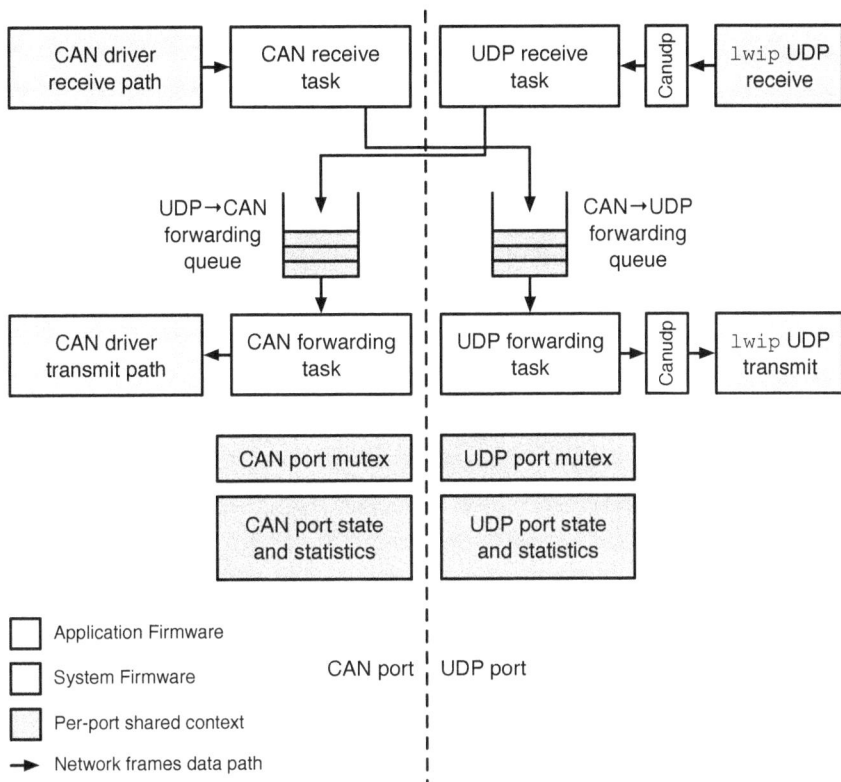

Figure 16.3 Gateway application firmware architecture (tasks, synchronization objects, and shared data structures) in this case study.

library—in the UDP port, due to its higher complexity with respect to the CAN port. For this reason, in the description that follows we focus only on the CAN port, while Section 16.2 describes the Canudp library in more detail. For each gateway port there are two tasks, called the *receive* and the *forwarding* task. The receive task executes the loop shown in Figure 16.4 until it is asked to terminate. Within the loop, it calls the CAN device driver, which is part of the system firmware, and waits until a CAN frame arrives, an error occurs, or a timeout expires. The next action it performs depends on the outcome of the call:

- If the receive task was able to receive a CAN frame successfully, it pushes the frame into the CAN→UDP forwarding queue, which belongs to the UDP port. This is done by the rx_handle_alerts() function shown in Figure 16.5 when it calls udp_port_enqueue().
- If an error occurred, the receive task determines whether or not it can recover from it. If the error is unrecoverable, the task informs the other port components and terminates. Otherwise, it logs the error and continues.

```
static void rx_task(
    void *arg                      /**< Raw task argument from FreeRTOS */
    )
{
    struct Can_port_ctx *ctx = (struct Can_port_ctx *)arg;
    enum Can_port_st st = CAN_PORT_ST_OK;

    ESP_LOGI(TAG, "Receive task started");

    if(twai_reconfigure_alerts(ALERTS_ALL, NULL) != ESP_OK
       || twai_start() != ESP_OK)
    {
        ESP_LOGW(TAG, "could not configure alerts or start the controller");
        st = CAN_PORT_ST_ESP;
    }

    else
    {
        while(st == CAN_PORT_ST_OK)
        {
            uint32_t alerts;
            esp_err_t esp_st =
                twai_read_alerts(&alerts, pdMS_TO_TICKS(CAN_HEARTBEAT_MS));

            switch(esp_st)
            {
            case ESP_OK:
                st = rx_handle_alerts(ctx, alerts);
                break;

            case ESP_ERR_TIMEOUT:
                /* Only check for termination */
                st = CAN_PORT_ST_TIMEOUT;
                break;

            default:
                st = CAN_PORT_ST_ESP;
                break;
            }

            if(st != CAN_PORT_ST_FREERTOS)
            {
                /* Update statistics and check for termination */
                ...
            }

            /* Local error recovery */
            if(is_recoverable(st))  st = CAN_PORT_ST_OK;
        }

        if(twai_stop() != ESP_OK)  st = CAN_PORT_ST_ESP;
    }

    report_task_term(ctx, st);
    vTaskDelete(NULL);
}
```

Figure 16.4   Gateway, main loop of the CAN receive task.

- The receive task does not perform any specific action when a timeout occurs, but it still checks if the port is shutting down and terminates if asked to, as described next.

```
static enum Can_port_st rx_handle_alerts(
    struct Can_port_ctx *ctx,    /**< CAN port context */
    uint32_t alerts              /**< Alerts from the CAN controller */
    )
{
    enum Can_port_st st = CAN_PORT_ST_OK;

    /* Log CAN errors but otherwise ignore them */
    if(alerts & ALERTS_ERR)
        ESP_LOGW(TAG, "error alerts: 0x%lx", alerts);

    if(alerts & ALERTS_RX)
    {
        twai_message_t twai_msg;
        esp_err_t esp_st = ESP_OK;
        while(st == CAN_PORT_ST_OK
            && (esp_st = twai_receive(&twai_msg, 0)) == ESP_OK)
        {
            /* Convert twai_message_t into a Canudp_msg */
            struct Canudp_msg msg;
            uint32_t msg_type = 0;

            if(twai_msg.extd)  msg_type |= CANUDP_MSG_TYPE_EXT;
            else  msg_type |= CANUDP_MSG_TYPE_BASE;

            if(twai_msg.rtr)  msg_type |= CANUDP_MSG_TYPE_RTR;
            canudp_msg_set_type(&msg, msg_type);
            canudp_msg_set_id(&msg, twai_msg.identifier);
            canudp_msg_set_dlc(&msg, twai_msg.data_length_code);

            for(int i=0, len=canudp_msg_length(&msg); i<len; i++)
                canudp_msg_set_data(&msg, i, twai_msg.data[i]);

            if(udp_port_enqueue(ctx->udp_ctx, &msg) != UDP_PORT_ST_OK)
            {
                ESP_LOGW(TAG, "forwarding failed");
                st = CAN_PORT_ST_FW;
            }
        }

        /* Report a CAN receive error only if esp_st says so */
        if(esp_st != ESP_OK && esp_st != ESP_ERR_TIMEOUT)
        {
            ESP_LOGE(TAG, "error receiving CAN message");
            st = CAN_PORT_ST_ESP;
        }
    }

    return st;
}
```

Figure 16.5   Gateway, CAN message handling.

Before waiting for the next CAN message to arrive, the receive task checks whether or not it should terminate. To do this, it queries the CAN *port context*, a per-port data structure that holds all the information pertaining to the port and is shared among all tasks and functions that operate on the port. Moreover, each port context contains a mutual exclusion semaphore (called mutex in the figure and in the code) that grants exclusive access to the context itself. In particular, the port context contains a state variable that can assume the following values:

Stopped: This is the initial state of the port and corresponds to the CAN_PORT_ STATE_STOPPED enumerated value in the source code. It indicates that no tasks are currently associated with the port, and hence, no tasks are using it. In this state, the context variable n_tasks, which counts how many tasks are associated with the port, is zero.

Running: This is the state of the port when it is operating normally and corresponds to the CAN_PORT_STATE_RUNNING enumerated value. In this state, n_tasks is 2, because both the receive and the forwarding tasks are active and are using the port.

Stopping: The port may be brought into this state when we wish to stop it normally, for instance, when we would like to shut the gateway down. It corresponds to the CAN_PORT_STATE_STOPPING enumerated value. The n_tasks variable counts how many tasks are still using the port.

Error: The receive task brings the port into this state, immediately before terminating, when it detects an unrecoverable error. It is indicated by the CAN_PORT_STATE_ERROR enumerated value in the code. Also in this case, the n_tasks variable counts how many tasks are still using the port. Unlike *Stopping*, being in the *Error* state implies an abnormal termination.

Given this information, the receive task may check for termination by querying the state variable. If its value is either *Stopping* or *Error*, it should terminate after decrementing n_tasks by one. Similarly, if the receive task determines it must terminate due to an unrecoverable error, it sets state to *Error*, decrements n_tasks by one, and terminates. This action also triggers the termination of the forwarding task, which performs exactly the same check.

From the description above it becomes evident that the simplest way to shut down a port normally is to set its state to *Stopping* and then wait until n_tasks goes down to zero, to ensure that no tasks are using it anymore, before deallocating the port context. In fact, this is exactly what the API that stops and destroys a port does. The corresponding code is shown in Figure 16.6.

More sophisticated approaches that avoid polling n_tasks repeatedly are of course possible. For instance, the port context could be augmented with a condition variable (explained in Chapters 5 and 7) that the port tasks signal upon termination and the port stop/destruction function passively waits upon while stopping the port.

However, taking into account that FreeRTOS does not directly provide condition variables and port shutdown is not performance-critical at all in this case study, the polling-based method has been chosen to avoid adding undue complexity to the discussion. As always, the operations just described, except for the timed wait, must be performed after gaining exclusive access to the port context by means of its mutex, to avoid race conditions. Figures 16.7 and 16.8 shows how the receive task reports its impending termination and checks for termination requests, respectively. The fragment of code shown in Figure 16.8 is embedded in the receive loop listed in Figure 16.4.

The structure of the forwarding task is similar, and in some ways complementary, to that of the receive task. Namely, as shown in Figure 16.9, it extracts a message from the UDP→CAN forwarding queue—the CAN-side counterpart of the

```
enum Can_port_st can_port_stop(
    struct Can_port_ctx *ctx     /**< CAN port context */
    )
{
    enum Can_port_st st = CAN_PORT_ST_TERM;
    int attempt = 0;
    while(st == CAN_PORT_ST_TERM && attempt++ < MAX_STOP_ATTEMPTS)
    {
        if(xSemaphoreTake(ctx->mutex, portMAX_DELAY) != pdTRUE)
            st = CAN_PORT_ST_FREERTOS;
        else
        {
            /* Ask tasks to terminate */
            if(ctx->state == CAN_PORT_STATE_RUNNING)
                ctx->state = CAN_PORT_STATE_STOPPING;

            /* Check whether they did or not */
            if(ctx->n_tasks == 0)
            {
                ctx->state = CAN_PORT_STATE_STOPPED;
                st = CAN_PORT_ST_OK;
            }

            if(xSemaphoreGive(ctx->mutex) != pdTRUE)
                st = CAN_PORT_ST_FREERTOS;

            vTaskDelay(pdMS_TO_TICKS(CAN_HEARTBEAT_MS));
        }
    }

    /* MAX_STOP_ATTEMPTS exhausted without being able to terminate the tasks
       associated with the context.  Report an error.
    */
    if(st == CAN_PORT_ST_TERM)  st = CAN_PORT_ST_STOP;
    ESP_LOGI(TAG, "port_stop status %d after %d attempts",
            st, attempt);
    return st;
}
```

Figure 16.6 Gateway, CAN port stop.

CAN→UDP queue mentioned previously—and schedules it for transmission by calling the appropriate function of the CAN driver. The forwarding task performs these operations in a loop until it is asked to terminate or an unrecoverable error occurs. It carries out the termination check exactly as the receive task does and reacts to errors in the same way.

Besides the mutex, the message queue, and the task-related state variables just described, the CAN port context also contains some counters that hold statistical information about the port. More specifically:

- The receive tasks counts how many messages it received from the CAN driver (rx counter) and how many errors the CAN driver reported (rx_err counter). Moreover, it also keeps track of the number of messages it successfully pushed into the CAN→UDP forwarding queue of the UDP port (fw counter) and how many errors it encountered while doing so (fw_err counter).

```
static void report_task_term(
    struct Can_port_ctx *ctx,    /**< CAN port context */
    enum Can_port_st st          /**< Status code */
    )
{
    if(st != CAN_PORT_ST_FREERTOS)
    {
        if(xSemaphoreTake(ctx->mutex, portMAX_DELAY) != pdTRUE)
            /* Unable to lock the context; no way to report the error */
            ESP_LOGE(TAG, "mutex lock failed");

        else
        {
            if(st != CAN_PORT_ST_TERM)  ctx->state = CAN_PORT_STATE_ERROR;
            ctx->n_tasks--;

            ESP_LOGI(TAG, "task terminated, %d tasks left", ctx->n_tasks);
            if(xSemaphoreGive(ctx->mutex) != pdTRUE)
                ESP_LOGE(TAG, "mutex unlock failed");
        }
    }
}
```

Figure 16.7   Gateway, reporting impending task termination.

```
/* Update statistics and check for termination */
if(xSemaphoreTake(ctx->mutex, portMAX_DELAY) != pdTRUE)
    st = CAN_PORT_ST_FREERTOS;
else
{
    if(st != CAN_PORT_ST_TIMEOUT)
        switch(st)
        {
        case CAN_PORT_ST_OK:
            ctx->stats.rx++;
            break;

        case CAN_PORT_ST_FW:
            ctx->stats.fw_err++;
            break;

        default:
            ctx->stats.rx_err++;
            break;
        }

    if(ctx->state != CAN_PORT_STATE_RUNNING)
        st = CAN_PORT_ST_TERM;

    if(xSemaphoreGive(ctx->mutex) != pdTRUE)
        st = CAN_PORT_ST_FREERTOS;
}
```

Figure 16.8   Gateway, termination check.

- The forwarding task counts how many CAN messages it transmitted successfully (tx counter) and how many transmission errors the CAN driver
  signaled (tx_err counter).

```
static void fw_task(void *arg)
{
    struct Can_port_ctx *ctx = (struct Can_port_ctx *)arg;
    enum Can_port_st st = CAN_PORT_ST_OK;

    ESP_LOGI(TAG, "Forwarding task started");

    while(st == CAN_PORT_ST_OK)
    {
        struct Canudp_msg msg;
        BaseType_t rtos_st =
            xQueueReceive(
                ctx->udp_to_can_fw, &msg, pdMS_TO_TICKS(CAN_HEARTBEAT_MS));

        /* Check for termination */
        if(xSemaphoreTake(ctx->mutex, portMAX_DELAY) != pdTRUE)
            st = CAN_PORT_ST_FREERTOS;
        else
        {
            if(ctx->state != CAN_PORT_STATE_RUNNING)
                st = CAN_PORT_ST_TERM;

            if(xSemaphoreGive(ctx->mutex) != pdTRUE)
                st = CAN_PORT_ST_FREERTOS;
        }

        /* Transmit the message only if we did not receive a termination
           request, we did not encounter other errors, and we successfully
           extracted a message from the forwarding queue.
        */
        if(st == CAN_PORT_ST_OK && rtos_st == pdTRUE)
        {
            uint32_t msg_type = canudp_msg_type(&msg);
            twai_message_t twai_msg;
            twai_msg.extd = ((msg_type & CANUDP_MSG_TYPE_EXT) != 0);
            twai_msg.rtr = ((msg_type & CANUDP_MSG_TYPE_RTR) != 0);
            twai_msg.identifier = canudp_msg_id(&msg);
            twai_msg.data_length_code = canudp_msg_dlc(&msg);

            for(int i=0, len=canudp_msg_length(&msg); i<len; i++)
                twai_msg.data[i] = canudp_msg_data(&msg, i);

            /* No need for mutual exclusion when calling twai_transmit()
               because the CAN port has its own mutex.
            */
            esp_err_t esp_st = twai_transmit(&twai_msg, portMAX_DELAY);

            /* Update statistics */
            ...
        }
    }

    report_task_term(ctx, st);
    vTaskDelete(NULL);
}
```

Figure 16.9   Gateway, CAN forwarding task.

- Finally, the CAN port enqueue for transmission function, which is the function invoked by the UDP port to push a message into the UDP→CAN forwarding queue, counts how many times it failed because the queue was

completely full (`tx_drop` counter). Unlike other kinds of forwarding error, this condition has a specific significance because it corresponds to a gateway overload. Its value is equal to the `fw_err` counter of the UDP port, except when other kinds of error occur.

The interposition of forwarding queues between the two ports, together with the additional receive and transmit queues that device drivers may also implement, absorbs and mitigates the effect of sporadic traffic bursts. A more serious, and possibly permanent, overload condition may still arise if one of the gateway ports is unable to transmit for a significant amount of time. For instance, on the CAN side this may occur when higher-priority traffic generated by other nodes saturates the bus and prevents the gateway from transmitting. Last, but not least, given the severe imbalance between the CAN bit rate (at most 1 Mb/s) and the Ethernet bit rate (at least 10 Mb/s and likely 100 Mb/s) an overload may also occur when the messages that the gateway receives from the UDP port systematically exceed what the CAN bus is able to accommodate.

Given that most message queue implementations, including the one provided by FreeRTOS, do provide a *blocking* operation that waits as long as necessary before inserting a message into the queue if the queue is full, making use of this capability would seem appealing at first sight to address or, at least, mitigate the overload issue, instead of just dropping messages altogether. However, the gateway must steer away from it to avoid deadlock, as described in Section 16.3.

Going back to the error counters, it is expected that all of them remain at zero under normal circumstances and the number of messages received by the CAN port is equal to the number of messages forwarded to the UDP port. In turn, this number should also be equal to the number of messages transmitted by the UDP port itself. The number of messages transmitted by the CAN port depends, instead, on the incoming traffic from the UDP side of the gateway. It is therefore unrelated to the number of messages handled by the CAN-side receive path.

It is worth remarking that, although the CAN port can recover from most errors, it is still crucial to detect and count them all, because they lead to the loss of messages while they are traveling through the gateway. As a consequence, these messages will never be forwarded and will never appear on the other gateway port.

At the same time, the presence of multiple distinct error counters along the path that each message follows while being forwarded helps identify the underlying reason for message drops. Considering a CAN message being forwarded to the UDP port, a CAN-side *receive error* indicates an issue with the CAN bus, the CAN controller, or its device driver. A *forwarding error*, instead, is most likely due to an overload or other issues within the UDP port. The *transmission error* counter helps telling apart errors internal to the gateway, counted by the forwarding error counter just mentioned, and UDP connectivity errors.

The two gateway ports provide virtually the same API, to the point that, when using an object-oriented language, seeing and treating them as specializations of an abstract class "gateway port" would be rather convenient. For this reason, as we did previously, we will discuss only the CAN-port API in detail here. The CAN port exports the following functions:

- Port initialization function:

```
enum Can_port_st can_port_init(struct Can_port_ctx *ctx);
```

This function initializes the context of a port, that is, a data structure that contains all the information the port requires to work. Its only argument ctx points to this data structure, which must have been pre-allocated by the caller. Besides preparing the CAN controller and device driver for use, it creates the mutual exclusion semaphore associated with the port and its forwarding queue. Moreover, it clears port statistics. However, for reasons that will become clearer later, the port initialization function does *not* start the receive and forwarding tasks yet and leaves the port in the **Stopped** state. All functions operating on a port context return an enumerated value that informs the caller about the outcome of the operation they were requested to perform. Any value other than CAN_PORT_ST_OK indicates an error. After a successful initialization, all the other functions that take a pointer to a port context as argument can safely be invoked on ctx.

- Port start function:

```
enum Can_port_st can_port_start(
    struct Can_port_ctx *ctx, struct Udp_port_ctx *udp_ctx);
```

This function binds the CAN port context ctx with the UDP port context udp_ctx and starts the tasks associated with ctx. After a successful start, the CAN port tasks may use the UDP port API on udp_ctx at any time. It is therefore crucial that the context referenced by udp_can has already been initialized before making this call. This is the main reason why the port initialization and start functions have been kept separate. Merging them into one single function would have made it impossible to ensure that a port were properly initialized before making its context known to the other port, or vice versa. As a consequence, there would have been an unsafe time window in which one port could have invoked the API of the other port on an uninitialized context. The port start function returns an error indication when asked to start a port that is already running or when it detects an unrecoverable error.

- Port stop function:

```
enum Can_port_st can_port_stop(struct Can_port_ctx *ctx);
```

This function stops the port corresponding to port context ctx, that is, it asks the tasks associated with it to terminate. To guarantee that no more API calls referring to the UDP port context passed to the CAN port start function (through its udp_ctx argument) are made after this function returns, it waits until it confirms that no tasks associated with ctx remain alive. Thus, the caller may safely destroy that UDP port context after this function returns, provided the UDP port has been stopped, too. The caller may also destroy the port context referenced by ctx if it is able to ensure that the other port is no longer using it. This can be accomplished by stopping both ports before destroying either of them and is yet another reason for keeping the port initialization/destruction functions completely separated from, and independent of, the port start/stop functions. An added benefit is that, since neither the start nor the stop function resets port statistics, a port can be repeatedly started and stopped without losing them. The port stop function has no effect and returns immediately if the port is already stopped.

- Port destruction function:

```
enum Can_port_st can_port_destroy(struct Can_port_ctx *ctx);
```

This function stops the port as `can_port_stop()` does and then destroys the given port context, freeing all resources allocated to it except for the port context data structure itself. After this function returns successfully, the caller may safely free the data structure as well.

- Port enqueue for transmission function:

```
enum Can_port_st can_port_enqueue(
    struct Can_port_ctx *ctx, const struct Canudp_msg *msg);
```

This function inserts the message referenced by msg into the UDP→CAN forwarding queue associated with the CAN port context ctx. The struct Canudp_msg data type is defined by the Canudp library (see Section 16.2) and provides a uniform way to refer to a CAN message, regardless of whether it comes from the CAN or the UDP port.

- Port state and statistics:

```
enum Can_port_st can_port_stats(
    struct Can_port_ctx *ctx,
    enum Can_port_state *state, struct Port_stats *stats);
```

This function fills the enumerated value referenced by state and the data structure referenced by stats with the state and statistics of port ctx, respectively. The port state can be CAN_PORT_STATE_STOPPED, CAN_PORT_STATE_RUNNING, CAN_PORT_STATE_STOPPING, or CAN_PORT_STATE_ERROR, corresponding to the abstract port states discussed previously. The struct Port_stats data type is common to both CAN and UDP ports because both kinds of port collect the same statistical information. Its members correspond to the counters also described previously. This function can be called periodically by a GUI or other supervisory tasks to monitor the state of the port and detect whether it stopped working completely due an unrecoverable internal error. Comparing port statistics retrieved at two different points in time reveals the amount of traffic the port handled in between, as well as the number and nature of any recoverable errors it encountered.

The corresponding UDP port functions are:

- ```
  enum Udp_port_st udp_port_init(
      struct Udp_port_ctx *ctx,
      const struct Canudp_config *config);
  ```

- ```
  enum Udp_port_st udp_port_start(
      struct Udp_port_ctx *ctx, struct Can_port_ctx *can_ctx);
  ```

- ```
  enum Udp_port_st udp_port_stop(struct Udp_port_ctx *ctx);
  ```

- ```
  enum Udp_port_st udp_port_enqueue(
      struct Udp_port_ctx *ctx, const struct Canudp_msg *msg);
  ```

- ```
  enum Udp_port_st udp_port_destroy(struct Udp_port_ctx *ctx);
  ```

- ```
  enum Udp_port_st udp_port_stats(
      struct Udp_port_ctx *ctx,
      enum Udp_port_state *state, struct Port_stats *stats);
  ```

As it can be seen, the only difference is that the UDP port initialization function `udp_port_init()` takes the additional argument `config`. It points to a data structure that contains the Canudp library configuration information, to be discussed in Section 16.2.

16.2   THE CANUDP LIBRARY

The Canudp library provides a portable way of broadcasting UDP datagrams that encapsulate a CAN frame on a local network interface. Symmetrically, it also enables library users to wait for the arrival of one of these messages and extract the CAN frame it contains. To guarantee proper communication among heterogeneous interconnected hosts, each with a possibly different word size and endianness, the UDP datagram payload contains data items that have a host-independent size and use the standard network byte order. The only underlying assumptions are that the library is compiled with the GNU C Compiler `gcc` on all hosts and these hosts all support 8-bit bytes as well as 32-bit unsigned integers.

The library also provides setters and getters that operate on the various fields of the data structure representing a CAN frame, the `struct Canudp_msg`, and convert between the native host representation and the representation used in the UDP payload as needed. Both the network interface name and the UDP port to be used can be configured at will. Moreover, the library supports simultaneous communication via multiple network interfaces and UDP ports, although the gateway uses only one at the moment. Discussing the design and making available the source code of the Canudp library has a twofold goal:

1. Show how to broadcast UDP datagrams containing information encoded in a host-independent way and let library users have access to this functionality by means of a simple API.
2. Illustrate how a careful design, especially regarding the subset of the system API the library is going to use, improves portability to the point that the same code can be compiled on systems that are very far from each other in terms of features and capabilities.

In this particular case, the Canudp library works equally well on Linux and its in-kernel TCP/IP protocol stack, and on ESP32-based systems that use FreeRTOS as operating system and lwIP as protocol stack. Since both systems provide the *Sockets API* described in Chapter 9, the only non-portable part of the library consists of the function that retrieves the IP address information of a certain network interface identified by name. This is because the functions to be used for this are not part of the aforementioned Sockets API, and hence, they vary from one system to another. Overall, the system-specific part of the Canudp library consists of 94 lines of code for the ESP32 and 166 lines of code for Linux, about 10% of the total. This percentage will likely decrease should the library grow bigger with new high-level functions.

The main data type defined by the Canudp library and used by both gateway ports is the `struct Canudp_msg`, a host-independent representation of a CAN frame:

```
struct __attribute__((packed)) Canudp_msg {
    uint32_t type;
    uint32_t id;
    uint8_t dlc;
    uint8_t data[CANUDP_MAX_DATA];
};
```

Its members represent the type of CAN frame (`type`), its identifier (`id`), its Data Length Code or DLC (`dlc`), and its payload (`data`). The use of exact-width integer data types, like `uint32_t`, ensures that they all have the same width and representation on all hosts that support them. This does not guarantee any specific endianness of multi-byte integers, but this is taken care of by getters and setters, as explained previously, with the help of the `htonl()` and `ntohl()` standard library functions. They convert a 32-bit integer from host to network byte order and vice versa.

Generally speaking, getters and setters are very straightforward, short functions. They have been declared `inline` and put directly in the library header for better performance. As an example, here are the ones for the `id` member:

```
static inline uint32_t canudp_msg_id(
    const struct Canudp_msg *msg)
{
    return ntohl(msg->id);
}

static inline void canudp_msg_set_id(
    struct Canudp_msg *msg, uint32_t id)
{
    msg->id = htonl(id);
}
```

The GCC-specific `packed` attribute in the data type definition prevents the compiler from inserting a machine-dependent amount of padding space between members. This would improve code performance, but most likely prevent meaningful network communication with other hosts. More sophisticated and even more portable ways to achieve network interoperability do exist, for instance, Google protocol buffers [35], but they were deemed too complex to be included in the case study.

The main functions of the Canudp library are:

- Initialization of a new library context:

```
enum Canudp_st canudp_open(
    struct Canudp_ctx *ctx,
    const struct Canudp_config *config,
    const struct timeval *receive_timeout);
```

This function initializes the context data structure referenced by `ctx` and allocated by the caller, filling it with all the information the library needs to send and receive UDP datagrams on a certain network interface and UDP port, as specified by the configuration referenced by `config`. After this function returns successfully, `ctx` may be passed as an argument to the other library functions. The return value of this function, as for all the other library functions, reflects the outcome of the call. Any value other than `CANUDP_ST_OK` indicates an error.

Internally, the `canudp_open()` function calls the non-portable function `canudp_fill_addr_config()` to retrieve the unicast and network-local broadcast address of the interface mentioned by `config`. Based on this information, the function opens two sockets, which will be used to transmit and receive UDP datagrams, respectively.

The inbound socket is permanently bound, by means of the `bind()` function, to the network-local broadcast address. To prevent the datagram receive function `canudp_receive()` from waiting indefinitely for the arrival of a message, this function also sets the receive timeout of the inbound socket to the value specified by the `receive_timeout` argument.

Symmetrically, the outbound socket is permanently connected to the network-local broadcast address. The local socket address, the so-called *ephemeral* address chosen by the system, is retrieved and stored into the context referenced by `ctx`. In this way, the library can identify datagrams sent by the same context and filter them out while receiving, thus preventing a context from receiving back its own messages.

- Closure of a library context:

```
enum Canudp_st canudp_close(
    struct Canudp_ctx *ctx);
```

This function frees the system resources assigned to the context referenced by `ctx`. After this function returns successfully, the caller may free the context data structure. Internally, the function simply closes the sockets associated with `ctx` by means of the `close()` function.

- Datagram transmission:

```
enum Canudp_st canudp_transmit(
    struct Canudp_ctx *ctx,
    const struct Canudp_msg *msg);
```

This function asks the underlying protocol stack to transmit the CAN frame held by the data structure referenced by msg, using context ctx. It is basically a wrapper around the socket send() function, invoked on the outbound socket associated with the context.

- Datagram reception:

```
enum Canudp_st canudp_receive(
    struct Canudp_ctx *ctx,
    struct Canudp_msg *msg);
```

This function waits for the arrival of a datagram by calling the recvfrom() function on the inbound socket associated with ctx and discarding datagrams sent by ctx itself. As explained previously, the distinction between datagrams sent by ctx from datagrams sent by others is based on comparing the source address of the incoming datagram with the ephemeral address assigned by the system to the outbound socket of ctx. If no datagrams arrive from ctx within the maximum amount of time specified by the receive_timeout argument provided when the context was opened, this function returns CANUDP_ST_TIMEOUT, a timeout indication.

The only non-portable function, which has the same signature but different implementations on Linux and the ESP32 is:

```
enum Canudp_st canudp_fill_addr_config(
    const struct Canudp_config *config,
    struct Canudp_addr_config *addr_config);
```

It retrieves the unicast and the network-local address assigned to the interface named by the configuration referenced by config and stores them into the data structure indicated by addr_config. The implementation depends on the underlying operating system and protocol stack:

- On Linux, it is based on the SIOCGIFADDR, SIOCGIFNETMASK, and SIOCGIFBRDADDR I/O control operations, invoked by means of the ioctl() function.
- On the ESP32, the same functionality is provided by the esp_netif_get_ip_info() function.

16.3   DEADLOCK ANALYSIS

Chapter 4 discussed *deadlock*, a pathological condition that may affect any system where *processes*, or *tasks*, may acquire exclusive use of some generic *resources* and then wait for other resources—possibly held by other tasks—to become available. This is a very broad definition and it should come to no surprise that, even in all

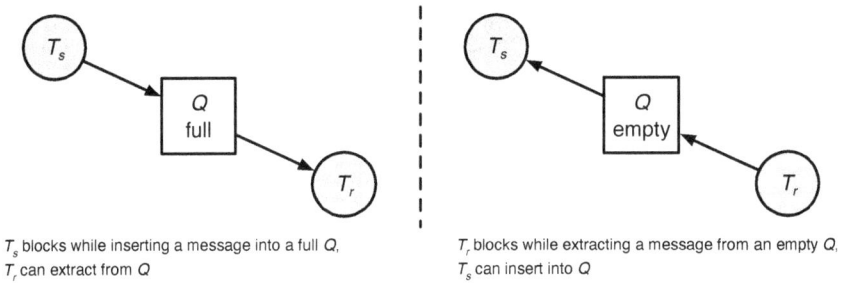

T_s blocks while inserting a message into a full Q, T_r can extract from Q

T_r blocks while extracting a message from an empty Q, T_s can insert into Q

Figure 16.10 Resource allocation graph fragments for an empty or full message queue.

its simplicity, the gateway being designed here still falls into it. As a consequence, a careful deadlock analysis is necessary, unless we are ready to accept that the gateway might inexplicably lock up in certain circumstances.

To reason about deadlock by means of the resource allocation graph defined in Section 4.3 it is important to properly identify what the tasks and resources of the gateway are. First of all, there are four tasks, namely, the two pairs of receive and forwarding tasks presented in Section 16.1 and shown in Figure 16.3. In other words, each port has two tasks, a receive and a forwarding task, associated with it. These two tasks need to gain exclusive access to the context of their own port while they operate, for instance, to update port statistics.

Moreover, each port has an enqueue for transmission function, which also needs exclusive access to the port context of its own port for the same reason. Since this function is called by the receive task of the opposite port, it indirectly grants this same exclusive access to that task as well.

Then, two exclusive-use resources that are clearly part of the gateway are the two port contexts protected by their mutual exclusion semaphores, one for each port. However, there are more. They are related to the way forwarding queues and, more generally, message queues, interact with the tasks that try to insert or extract messages from them when they are full or empty, respectively. More specifically:

- As long as a message queue Q is neither empty nor full, insertions and extractions do not block the calling tasks and may not create any arcs in the resource allocation graph.
- When Q is *full*, the insertion of a message normally blocks the calling task T_s until space becomes available, unless the task asked for a non-blocking operation. In terms of the resource allocation graph, we can model this wait with a *request* arc that goes from T_s to the "Q full" resource, as shown on the left side of Figure 16.10. The fact that a single other task T_r has the capability to extract a message from Q and unblock T_s can then be modeled as an *ownership* arc that goes from "Q full" to T_r.
  In the gateway, for instance, the CAN→UDP forwarding queue may be full while the CAN receive task performs a blocking insertion on it. Since the

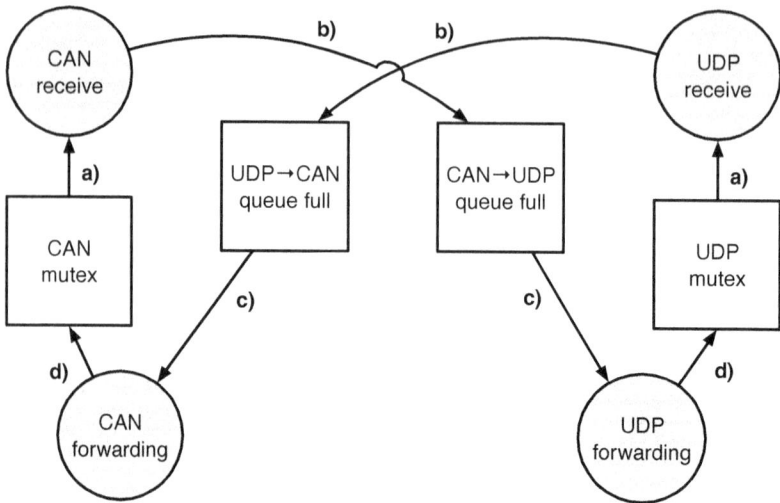

Figure 16.11 Resource allocation graph for the whole gateway when both forwarding queues are full.

UDP forwarding task is the only one responsible for extracting from it, the corresponding resource allocation graph has an arc that goes from the CAN receive task to the "CAN→UDP queue full" resource, and another arc from this resource to the UDP forwarding task.

- Symmetrically, when Q is *empty* and a task T_r attempts to extract a message from it, it blocks until some other task inserts a message into Q. The task T_s that can insert such a message owns the resource "Q empty" in this case. Hence, in the resource allocation graph there will be an arc from T_r to "Q empty" and an arc from "Q empty" to T_s, as shown on the right side of Figure 16.10.

  In the gateway this happens when the UDP forwarding task is waiting on an empty CAN→UDP forwarding queue and the CAN receive task is the only task that is capable of unblocking it by inserting a message into that queue.

Since a message queue cannot be empty and full at the same time, unless it is a rendezvous queue (see Section 6.3), the two graph fragments shown in the figure cannot appear together in the same resource allocation graph.

Under the assumptions just discussed, when both the CAN→UDP and the UDP→CAN forwarding queues are full, we can draw the resource allocation graph shown in Figure 16.11 for the whole gateway. The meaning of the arcs labeled **a)** through **d)** in the figure is:

a) The receive tasks hold their own port mutex while operating on the other port to insert a message in its forwarding queue.

b) Both receive tasks are blocked; they are waiting to insert a message into the forwarding queue of the other port because it is full.

c) The two forwarding tasks are the only tasks able to extract a message from the forwarding queue of their port.

d) The forwarding tasks need to acquire the mutex of their port while they iterate and are currently waiting for the mutex to be released.

Together, all these arcs, along with the tasks and resources they link, form a *cycle*. As explained in Section 4.3, this implies the gateway is deadlocked. To better convince ourselves this is really the case, we may describe the scenario that the resource allocation graph represents in informal terms.

As described previously, we are in a situation in which both forwarding queues are full. Depending on their size, this scenario may be more or less remote, but it is important to remark that we cannot rule it out completely. Both receive tasks have received a message successfully and are trying to insert it into the forwarding queue of the other port while holding the mutex of their own port. They are both blocked because both queues are full, in agreement with our hypotheses.

None of the forwarding tasks that, in theory, would have the ability to unblock the receive tasks (by extracting a message from the message queues they are blocked on) can proceed. They are both blocked because they are waiting to acquire exclusive access to their port context. They may need to do this, for instance, to update port statistics depending on the outcome of the previous operation they performed. As a result, all tasks stay blocked for an indefinite amount of time, thus rendering the gateway inoperable.

Fortunately, the theory developed in Chapter 4 helps us solve this problem. Namely, we know we can prevent the deadlock if we can ensure that at least one of the arcs that are part of the cycle in the resource allocation graph may never be established. Although there is no easy way to prevent arcs of type **c)** and **d)** from forming, because they are deeply rooted in the design principles on which the gateway is based, it is still possible to work on arcs of type **a)** and **b)**. In particular:

- We can rule out the existence of arcs of type **a)** if we implement the receive tasks so that they do not operate on the other port while they are holding exclusive access to their own port context. This choice does not impose any undue constraint on the design because, by definition, any forwarding decision the receive tasks may have to make—and for which they may need to consult the context of their own port—has already been made before they interact with the other port. At the same time, the receive tasks are still able to update port statistics based on the outcome of the forwarding request they made, by acquiring their port mutex again after the request has been completed, either successfully or unsuccessfully.

- Arcs of type **b)** may exist only if queue insertions are *blocking* when the target queue is full. If we make these operations return an error, instead, no arcs of this type will ever be part of the resource allocation graph. As discussed previously, overfilling a forwarding queue is a symptom that the gateway is unable to sustain the traffic it has to handle, not because of a sporadic burst, but due to a persistent overload. Under this condition, it makes sense for the gateway to start dropping packets as soon as it detects the overload, rather than blocking while trying to forward them and potentially exacerbating the issue.

In this scenario, the first option is as feasible and appealing as the second. However, based on the symmetry shown in Figures 16.3 and 16.10, we can infer that an analogous resource allocation graph, involving empty instead of full queues and also containing a cycle, exists. This graph describes the condition in which both forwarding tasks are waiting to receive a message from their respective forwarding queue while they hold the mutex of their own port. At the same time, the two receive tasks cannot proceed because they are waiting to acquire the mutex of their port before they can insert any messages into the forwarding queues.

Although, in principle, the two approaches to preventing the deadlock are still practicable also in this case, suppressing arcs of type **b)** would be quite inefficient from the implementation point of view. This is because it would make the system non-blocking in the common case in which the gateway is idle and its forwarding queues are empty. In turn, this would lead to a waste of CPU resources and, were task priorities and queue polling intervals not chosen wisely, to a possible starvation of the tasks with a priority lower than the priority of the receive tasks.

As a consequence, the gateway implementation prevents arcs of type **b)** when the queues are full, because it is more meaningful from the application point of view, but it prevents arcs of type **a)** when the queues are empty, for the reasons just described. All in all, this is a typical example of two solutions that, although equivalent from the theoretical point of view, may lead to different and more or less convenient outcomes when applied to different practical scenarios.

16.4   LINUX TEST PROGRAM

The Linux test program transmits and receives UDP datagrams encapsulating a CAN frame exactly as the gateway does. In fact, they both use the same library—the Canudp library described in Section 16.2—to do so. The main difference is that the gateway is an embedded device that neither generates traffic on its own nor prints out in any way the data flowing through it. In other words, it merely "moves" messages from one of its ports to the other. On the contrary, the Linux test program is a standalone command-line tool that can either wait for messages to arrive and dump them in human-readable form or generate traffic as instructed by the command-line options and arguments it receives.

As just described, the Canudp library is in charge of the bulk of what the test program does. Besides that, most of its source code consists of a set of functions that parse the command line and extract from it the information the tool needs. As an example, Figure 16.12 shows how the main command-line parsing function is organized. Based upon that information, the -t command-line option in particular, the tool first of all decides whether it should receive or transmit.

When receiving, the tool repeatedly calls the `canudp_receive()` function from `exec_receive()` and prints out the messages it receives on `stderr`. As shown in

```
static int parse_options(
    int argc,                    /**< Argument count */
    char * const argv[],         /**< Argument vector */
    struct Command_line *command_line /**< Structure to be filled */
    )
{
    const struct Command_line def_command_line = {
        .canudp_config.if_name = DEFAULT_IF_NAME,
        .canudp_config.udp_port = DEFAULT_PORT,
        .ff = FF_BASE,
        .rtr = false,
        .op = OP_RECEIVE
    };
    int st = 0;
    int opt;

    *command_line = def_command_line;
    while((opt = getopt(argc, argv, "i:p:ert")) != -1 && st == 0)
    {
        switch(opt)
        {
        case 'i':
        {
            /* Interface name */
            strncpy(
                command_line->canudp_config.if_name, optarg,
                CANUDP_IFNAMSIZ);
            command_line->canudp_config.if_name[CANUDP_IFNAMSIZ-1] = '\0';
            break;
        }

        case 'p':
        {
            /* Port number */
            if(parse_int(
                    optarg, 0, 65535, &command_line->canudp_config.udp_port))
            {
                fprintf(
                    stderr, "%s: invalid port number '%s'; using default.\n",
                    argv[0], optarg);
            }
            break;
        }

        case 't':
            command_line->op = OP_TRANSMIT;
            break;

        /* ... More options omitted... */
        case '?':
            st = -1;
            break;
        }
    }

    /* ... Error handling ... */

    return st;
}
```

Figure 16.12 Linux test program, command-line options handling.

```
static int exec_receive(
    struct Canudp_ctx *ctx                    /**< Canudp context */
    )
{
    enum Canudp_st st;
    struct Canudp_msg msg;
    while((st = canudp_receive(ctx, &msg)) == CANUDP_ST_OK)
    {
        /* This is not a switch because multiple flags may be set */
        uint32_t type = canudp_msg_type(&msg);
        if(type & CANUDP_MSG_TYPE_BASE) fprintf(stderr, "B");
        if(type & CANUDP_MSG_TYPE_EXT)  fprintf(stderr, "E");
        if(type & CANUDP_MSG_TYPE_RTR)  fprintf(stderr, "R");
        if(type & CANUDP_MSG_TYPE_ERR)  fprintf(stderr, "*");

        fprintf(stderr, " 0x%" PRIx32 " [%" PRIx8 "] ",
                canudp_msg_id(&msg), canudp_msg_dlc(&msg));

        int len = canudp_msg_length(&msg);
        for(int i=0; i<len; i++)
            fprintf(stderr, " %02" PRIx8,
                    canudp_msg_data(&msg, i));

        fprintf(stderr, "\n");
    }

    if(st == CANUDP_ST_TIMEOUT)
        fprintf(stderr, "(Timeout receiving from UDP)\n");
    else
        fprintf(stderr, "canudp_receive: error %d\n", st);
    return (st != CANUDP_ST_TIMEOUT) ? EXIT_FAILURE : EXIT_SUCCESS;
}
```

Figure 16.13 Linux test program, receive loop.

```
static int exec_transmit(
    struct Canudp_ctx *ctx,                   /**< Canudp context */
    struct Canudp_msg *msg        /**< Message to be transmitted */
    )
{
    enum Canudp_st st = canudp_transmit(ctx, msg);
    if(st != CANUDP_ST_OK)
        fprintf(stderr, "canudp_transmit: error %d\n", st);
    return (st != CANUDP_ST_OK) ? EXIT_FAILURE : EXIT_SUCCESS;
}
```

Figure 16.14 Linux test program, transmit function.

Figure 16.13, `exec_receive()` returns to the caller and the test program terminates when `canudp_receive()` times out without receiving any message from the network. In another part of the code, now shown here, the timeout interval is set to 30 s. When asked to transmit, the tool fills a `struct Canudp_msg` with the information the user provided on the command like and calls the `exec_transmit()` function to transmit it, as shown in Figure 16.14.

16.5 SUMMARY

This chapter presented a case study involving a gateway, that is, a device that interconnects multiple networks, possibly based on different underlying technologies and protocols, and enables their nodes to seamlessly exchange information through it. Although the application by itself is fairly simple, it provided a convenient way to illustrate how this kind of embedded software is designed and implemented.

At the same time, the case study gave us the opportunity to stress that, no matter how simple or complex a software project is, it is still important to put the theoretical knowledge described in the previous chapters of this book to good use. As an example, we showed that a careless design may lead to a deadlock. Moreover, it turned out that two approaches to prevent deadlock, although equally good from the theoretical point of view, had very different side effects on gateway performance and CPU usage.

16.6 EXERCISES

EXERCISE 1

The functions that stop a gateway port, `can_port_stop()` and `udp_port_stop()`, contain a polling loop. In it, they query the port context variable `n_tasks`, which keeps track of how many tasks are still associated with the context. The functions exit from the loop and return to the caller when `n_tasks` goes down to zero. The loop iterates every `CAN_HEARTBEAT_MS` and `UDP_HEARTBEAT_MS` milliseconds, respectively. This approach is very easy to implement, but it has two main shortcomings:

- It does not react to task termination immediately. Namely, up to `CAN_HEARTBEAT_MS` or `UDP_HEARTBEAT_MS` millisecond may elapse between task termination and detection.
- It is based on an active, rather than passive, wait. Besides wasting CPU time, this shortcoming also prevents lowering `CAN_HEARTBEAT_MS` or `UDP_HEARTBEAT_MS` to improve reactivity because this would further increase CPU load.

Rewrite the termination detection logic to use a passive wait.

Hint: A condition variable within a monitor (Chapter 5) would enable a terminating task to detect whether it is the last one to terminate and notify the gateway stop function accordingly, thus solving the problem. Although FreeRTOS does not by itself provide monitors, they can be implemented based on semaphores. The POSIX version of condition variables (Chapter 7) may be a starting point for the implementation. The solution to Exercise 1, Chapter 8 may also turn out to be useful.

EXERCISE 2

The tasks associated with gateway ports ensure that any passive wait for external events they perform may not last for an indefinite amount of time, so that they can check for a termination request and honor it. Besides this "voluntary" termination, FreeRTOS provides two other ways to terminate a task:

- vTaskDelete(), which immediately deletes a task, regardless of its state.
- xTaskAbortDelay(), which forces a task to leave the blocked state and become ready, thus terminating any wait it may be performing.

Discuss advantages and disadvantages of these alternative approaches to task termination with respect to the method used by the gateway.

Hint: Consider the effect that immediate task termination may have on any mutually exclusive resources the task may hold. Moreover, think about the side effects that forcibly terminating a wait may have, especially when the wait is performed deep within a library that may or may not have been designed with this in mind.

17 Real-Time, High-Performance Data Acquisition

In this chapter we use the RedPitaya STEMlab 125-14 board (referred below as RP board) to develop real-time applications. Red Pitaya is a project intended to be alternative for many expensive laboratory measurement and control instruments. The RP board mounts two 125 MS/s radiofrequency (RF) inputs and two 125 MS/s RF outputs with 50 MHz analogue bandwidth and 14-bit analog-to-digital (ADC) and digital-to-analog (DAC) converters. The ADC converters are connected to a Zynq System on Chip (SoC), that integrates a dual core ARM Processor with a Field Programmable Gated Array (FPGA). This combination provides an extremely powerful tool for fast digital processing and control. Very fast data processing can in this way be carried out by the FPGA component, leaving more complex computation to the processor, intimately connected to the FPGA hardware. A variety of applications have been developed for the RP board, such as Oscilloscope and Signal generator, Spectrum analyzer, Bode analyzer, Logic analyzer, etc. and are freely available at the RedPitaya site `https://redpitaya.com/applications-measurement-tool`. These applications hide every implementation detail and users interact with them via the board's network interface. A Web server running on the RP board exports application interfaces that can be used via a Web browser. Alternatively, MATLAB, LabVIEW and Python interfaces are available exporting the RedPitaya functionality on a host computer connected to the RP board via the network. For the above reasons, the RP board is an extremely useful tool in laboratories for measurement and learning. In this chapter, however, we adopt a different approach, that is, developing a real-time application from scratch, without relying on any pre-built tool. In this way, we shall discover how to configure and program the FPGA and how to connect it to the Processor in the SoC system. The final step will be the development of a Linux driver connecting user applications to the underlying hardware. Not all the above steps will be presented here in detail because topics such as FPGA programming are outside the scope of this book. We shall concentrate more on the development of the Linux driver in a real-time context in order to show how it can be exploited in a real-time application such as feedback control. Nevertheless, we shall provide an overview of every involved component, addressing the interested reader toward specific literature for the topics not fully covered here.

Our sample application, called `rp_adc_dac`, will collect data from a high speed 2 channels ADC converter and write output data to the high speed 2 channels Digital to Analog (DAC) converter. Both ADC and DAC converters are hosted on the RP

DOI: 10.1201/9781003593416-17

355

board and are connected to some pins of the Zynq SoC device, also hosted on the RP board. Reading from ADC channels and writing to DAC channels will be made possible via the **read**() and **write**() operations performed over a Linux driver. Real-time application, such as feedback control, can thus be implemented by programs executed in user mode and running on the RP board itself.

The three main steps in the development of our application, and more in general in any SoC application are the following:

1. The development of the FPGA application interfacing the hardware and carrying out possible computation at a speed that is not achievable using the processor. In our application, the FPGA will be used to move data from the ADC inputs, perform subsampling based on the desired data sampling rate, handle start and stop commands and enqueue data on a First-In-First-Out (FIFO) queue that will dequeued by read accesses performed by the driver code.

2. The development of a software driver accessing memory locations exported by the FPGA and providing the required synchronization, i.e. waiting for the availability of a new ADC sample during ADC data readout. Linux is the running operating system (OS) for the RP board and therefore a Linux driver is presented here. Observe that other approaches are possible when developing a SoC application. A different operating systems such as FreeR-TOS may be chosen or even no OS at all. This latter configuration, called *Bare Metal* is adopted on many microcontrollers, where the running software basically executes a control loop, reading from input devices, performing computation and writing to output devices.

   The Linux driver used in this application will be presented in detail and it can be used as a template for the development of other drivers on this and other boards hosting a SoC running Linux.

3. The development of user programs for real-time applications. Here, we shall present and discuss a very simple program, just reading ADC samples and outputting them via the DAC converter. With the help of an oscilloscope it is then possible to measure the delay between the input and output waveform, that is, the time required by the system to read the sample pass it to the program, and write it back. We shall discuss different driver configurations and their effect on delay and throughput.

17.1   FPGA DEVELOPMENT

A field-programmable gate array (FPGA) is an integrated circuit that can be repeatedly programmed after manufacturing. A FPGA is composed of programmable logic blocks with a connecting grid that can be programmatically configured to interconnect with other logic blocks to perform various digital functions. A FPGA configuration is generally written using a hardware description language (HDL). Several HDL languages are available for FPGA programming such as VHDL and Verilog. Even if a program written in an HDL language may looks at a first glance similar to a normal program for a CPU, there is a basilar difference between the twos. Indeed, CPUs are general-purpose processors that can perform a wide range of tasks, while FPGAs can be considered highly specialized processors that can be programmed to perform specific tasks with great efficiency. FPGA programming consists of creating a logic circuit that will perform a requested algorithm. However, the building blocks of this algorithms are not variables, conditions and operations to be performed, rather, logic gates, adders, registers and multiplexers. Under this perspective we can better understand what are the operations carried out by the FPGA in our application. We need first to take the samples produced by the ADC circuitry hosted on the RP board along with the Zynq chip. Digitalized samples are transferred via a given digital protocol that must be implemented by the FPGA logic in order to successfully transfer samples. ADC samples are then transferred to other blocks for (1) enable or disable the flow of samples based on a register (`command_register`) defined in the FPGA whose address is then exported to the CPU (we describe this later in more detail); (2) downsample ADC samples by dividing the data acquisition 125MHz clock by the content of another register (`decimation_register`); (3) transfer the downsampled stream to the head of a FIFO queue whose tail is read via memory access performed by the processor.

In order to safely transfer data from one block to another within the FPGA, a simple communication bus, called `AXI bus`, is implemented in the FPGA logic. In its simplest configuration the AXI bus defines a number of data lines, bringing the bits of the data being transferred and an enable line. Data are transferred when the enable line is high. Normally the enable line is maintained high during one clock cycle triggering in this way data transfer.

In summary, programming the FPGA consists in developing the involved blocks using an HDL language such as VHDL or Verilog, and then assembling them to reach the desired functionality. This process is greatly simplified by two important factors:

1. The availability of powerful graphical tools that allow visually assembling blocks in a FPGA project. For the Zynq SoC the **Vivado** Integrated Development Environment (IDE) is used to develop and assemble FPGA components. The complete project described in this chapter has been developed in Vivado and its complete block design is shown in Figure 17.1. The single blocks will be briefly introduced in the following.

2. The availability of a large number or ready-to-use blocks for a variety of function. Intellectual property (IP) cores are reusable HDL component in FPGA that can be integrated into complete implementations using design tools such as Vivado (from AMD) or Platform Designer (from Intel). Many IP cores are freely available, others, carrying out more complex functions, must be purchased.

For our `rp_adc_dac` application, the steps described below have been implemented starting from an empty Vivado project . Some of the used blocks have been taken from the Vivado toolkit, others from the RedPitaya community and others

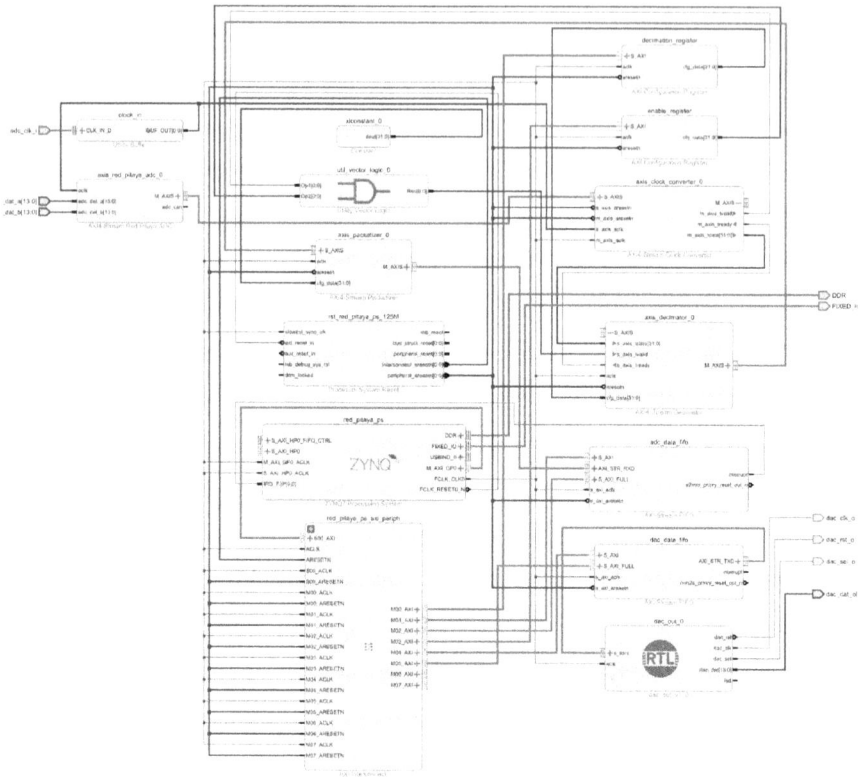

Figure 17.1   The complete Vivado project

developed ad hoc using the VHDL language. As stated before we shall not describe in detail the blocks implementation, but the whole project, including the source code for the blocks developed ad hoc and from the RedPitaya community, is available on GitHub (https://github.com/minimap-xl/RTOS_Book). In that repository all the required material and detailed instructions are available to replicate this project only using a RedPitaya board with Ubuntu Linux available at https://redpitaya. readthedocs.io/en/latest/quickStart/SDcard/SDcard.html without the need of installing Vivado. If you are willing to replicate the steps described below in FPGA construction you can install Vivado on your system and take the complete project from GitHub, but this is not required to run the presented application on your RP board.

The first step, starting from an empty Vivado project, is to instantiate the Zynq Processing System, that is, the logic required to connect the FPGA to the ARM processor hosted on the Zynq SoC. In addition to this block two more system blocks are

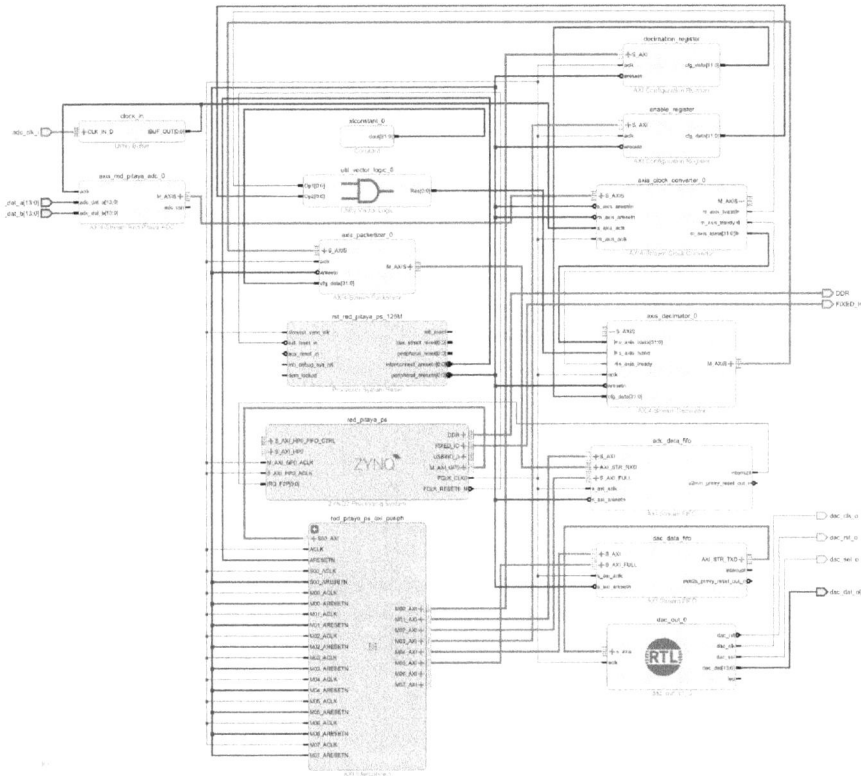

Figure 17.2   The Zynq components in the Vivado project

requited for our project (see Figure 17.2): the `Processor System Reset` to handle
the system reset logic and the `AXI Interconnect` block bringing the required logic
to connect the used registers (in our case the Decimator Register and Enable Register) and the FIFO queues (one for ADC readout, the other for writing to DAC) to
the internal processor bus so that they can be programmed via software using proper
addresses in the physical addressing space.

After the system components have been instantiated in the Vivado project,
we proceed by instantiating two configuration registers (`enable_register` and
`decimation_register`). The implementation of the configuration registers is written in Verilog and has been made available by the RedPitaya community. These
registers are connected to the AXI Interconnect component and export a number of
bits (32 for the decimation register and 1 for the enable register) to be used in the
FPGA logic. Luckily, most interconnection is automatically generated by the Vivado that is smart enough to recognize when the instantiated components need to be

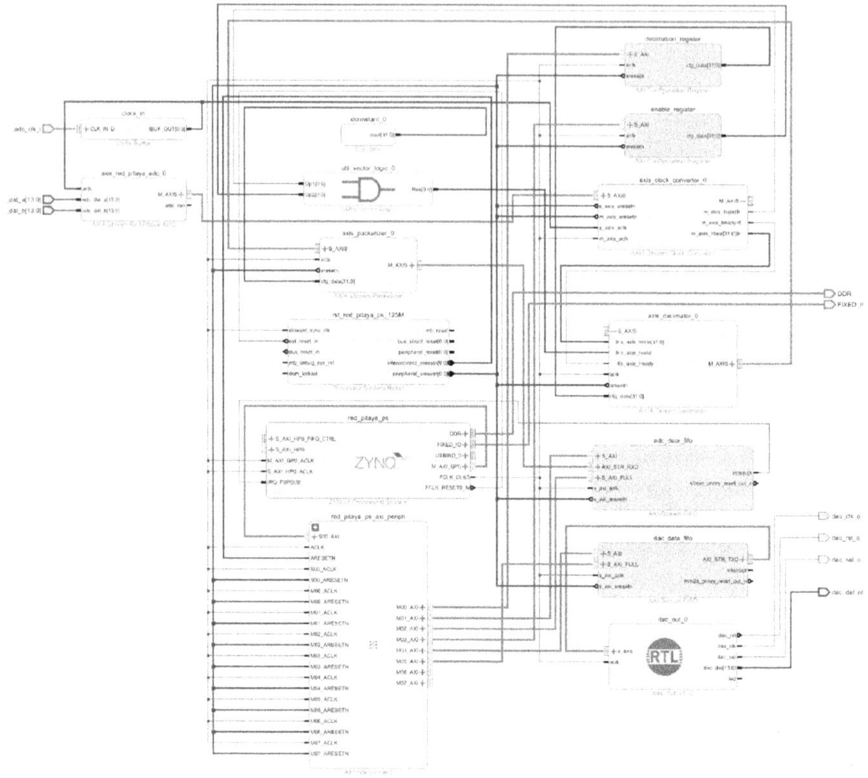

Figure 17.3 The registers and the FIFO queues in the Vivado project

interconnected with the processor bus. The used FIFO queues are freely available in the Vivado design suite (AXI Streaming FIFO) and implement the required logic for enqueueing incoming data arriving from the FPGA logic, i.e. from the ADC converter, to be read by the software via read operations (`adc_data_fifo`), or from the program, via write operation, toward the FPGA for DAC (`dac_data_fifo`)

The registers and the FIFO queues in the Vivado project are shown in Figure 17.3

The ADC logic is carried out by three blocks in the Vivado project. The physical inputs, represented by the three input connectors `adc_clk_i` (1 bit) `dat_a` (14 bits) and `dat_b` (14 bits), shown on the left of Figure 17.1, bring the ADC clock (125 MHz) and the 14 converted bits for channel A and B, respectively. The buffered clock is given as input to block `axis_redpitaya_adc` that converts the incoming data stream into an equivalent stream over the AXI bus. Block `axi_clock_converter` provides the adaption between the external ADC clock and the internal Zynq clock. Even if the two clocks have the same frequency (125 MHz), they represent two different clock domains and their combined usage without precautions may produce metastability in digital circuitry. Metastable states give intermediate values other than high or low or 1 or 0, leading to logical incorrectness. The condition of metastability in digital circuits propagates errors to the remaining part of the circuit and as such must be avoided as far as possible. Several techniques exist to reduce the risk of metastability to a minimum, and in practice pre-built blocks, like `axi_clock_converter`, are used to move from one clock domain to another. The resulting data flow, sent over the output AXI bus of block `axi_clock_converter`, brings sampled ADC data (two 14 bit channel, sign extended to 16 bits and packed in a 32 bits word) at the ADC sampling speed of 125MHz. This means that the data strobe represented by the AXI enable line has a frequency of 125 MHz, corresponding to the internal frequency of the Zynq FPGA.

We need now to perform two operations over to ADC samples stream before sending it to the FIFO queue. The first operation enables or disables ADC data readout under software control. This is achieved by writing 1 (enable) or 0 (disable) in the 1 bit register `enable_register` (actually, the register are accessed as 32 bit words, so in this case we are interested in the least significant bit). The output bit of `enable_register` is put in AND with the enable output line of block `axi_clock_converter`. In this way, when the output bit of `enable_register` is 0, no data strobe is passed to the following block, so inhibiting data transfer. If the output bit is 1, ADC data are regularly transferred to the following block.

The second operation is ADC data downsampling (software would be never able to sustain a data flow rate of 125 MHz). The downsampling rate is represented by the 32 output bits of register `decimation_register` provided as input to block `axis_decimator`. This block internally performs data downsampling and therefore the strobe rate of the output AXI bus becomes 125/N MHz, where N is the content of `decimation_register`. Observe that the output AXI bus is in general graphically displayed by Vivado as a thick line. When needed (in our case between `axi_clock_converter` and `decimation_register`) it can be displayed in detail thus allowing manipulating the single lines.

Finally, the resulting ADC samples stream is made readable by the software via the `adc_data_fifo` block, after passing the `axi_packetizer` block that handles another control line (`tlast`) of the AXI bus as expected by `adc_data_fifo`. `tlast` indicates the end of a block of data when block transfer is performed on the AXI bus. In our case the block trivially corresponds to the single 32 bit word so that `tlast` is issued at every sample transfer.

The use of a FIFO queue for data readout is necessary in order to account for possible transient delays on the processor side. The software must of course be able to read data with an average rate corresponding to the downsampled data rate in order to avoid FIFO overflow, but the processor may be temporarily busy (e.g. when serving an interrupt) and not ready to immediately accept an incoming sample. `adc_data_fifo` is then connected to the `AXI interconnect block` to interface to the Processor bus. `adc_data_fifo` has also an interrupt output that is connected to the Zynq processor. An interrupt request will be then generated when a sample is

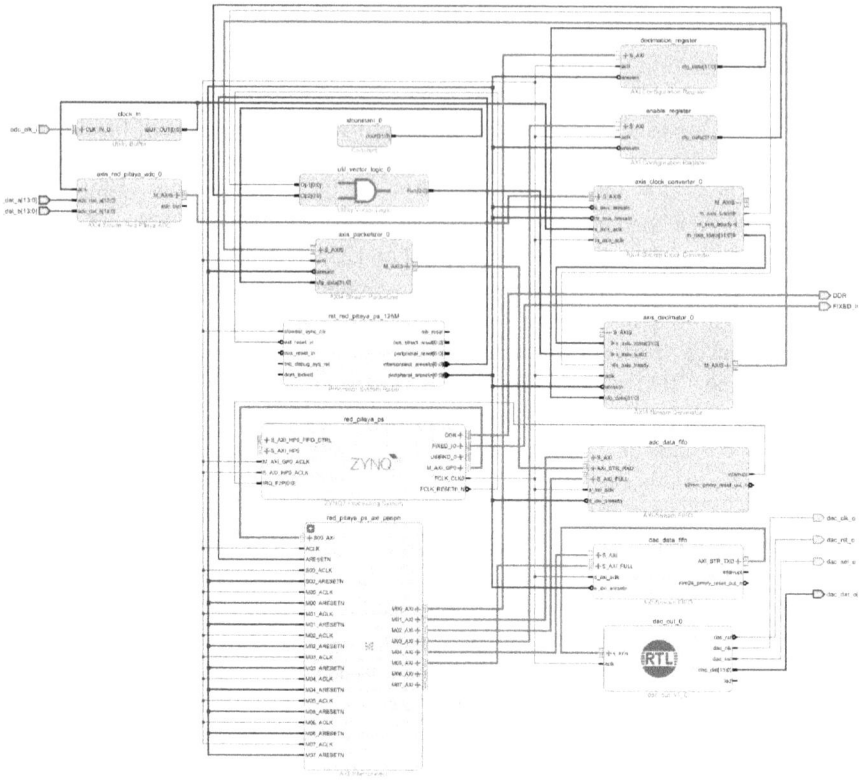

Figure 17.4   The chain of blocks involved in adc data acquisition.

present in the FIFO queue, and it will be used by the software driver to synchronize data readout. The whole chain of involved blocks in the data acquisition process is shown in Figure 17.4.

The management of DAC conversion is carried out by blocks dac_data_fifo and dac_out. dac_data_fifo implements a FIFO queue from software to the FPGA. Data samples (32 bits words packing two 14 bit DAC channels) written by the software via memory write operations on the FIFO are then made available over the output AXI bus. In order to synchronize output data flow to the DAC conversion rate, an internal line of the AXI bus, called ready, is set by the receiver dac_out when it is ready to accept a new data sample. Block dac_out will carry out the required logic for sending the data sample to the DAC device via the FPGA outputs dac_wrt_o, dac_ck_o, dac_rst_o, dac_set_o, and dac_dat_o, shown on the right of Figure 17.1. We won't enter into the details of the ADC and DAC logic here, but the interested reader can find the HDL source code of both blocks (axis_red_pitaya_adc and dac_out) in the GitHub repository.

The FPGA project can now be compiled by Vivado and a `bitstream.bin` file is produced. The bitstream defines the internal FPGA configuration to achieve the functionality as specified by the Vivado project. The bitstream file for the `rp_adc_dac` project is available in the GitHub repository and therefore it is not necessary to have Vivado installed to run the `rp_adc_dac` project. The FPGA of the RP board can be dynamically configured directly from a terminal connected to the board (either a terminal connected via USB or a Secure Shell remote session) via the following command:

```
fpgautil -b bitstream.bin
```

After having briefly introduced the FPGA architecture of our `rp_adc_dac` project, we concentrate now on how this systems interfaces to software. In particular, we shall now see at what addresses a program running on the Zynq ARM processor can access the registers and the FIFO queues. The memory addresses of the components declared in a project are automatically assigned by Vivado. In particular, a single address is assigned to registers `enable_register` and `decimation_register` and the value of those registers can be assigned and read back via a write and a read instruction at the corresponding address, respectively. The management of the FIFO queues is a bit more complicated, and two addresses are assigned to every queue instance. Starting at the two addresses, a set of internal register is exported by every queue so that the queue can be properly configured and used. In our project we shall use the following internal register for the ADC FIFO queue:

- **Receive Data FIFO Occupancy** returning the number of items currently in the FIFO.
- **Receive Data FIFO** returning and dequeueing the first item in the queue upon a read instruction.
- **Interrupt Enable Register** used to enable interrupt sources (in our case only data availability).
- **Interrupt Status Register** used in the interrupt management for reading and resetting the active interrupts.

The internal registers used for the DAC FIFO queue are the following:

- **Transmit Data** where data is written in FIFO.
- **Transmit Data Length** used to trigger FIFO output toward the DAC.

Note that the assigned addresses that are shown in the Vivado interface are *physical addresses*, and you may recall from Chapter 2 that physical addresses cannot be used directly on a system with virtual memory as Linux is. However, as we shall see, we don't need to explicitly use those numbers in order to let our driver communicate with registers and FIFO queues.

At this point we have all the required elements for interfacing a software program to the FPGA hardware. ADC data configuration is performed by writing the requested decimation to the address corresponding to `decimation_register`; ADC is started and stopped under program control by writing to the address corresponding to `enable_register`; incoming ADC data can be inspected and accessed reading the memory locations corresponding to the `Receive Data FIFO Occupancy` and `Receive Data FIFO` registers, respectively; data samples can be written to the DAC by writing at the memory locations corresponding to `Transmit Data` and `Transmit Data Length` registers. However, it is not possible to write a program directly handling the memory accesses because reading and writing to physical devices is not allowed in user mode. For this reason we need to develop a *software driver* as explained in the next section.

Before discussing the development of the Linux software driver we need another piece of information, i.e. to know how Linux discovers connected hardware devices. We have already seen in Chapter 2 that devices present themselves to the processor via a set of physical addresses over the bus. Clearly, such addresses cannot be hardwired in the hardware device, otherwise memory clashes may occur when combining different devices on different computer architectures. In most architectures devices are discovered during the boot of the system. This is the case of the devices mapped over the PCI or PCI-e bus, where a set of fixed configuration addresses uniquely define the connected devices. When a PCI device is discovered, a negotiation between the processor and the device itself occurs, and the set of physical addresses for the registers of that device is assigned by the OS. At that point the device is configured and becomes active, and at the same time the OS is aware of the physical addresses of its registers. This is however not the case for the Zynq architecture, and more in general for SoC systems, where the devices, or at least their interfaces such as the ADC and DAC devices in the RP board, are intimately connected to the processor in the same chip. In this case a static definition, called `device tree`, is defined. During boot the device tree is parsed so that the OS can know what are the connected devices as well as the physical addresses of their registers, the used interrupts and the associated DMA controllers.

When developing a Zynq application the device tree is generated by the Vivado tool based on the memory mapping information for that project. For the `rp_adc_dac` project the device tree will include the memory interfaces of `enable_register`, `decimation_register`, `adc_data_fifo`, and `dac_data_fifo`, in addition to other system devices. On the RP board, when booting, Linux searches the device tree definition in file

`/boot/dts/z10_125/devicetree.dtb`

and uses the information stored in that file to build an internal representation of the devices of the system. We shall see in the next section how the software driver interacts with Linux in order to discover the registers and FIFO interfaces so that it can configure and handle ADC and DAC data streams.

17.2   THE LINUX DRIVER

We are now ready to build the last component that will allow user programs read ADC data and write DAC data. In the previous section we have seen how the FPGA has been configured for our purpose and how the hardware interacts with the software. In summary we have now (1) one memory location for `enable_register` and `decimator_register`, respectively and (2) a set of memory locations, corresponding to a set of control registers, for `adc_data_fifo` and `dac_data_fifo`, respectively.

The driver will provide a way for programs to control the registers and to read and write from the ADC and DAC FIFOs, respectively. In particular, we shall develop the device driver for a *character device*. A character device is an abstraction provided by Linux to support devices that can be read from or written to with relatively small data transfers in the order of few bytes. They are abstracted as files in the file system and are accessed via a limited set of functions (`read()`, `write()`, `ioctl()`, ...). In our application it makes sense to use `read()` for reading ADC samples and `write()` for writing DAC samples. It is less intuitive how dealing with the configuration, that is, how giving appropriate values to `enable_register` and `decimation_register`. For this purpose we shall use the `ioctl()` function, taking two arguments, where the first integer argument defines the target register and data direction, and the second argument is the address of a user variable containing the value to be written or receiving the value to be read.

    The structure of a device driver in Linux is dictated by a set of rules to which the driver program has to adhere in order to be accepted by the OS when the driver is loaded via the `insmod` shell command. The argument to the `insmod` command is the file containing the compiled driver code, in our case `rp_adc_dac.ko` (ko represents the classical extension of the compiled driver code in Linux), whose source is in file `rp_adc_dac.c`.

The key structure in the driver is listed below:

```
static struct file_operations fops = {
.read = device_read,
.write = device_write,
.open = device_open,
.release = device_release,
.mmap = device_mmap,
.llseek = device_lseek,
.unlocked_ioctl = device_ioctl,
.poll = device_poll,
};
```

This structure is used by Linux to map the set of routines exported by all device drivers onto the specific routines for this driver. The names of the fields correspond to the routines defined by Linux and their assigned values are the addresses of the specific routines for this driver. In our case we are only interested in `read` and `write` for reading ADC samples and writing DAC samples, `open` and `release` invoked when this device is open and closed by a user program, respectively, `poll` used to synchronize ADC readout and `ioctl` to read and write the configuration registers. The other routines are dummy in our applications.

Another important structure in the driver is the private information listed below:

```
struct rp_adc_dac_dev {
/* kernel internal structure sed to discover Zynq resources */
        struct platform_device *pdev;
```

```
/* IRQ number for adc data fifo interrupt */
      int irq;

/* Kernel internal structure to represent character devices */
      struct cdev cdev;

/* Pointers to enable register, decimator register,
   adc data fifo and dac data fifo */
      void * iomap_decimation_register;
      void * iomap_enable_register;
/* For each FIFO two pointers are retrieved.
   These pointers address the internal FIFO IP registers */
      void * iomap_adc_data_fifo;
      void * iomap1_adc_data_fifo;
      void * iomap_dac_data_fifo;
      void * iomap1_dac_data_fifo;

/* asynchronous read flag */
      int async;

/* Actual number of samples in adc data fifo */
      u32 fifoSamples;

/* Wait queue used to synchronize readout */
      wait_queue_head_t readq;

/* Interrupt generation flag */
      int fifoHalfInterrupt;
}
```

This structure contains all the meaningful specific information for each instance of the device. Note that in our case we have only one instance of the rp_adc_dac device in the system and therefore we store this structure in variable staticPrivateInfo, statically allocated. We shall enter into the detail of every field in the remaining of this section, but it is worth discussing now the void* pointers declared in staticPrivateInfo. These pointers will be mapped onto the actual physical addresses of the decimation and enable registers (iomap_decimation_register and iomap_enable_register, respectively) and to the two base addresses for the used ADC and DAC FIFOs (recall that two addresses have been exported for every FIFO queue in the FPGA). As Linux uses virtual memory addresses, it is not possible to access directly the physical addresses of those hardware components in the FPGA, but the OS will configure the corresponding Page Table entries so that the virtual addresses specified by the pointer contents are mapped to the correct physical addresses. So, reading and writing to iomap_decimation_register and iomap_enable_register means reading and writing directly the target registers. For the FIFOs, the addresses will represent the base address for the used FIFO registers required to dequeue and enqueue data samples.

In addition to the routines defined in the structure file_operations a couple of other routines must be defined in the device driver, in particular init and exit routines, that are invoked by the OS when the driver is loaded (via command insmod) and unloaded (via command rmmod), respectively. The code below refers to routine rp_adc_init that is declared as the init routine via the macro module_init(rp_adc_dac_init)

```
static int __init rp_adc_dac_init(void)
{
        int err, devno;
        dev_t newDev;
```

```
/* Allocate a range of char device numbers.
   The major number will be chosen dynamically,
   and returned (along with the first minor number) in newDev */
        err = alloc_chrdev_region(&newDev, 0, 1, DEVICE_NAME);
        if(err < 0)
        {
                printk ("alloc_chrdev_region failed\n");
                return err;
        }
```

```
/* Major device number */
        id_major = MAJOR(newDev);
```

```
/* Initialize cdev. The struct cdev is the kernel's internal */
/* structure that represents char devices. */
        cdev_init(&staticPrivateInfo.cdev, &fops);
```

```
/* Set required cdev fields */
        staticPrivateInfo.cdev.owner = THIS_MODULE;
        staticPrivateInfo.cdev.ops = &fops;
```

```
/* Prepare Major and minor ids. Major id is returned by macro MAJOR
   Minor id is set to 0 because only one device instance
   is defined in the system. */
        devno = MKDEV(id_major, 0);
```

```
/* add the char device to the system */
        err = cdev_add(&staticPrivateInfo.cdev, devno, 1);
        if(err < 0)
        {
                printk ("cdev_add failed\n");
                return err;
        }
```


```
/* create a struct class pointer */
        rp_adc_dac_class = class_create(THIS_MODULE, DEVICE_NAME);
        if (IS_ERR(rp_adc_dac_class))
        {
                return PTR_ERR(rp_adc_dac_class);
        }
/* create a device and registers it with sysfs */
        device_create(rp_adc_dac_class, NULL, devno, NULL, DEVICE_NAME);
```

```
/* Initialize the wait queue that will be used for synchronous read */
        init_waitqueue_head(&staticPrivateInfo.readq);
```

```
/* platform driver register will trigger invocation of specific probe
   routines based on the device tree definition. Here, it will be used
   to map register (enable register and decimationa register)
   and fifo addresses (adc data fifo and dac data fifo)
   the passed struct platform driver structure defines the names of the
   the specific probe routines (rp adc dac probe when loading module,
   rp adc dac remove when unloading module) */
        return platform_driver_register(&rp_adc_dac_driver);
}
```

The first action performed by the init routine is the allocation of a major and a set of minor device numbers. Linux device instances are characterized by a *major device number* and a *minor device number*. The major number identifies the type of the device and the minor number is used within the device driver to distinguish between multiple devices of same type. In our case we have only one instance of the rp_adc_dac device, so the only minor number in our case is 0, but other devices may be replicated in different instances (e.g. serial ports). Routine alloc_chrdev_region() is provided by Linux to dynamically allocate a device major number along with a range of minor numbers. The major number, used in subsequent calls, is then derived from the returned dev_t structure via the macro MAJOR. Afterward, the cdev structure, that is the kernel's internal structure that represents this char device, is initialized calling cdev_init() and passing a preallocated cdev structure in the private information (staticPrivateInfo). After setting a couple of required cdev fields, the specific device instance, represented by the returned major number and by 0 as minor number, is added to the system by mean of cdev_add(). To complete the device registration procedure, routines class_create() and device_create() are called afterward. At this point the device instance is visible in the Linux file system in the /dev directory (/dev/rp_adc_dac) along with the other devices of the system.

Routine init_waitqueue_head() is then called to initialize an internal monitor that will be used in synchronous read to wait the availability of a new sample (more on that below).

Finally, routine platform_driver_register() is called to map physical addresses against the pointers declared in staticPrivateInfo field. There is a bit of magics behind the argument rp_adc_dac_driver passed to platform_driver_register(), that is declared as follows:

```
static struct platform_driver rp_adc_dac_driver = {
.driver = {
.name   = MODULE_NAME,
.owner = THIS_MODULE,
.of_match_table = rp_adc_dac_of_ids,
},
.probe = rp_adc_dac_probe,
.remove = rp_adc_dac_remove,
};
```

This structure instructs the Linux kernel that this driver is interested to a particular set of hardware resources and that, when each of these resources are discovered during the driver initialization, routine rp_adc_dac_probe has to be called by the kernel. Conversely, when the driver is unloaded, routine rp_adc_dac_remove (in our case dummy) is called by the kernel to relinquish the resources when the driver is unloaded. The specification of the hardware of interest is defined by field of_match_table that is declared as follows:

```
static const struct of_device_id rp_adc_dac_of_ids[] = {
{ .compatible = "xlnx,axi-cfg-register-1.0",},
{ .compatible = "xlnx,axi-fifo-mm-s-4.1",},
{}
};
```

A list of .compatible fields characterizes the hardware specification. But from where these names come? You may recognize similar names for the registers and FIFO queues declared in the FPGA schema shown in Figure 17.1. Indeed the HDL implementation of those components provides the compatibility names that Vivado will use when building the device tree for declaring the memory locations exported by the FPGA to Linux. As there are two registers and two FIFOs declared in the FPGA, we expect that routine `rp_adc_dac_probe` will be called twice for the registers and twice for the FIFOs. At every call, a set of arguments will provide the necessary information for mapping the pointers in the `privateInfo` structure. The source code of routine `rp_adc_dac_probe` is listed below:

```
static int rp_adc_dac_probe(struct platform_device *pdev)
{
        u32 off;
        struct resource *r_mem;
        char deviceAddress[9];

/* The first 8 chars correspond to the hex address in FPGA */
        memcpy(deviceAddress, pdev->name, 8);
        deviceAddress[8] = 0;

/* Get information on the structure of the device resource
   (start address and end address), in order to map it in memory. */
        r_mem = platform_get_resource(pdev, IORESOURCE_MEM, 0);
/* align mapped memory region to page boundary */
        off = r_mem->start & ~PAGE_MASK;

        if(!strcmp(deviceAddress, "50000000" )) //Decimation Register
        {
/* Map memory region declared in FPGA onto Kernel address space */
                staticPrivateInfo.iomap_decimation_register =
                        devm_ioremap(&pdev->dev,r_mem->start+off,0xffff);
        }
        else if (!strcmp(deviceAddress, "60000000")) //Enable Register
        {
                staticPrivateInfo.iomap_enable_register =
                        devm_ioremap(&pdev->dev,r_mem->start+off,0xffff);
        }
        else if (!strcmp(deviceAddress, "43c20000")) //dac data fifo
        {
/* For fifo two memory regions are defined. They are mapped against
   iomap dac data fifo and iomap1 dac data fifo, respectively */
                staticPrivateInfo.iomap_dac_data_fifo =
                        devm_ioremap(&pdev->dev,r_mem->start+off,0xffff);
                r_mem = platform_get_resource(pdev, IORESOURCE_MEM, 1);
                off = r_mem->start & ~PAGE_MASK;
                staticPrivateInfo.iomap1_dac_data_fifo =
                        devm_ioremap(&pdev->dev,r_mem->start+off,0xffff);
        }
        else if (!strcmp(deviceAddress, "43c00000")) //adc data fifo
        {
                staticPrivateInfo.iomap_adc_data_fifo =
                        devm_ioremap(&pdev->dev,r_mem->start+off,0xffff);
                r_mem = platform_get_resource(pdev, IORESOURCE_MEM, 1);
                off = r_mem->start & ~PAGE_MASK;
                staticPrivateInfo.iomap1_adc_data_fifo =
                        devm_ioremap(&pdev->dev,r_mem->start+off,0xffff);
/* adc data fifo can generate an interrupt, so it must be registered */
                staticPrivateInfo.irq = setIrq(pdev);
        }
        else
        {
                printk("ERROR: Unexpected rp_adc_dac_probe call\n");
        }
        return 0;
}
```

A first problem for properly assigning the correct pointers during the invoca-
tions of `rp_adc_dac_probe` is to discern to what register of FIFO the current
call refers to. Here we use a convention adopted by Vivado in the generation of
the device tree, that is assigning a name to the resource that contains the physical address as a hex text. The physical addresses of every component are known
(they can be displayed by the *Address Editor* tab in the Vivado IDE). For example, 0x50000000 is the physical address of the decimation register. The resource
name is provided in the `name` field of the passed argument pdev and therefore
we can discover in this routine to which component every `rp_adc_dac_probe`
call refers to. The pointer assignment is then performed in two steps: first a
memory resource descriptor (`struct resource *r_mem`) is retrieved by calling
`platform_get_resource(pdev, IORESOURCE_MEM, 0)` and then the virtual address returned by `devm_ioremap()` is assigned to the corresponding pointer. Under
the hood, the Linux kernel has arranged a Page Table entry in the Page Table structure used to map the kernel address space onto the physical address space so that the
virtual address stored in the pointer gets translated into the correct physical address.
The pointer assignment for the FIFO pointers is similar, except that in this case two
addresses are retrieved for every FIFO instance. These addresses will represent the
base address for accessing the FIFO internal registers, as we shall see shortly. Finally, for the `adc_data_fifo` component we need to retrieve interrupt information
and to associate a interrupt routine with it. When configuring `adc_data_fifo` in
the Vivado FPGA project, the generation of interrupt has been enabled for that component in order to signal the availability of new samples in the queue. The fact that
an interrupt will be generated for that component has been then reported by Vivado
in the device tree in order to inform the kernel and the driver. The association of an
interrupt routine to that interrupt is carried out by routine `setIrq` shown below:

```
static int setIrq(struct platform_device *pdev)
{
        int res;
/* Get Interrupt Request number (IRQ) associated with this device */
        int irq = platform_get_irq(pdev,0);
/* Associate an Interrupt Service Routine (IRQ cb) to this IRQ
   The argument passed by theInterrupt Service Routine is the address
   of the device private data (staticPrivateInfo) */
        res = request_irq(irq, IRQ_cb, IRQF_TRIGGER_RISING ,"rp_adc_dac",
            &staticPrivateInfo);
        if(res)
                printk("rp_adc_dac: can't get IRQ %d assigned\n", irq);
        return irq;
}
```

Kernel routine `platform_get_irq` returns the interrupt request number for that
device, that is passed to `request_irq` to associate the driver routine IRQ_cb with
the generation of that interrupt.

At this point, the hardware has been configured by all the activities triggered by a single call to `platform_driver_register()` during the driver initialization. We can
concentrate now on the driver operations performed in run time, i.e. when the device
is open, read, written or configured. `device_open` is called when a user program

opens this device (recall that the device presents itself in the /dev directory). Its code is listed below:

```
/* Invoked by linux when a user program opens the rp adc dac device */
static int device_open(struct inode *inode, struct file *file)
{

/*Retrieve the device specific information from the passed inode
  structure using macro container'of. This is the standard way for
  retrieving private driver information, even if in this case this
  information is already available in staticPrivateInfo variable */
        struct rp_adc_dac_dev *privateInfo =
                container_of(inode->i_cdev, struct rp_adc_dac_dev, cdev);

/*copy the pointer to device private information to field private data
  of the passed file argument so that it can be retrieved
  by the following device methods */
        file->private_data = privateInfo;

/* Retrieve mode from the flags passed by the program to open() function */
        privateInfo->async = (file->f_flags & O_NONBLOCK);
/* Default behavior in synchronous mode is to generare an interrupt
   when a sample arrives in the empty adc data fifo.
   It can be later changed via ioctl call */
        if(!privateInfo->async)
        {
                setFifoFirstInterrupt(privateInfo);
        }
        return 0;
}
```

The first instruction retrieves the staticPrivateInfo structure. The way this is performed is rather tricky, but this is the standard in Linux drivers: the cdev structure that has been initialized in rp_adc_dac_init was allocated within the staticPrivateInfo structure (in our case statically allocated, otherwise dynamically allocated when more than one instance of the given device is supported). The same cdev structure is pointed by the passed inode structure passed as argument to device_open. Macro container_of returns the address of the structure containing that field, in our case staticPrivateInfo. Observe that our case is a simplified one, i.e. supporting only one instance of the rp_adc_dac device and therefore we could have accessed directly staticPrivateInfo (it is declared as a static variable). Nevertheless the method used in device_open is the classical one adopted to retrieve the private information and store its pointer in the field private_data of the passed struct file, that will be passed to the other run time driver routines.

Afterward, a flag of the privateInfo structure keeps track of the mode the device has been opened by the user program. Two read modes are supported by this driver: synchronous and asynchronous. When in synchronous mode, a read call suspends the calling program until an ADC sample has been enqueued in the FIFO and can be returned, otherwise (in asynchronous mode), read returns soon reporting in case that no samples have been read (recall that in a Linux device, the read() operation returns the number of bytes read). To handle the synchronous mode, an interrupt from the adc_data_fifo FPGA component is enabled. Two options are supported, i.e. generate the interrupt as soon as a new ADC sample has been enqueued in the empty adc_data_fifo, or to generate the interrupt when the number of samples exceeds a given threshold (16 kSamples in our application, half the FIFO size). The two options have different implications in latency and throughput that will be discussed later in this chapter. Routine `setFifoFirstInterrupt`, shown below along with other FIFO utility routines, defines the interrupt policy by writing an appropriate value in the Interrupt Enable Register of the FIFO.

```
/* Support Function for configuring interrupts in adc data fifo */
static void setFifoHalfInterrupt(struct rp_adc_dac_dev *dev )
{
        dev->fifoHalfInterrupt = 1;
/* Enable interrupt when fifo length exceeds 1024 samples
  (as configured in FPGA Interrupt enable register of
   the fifo is configured */

        Write(dev->iomap_adc_data_fifo,IER,0x00100000);
}

static void setFifoFirstInterrupt(struct rp_adc_dac_dev *dev )
{
        dev->fifoHalfInterrupt = 0;
/* Enable interrupt when the first sample arrives and the fifo is empty
   Interrupt enable register of the fifo is configured */
        Write(dev->iomap_adc_data_fifo,IER,0x04000000);
}

/* Support functionf for clearing adc data fifo and dac data fifo */
static void clearAdcFifo(struct rp_adc_dac_dev *dev )
{
/* Clear all pending interupts */
        Write(dev->iomap_adc_data_fifo,ISR,0xFFFFFFFF);
/* Disable all interrupts */
        Write(dev->iomap_adc_data_fifo,IER,0x00000000);
/* Reset fifo */
        Write(dev->iomap_adc_data_fifo,RDFR,0xa5);
}
```

Routine `device_release`, shown below, is called when the user program closes the `rp_adc_dac` device.

```
static int device_release(struct inode *inode, struct file *file)
{
/* retrieve private device info */
        struct rp_adc_dac_dev *dev = file->private_data;
/* If not found something went wrong */
        if(!dev) return -EFAULT;

/* Clear FPGA Enable Register to stop reception of adc samples */
        *((u32 *)dev->iomap_enable_register) = 0;
/* Clear adc data fifo */
        clearAdcFifo(dev);
/* Wake up possible threads waiting for synchronous data */
        dev->fifoSamples = 0;
        wake_up_interruptible(&dev->readq);
        return 0;
}
```

The `privateInfo` structure is accessed via the passed `private_data` pointer in the `file` argument passed in the call. The same operation is performed at the beginning of the remaining driver routines. This routine clears `adc_data_fifo` writing appropriate values in the FIFO registers (via routine `clearAdcFifo`) and finally it awakes any pending task possibly waiting for ADC data availability so that it can safely return.

Writing a new DAC sample is performed by the `device_write` driver routine whose code is shown below:

```
/* Support functions for writing fifo registers.
   The second argument is the offset in the register set */
static void Write(void *addr, enum AxiStreamFifo_Register op, u32 data )
{
        *(u32 *)(addr+op) = data;
}

static ssize_t device_write(struct file *filp, const char *buff,
        size_t len, loff_t *off)
{
        u32 val;
        struct rp_adc_dac_dev *dev;

/* writing ot DAC makes sense only one sample at a time */
        if(len != sizeof(u32))
        {
                return -EINVAL;
        }
/* Retrieve device private info */
        dev = (struct rp_adc_dac_dev *)filp->private_data;

/* Copy value from user address space */
        if(copy_from_user (&val, (void __user *)buff, sizeof(u32)))
        {
                return -EFAULT;
        }

/* Write val in dac data fifo and transmit it */
        Write(dev->iomap1_dac_data_fifo,0, val);
        Write(dev->iomap_dac_data_fifo,TLR, 4);
        return sizeof(u32);
}
```

The arguments passed to `write` by the user program are the pointer to the data to be written and its length in bytes. These arguments are the passed by the kernel to `device_write` via arguments `buff` and `len`, respectively. In our case writing of a single sample (packing two output channels) packed in a 4 byte word is only supported. The value is then retrieved from the passed pointer, but there is here a problem: the passed pointer refers to the address space of the calling program, which is different from the kernel address space (recall that the driver routines run in kernel mode). This is the reason why routine `copy_form_user` is used instead of a straight copy. Indeed the kernel has all the information (i.e. the page table contents) for properly mapping memory addresses. Once the value has been copied, it is written into `dac_data_fifo` and transmitted by writing in the appropriate FIFO registers.

ADC samples are read by user program via the driver's `read` routine that is mapped by the kernel onto driver routine `device_read` listed below.

```
/* Support functions for reading fifo registers */
static u32 Read(void *addr, enum AxiStreamFifo_Register op )
{
        return *(u32 *)(addr+op);
}
static ssize_t device_read(struct file *filp, char *buffer,
        size_t length, loff_t *offset)
{
        u32 i = 0;
```

```
          u32 actSamples;
/* Samples are 4 bytes and bring 2 16 bits ADC channels. */
          u32 requestedSamples = length/sizeof(u32);
/* Retrieve device private info */
          struct rp_adc_dac_dev *dev =
                    (struct rp_adc_dac_dev *)filp->private_data;
/* The passed buffer is considered a 32 bit array */
          u32 *b32 = (u32*)buffer;

/* Get the number of samples in the fifo via fifo register RDFO */
          dev->fifoSamples = Read(dev->iomap_adc_data_fifo,RDFO);
/* If no samples in fifo and synchronous mode, wait the sample
   availability event readq will be set by the interrupt routine
   triggered either by the availability of the first sample in the
   queue (halfInterrupt == 0) or the availability of 1024 samples in
   the queue () halfInterrupt == 1) */

          while(dev->fifoSamples == 0 && !dev->async)
          {
                    if(wait_event_interruptible(dev->readq,
                              (dev->fifoSamples > 0)))
                              return -ERESTARTSYS;
/* Read again the number of samples in the fifo
(more may have been arrived) */

                    dev->fifoSamples = Read(dev->iomap_adc_data_fifo,RDFO);

/* This will happen only if the device is closed meanwhile */
                    if(dev->fifoSamples == 0)
                    {
                              return 0;
                    }
          }

/* The fifo is 32k samples wide. It should never apporach its limit. */
          if(dev->fifoSamples > 32700)
          {
/* This can be considered an error condition, report it in the log file
   that can be read issuing dmesg command */
                    printk("rp_adc_dac Overflow!\n");
                    return -EFAULT;
          }
/* There could be more samples in fifo than those requested in read() */
          actSamples = (dev->fifoSamples > requestedSamples)?
                    requestedSamples:dev->fifoSamples;
          for(i = 0; i < actSamples; i++)
          {
/* Read the current samples (actually tow channels) from fifo register */
                    u32 currSample = Read(dev->iomap1_adc_data_fifo,RDFD4);
/* Copy the sample in the user buffer. A direct copy cannot work
   because the kernel address space is different from the user space.
   put user performeed the appropriate virtual address translation. */
                    put_user(currSample, b32++);
                    dev->fifoSamples--;
          }
/* Return the actual number of read samples, that may be 0
   for asynchronous operation. */
          return sizeof(u32)*actSamples;
}
```

After retrieving the private info structure from the passed `filp` argument in the usual way, the length of `adc_data_fifo` is checked in order to see if acquired samples are available. If there are samples, a further check is done to verify that they are not too much, i.e. the FIFO is almost full, indicating a very likely overflow. If the check passes, FIFO samples are read until either the FIFO has been emptied or the number of bytes to be read has been reached. Every read sample is copied in the user buffer via kernel routine put_buffer in order to account for the different addressing spaces. If there are no samples in the FIFO, the behavior of device_read depends on whether the device was open in synchronous or asynchronous mode. If the device has been open in asynchronous mode, device_read simply returns reporting that 0 samples have been read. Otherwise, the calling program is suspended until one or more samples are available in the FIFO (we shall discuss later the implication of these different policies). The suspension is performed by kernel routine wait_event_interruptible that suspends this process execution until both the wait queue readq used to synchronize readout is awakened and the condition (dev->fifoSamples > 0) holds. In this case the interrupt mechanism, enabled in the FPGA project for adc_data_fifo is used and the interrupt routine IRQ_cb, listed below, is called in response to the interrupt.

```
irqreturn_t IRQ_cb(int irq, void *dev_id)
{
/* The interrupt handler receives the device private info */
        struct rp_adc_dac_dev *dev =  dev_id;

/* Reset all pending interrupts and disable further interrupts */
        Write(dev->iomap_adc_data_fifo,ISR,0xFFFFFFFF);
        Write(dev->iomap_adc_data_fifo,IER,0x00000000);

/* Read from the fifo register RDFO the number of pending samples */
        dev->fifoSamples = Read(dev->iomap_adc_data_fifo,RDFO);
/* wake up possible threads waiting for data availability  */
        wake_up_interruptible(&dev->readq);

/* Re-enable interrupts */
        if(dev->fifoHalfInterrupt)
                Write(dev->iomap_adc_data_fifo,IER,0x00100000);
        else
                Write(dev->iomap_adc_data_fifo,IER,0x04000000);
        return IRQ_HANDLED;
}
```

The first action of the interrupt routine is to temporarily disable further interrupts from adc_data_fifo by writing in the appropriate FIFO registers. Then the actual number of samples is recorded in the private info structure and the wait queue awakened, possibly making a waiting process ready for execution again. Finally, interrupts (either first sample or threshold) are re-enabled.

Synchronization is also carried out by driver routine device_poll, invoked by the kernel when a user program invokes either poll() or select() to wait for one of a set of file descriptors to become ready to perform I/O.

```
static unsigned int device_poll(struct file *file,
        struct poll_table_struct *p)
{
        unsigned int mask=0;
        struct rp_adc_dac_dev *dev =
                (struct rp_adc_dac_dev *)file->private_data;
/* Poll or select cannot be called in asynchronous mode */
        if(dev->async)
        {
                return -EINVAL;
```

```
            }
/* Get the number of samples in the fifo via fifo register RDFO */
        dev->fifoSamples = Read(dev->iomap_adc_data_fifo,RDFO);
/* If no samples available AND synchronous mode, wait readq */
        if(dev->fifoSamples ==0 && !dev->async)
        {
                poll_wait(file,&dev->readq,p);
        }
/* Report in the poll mask successful input data availability */
        mask |= POLLIN | POLLRDNORM;
        return mask;
}
```

Finally, device routine `ioctl` is used to control the enable and decimation registers. More in general, in Linux device, routine `ioctl` is used all the operations that cannot fit in routine `read` or `write`. We have already seen in Chapter 2 how `ioctl` routines have been used extensively to acquire frames from a camera device. `ioctl` accepts two parameter, an integer parameter that specify the type of operation being requested and a generic pointer. The pointed data depends on the kind of operation specified, and in this case it simply refers to the 32 bit content of the register being read or written. Device `ioctl` is mapped by the kernel onto driver routine `device_ioctl` listed below:

```
static long device_ioctl(struct file *file, unsigned int cmd,
        unsigned long arg)
{
/* Retrieve device private data from the passed struct file */
    struct rp_adc_dac_dev *dev = file->private_data;

/* cmd corresponds to the passed command code */
    switch (cmd) {
/* Read content of decimation register
    The register address is mapped to iomap decimation register */
      case RP_ADC_DAC_GET_DECIMATION_REGISTER:
        {
/* The copy must span two different address spaces */
          if(copy_to_user ((void __user *)arg,
            dev->iomap_decimation_register, sizeof(u32)))
          {
            return -EFAULT;
          }
          return 0;
        }
/* Write content of decimation register */
      case RP_ADC_DAC_SET_DECIMATION_REGISTER:
        {
          if(copy_from_user (dev->iomap_decimation_register,
            (void __user *)arg, sizeof(u32)))
          {
            return -EFAULT;
          }
          return 0;
        }
/* Read content (only last bit significant) of enable register */
      case RP_ADC_DAC_GET_ENABLE_REGISTER:
        {
          if(copy_to_user ((void __user *)arg,
            dev->iomap_enable_register, sizeof(u32)))
          {
            return -EFAULT;
          }
          return 0;
```

```
      }
/* Write enable register. Used to start or stop data readout */
   case RP_ADC_DAC_SET_ENABLE_REGISTER:
   {
       if(copy_from_user (dev->iomap_enable_register,
          (void __user *)arg, sizeof(u32)))
       {
         return -EFAULT;
       }
       return 0;
   }
/* Enable interrupt when adc data fifo size exceeds 16k samples */
   case RP_ADC_DAC_FIFO_INT_HALF_SIZE:
   {
       setFifoHalfInterrupt(dev);
       return 0;
   }
/* Enable interrupt when the first sample is received by adc data fifo */
   case RP_ADC_DAC_FIFO_INT_FIRST_SAMPLE:
   {
       setFifoFirstInterrupt(dev);
       return 0;
   }
/* Invalid ioctl code */
   default:
      return -EAGAIN;
  }
}
```

Based on the passed code (codes are defined in the include file `rp_adc_dac.h`), the corresponding operation is performed. In addition to reading and writing register, two codes are used to set the interrupt generation when the fist sample is inserted in empty ADC FIFO or a threshold in the number of enqueued samples is exceeded.

We are now done with the driver. The complete source code is available on GitHub along with the files that are required to configure the RP board (FPGA bin file and devicetree). In the next section we shall demonstrate how it can be used in user programs and we shall discuss the implication in latency and throughput of the different device configurations.

17.3 USING THE DRIVER

We are now ready to put everything at work. Let's recall first the few steps required to set up the system:

1. Prepare the SD card with Ubuntu Linux that will be inserted in the board. Its content can be downloaded from `https://redpitaya.readthedocs.io/en/latest/quickStart/SDcard/SDcard.html`. In that page you can also find the instructions for flashing the SD card.
2. Insert the SD and connect the board to a terminal via the USB-C connector and to the network via the RJ45 network connector. The default network configuration is Dynamic Host Configuration (DHCP).
3. Power on the system and the boot procedure will be displayed on the connected terminal. The assigned IP address can be found with the command `ip a` and then you can connect remotely to the board via `ssh`.

4. Once you are connected to the board, copy files `bitstream.bin` and `devicetree.dtb` in your home directory and give the following commands:

   • `fpgautil -b bitstream.bin`
   • `cp devicetree.dtb /boot/dts/z10_125/devicetree.dtb`

   Reboot the system afterward. The first command configures the onboard FPGA and must be executed every time the rp board is restarted (you may put it in a startup shell script such as `.bashrc`). The second command copy the device tree configuration in the directory used by Ubuntu during bootstrap and is done once for all.

5. Compile the driver code. Observe that driver C code cannot be compiled straight as any other C program because it requires the Linux kernel configuration files. All the required packages are already available in the Red-Piteya Ubuntu distribution and there is no need for additional package installation. The Makefile listed below shall be used to build the driver code (`rp_adc_dac.ko`) from files `rp_adc_dac.c` and `rp_adc_dac.h`.

```
obj-m = rp_adc_dac.o
KVERSION = $(shell uname -r)
all:
make  CC=arm-linux-gnueabihf-gcc ARCH=arm \
KCFLAGS="-march=armv7-a -mcpu=generic-armv7-a" \
-C /lib/modules/$(KVERSION)/build M=$(PWD) modules
clean:
make -C /lib/modules/$(KVERSION)/build M=$(PWD) clean
```

6. Load the driver with the command

```
insmod rp_adc_dac.ko
```

You can check if everything was ok having a look at the kernel console via command dmesg. The driver code invoked during driver installation prints some information that looks like the following.

```
[ 2884.456097] Loading module rp_adc_dac
[ 2884.456121] MAJOR ID...236
[ 2884.456134] mknod /dev/rp_adc_dac c 236 0
[ 2884.456724] rp_adc_dac_probe  43c00000.axi_fifo_mm_s
[ 2884.456741] mem start: 43c00000
[ 2884.456750] mem end: 43c0ffff
[ 2884.456757] mem offset: 0
[ 2884.456834] IRQ: 38
[ 2884.456873] rp_adc_dac: got IRQ 56 assigned
[ 2884.461386] rp_adc_dac_probe  43c20000.axi_fifo_mm_s
[ 2884.461405] mem start: 43c20000
[ 2884.461414] mem end: 43c2ffff
[ 2884.461422] mem offset: 0
[ 2884.461645] rp_adc_dac_probe  50000000.axi_cfg_register
[ 2884.461660] mem start: 50000000
[ 2884.461669] mem end: 50000fff
```

```
[ 2884.461676] mem offset: 0
[ 2884.461851] rp_adc_dac_probe  60000000.axi_cfg_register
[ 2884.461864] mem start: 60000000
[ 2884.461872] mem end: 60000fff
[ 2884.461880] mem offset: 0
```

If you see it, everything worked and the driver is ready for use. The driver will also appear in /dev directory.

We shall now present a simple program that opens and configure the driver and then in a loop it reads the two ADC channels and send the acquired samples to the two DAC channels. This program can then be used as a template in a feedback application, performing control computation on the ADC data (data from sensors) and sending the results to the DAC channels (data to actuators). The code of the program is listed below. The program is also available in the Project GitHub repository.

```c
#include <sys/ioctl.h>
#include <fcntl.h>
#include <stdlib.h>
#include <string.h>
#include <errno.h>
#include <unistd.h>
#include <stdio.h>
#include "rp_adc_dac.h"

/* rp adc dac test program. This user program opens the rp adc dac
   device and enters a loop where ADC samples are sent back to the DAC.
   The program accepts as arguments:
   - The mode of operation: async, sync1, sync1024.
   - The sampling frequency
   - The buffer dimension   */
int main(int argc, char *argv[])
{
        enum   {
                ASYNC,  /* Asynchronous readout, no wait */
                SYNC1,  /* Synchronous read, first available sample */
                SYNC16K /* Synchronous read, 16k samples exceeded */
        } mode;
        int fd, reg, rb, freqDiv, freq, bufferDim, readSamples, i;
        unsigned int *dataBuf;
        unsigned int outData;
        short chan1, chan2;
        float fchan1, fchan2;
        if (argc != 4)
        {
                printf("Usage: adc_dac_main <sync|sync1|sync16k> \
                  <sampling frequency> <buffer dimension>");
                exit(0);
        }
        if(!strcmp(argv[1], "async"))
                mode = ASYNC;
        else if(!strcmp(argv[1], "sync1"))
                mode = SYNC1;
        else if(!strcmp(argv[1], "sync16k"))
                mode = SYNC16K;
        else
        {
                printf("Invalid mode: %s\n", argv[1]);
                exit(0);
        }
        sscanf(argv[2], "%d", &freq);
        freqDiv = 125000000/freq; //Base sampling frequency is 125Mhz
```

```
/* Get the buffer dimension in samples */
        sscanf(argv[2], "%d", &bufferDim);
        dataBuf = (int *)malloc(sizeof(int)*bufferDim);
/* Open the device */
        if(mode == ASYNC)
                fd = open("/dev/rp_adc_dac", O_NONBLOCK) ;
        else
                fd = open("/dev/rp_adc_dac", O_RDWR);
        if (fd < 0)
        {
                printf("Cannot open /dev/rp_adc_dac\n");
                exit(0);
        }

/* Set frequency decimation */
        reg = freqDiv;
        ioctl(fd, RP_ADC_DAC_SET_DECIMATION_REGISTER, &reg);

/* Check that the value has been set,
   just to make sure that the FPGA has been configured */
        reg = 0;
        ioctl(fd, RP_ADC_DAC_GET_DECIMATION_REGISTER, &reg);
        printf("Decimation: %d\n", reg);

/* Depending on the mode, define when the ADC FIFO
   generates an interrupt */
        if (mode == SYNC1)
                ioctl(fd, RP_ADC_DAC_FIFO_INT_FIRST_SAMPLE, NULL);
        if(mode == SYNC16K)
                ioctl(fd, RP_ADC_DAC_FIFO_INT_HALF_SIZE, NULL);

/* start data readout */
        reg = 1;
        ioctl(fd, RP_ADC_DAC_SET_ENABLE_REGISTER, &reg);
        while(1)
        {
                rb = read(fd, dataBuf, sizeof(int)*bufferDim);
                if(rb < 0)
                {
                        printf("ADC read failed\n");
                        exit(0);
                }
                readSamples = rb/sizeof(int);
/* The number of read samples can any value between 0 and bufferDim */
                for(i = 0; i < readSamples; i++)
                {
/* two 14 bit channels are packed in a 32 bit word
   Least significant bits encode Channel1 */
                        memcpy(&chan1, &dataBuf[i], sizeof(short));
                        chan2 = dataBuf[i] & (0xffff0000)>>16;
/* Conversion from raw data into voltage.
   Scale may depend on the RedPitaya configuration */
                        fchan1 = 30. * chan1/8192.;
                        fchan2 = 30 * chan2/8192.;

/* A control algorithm may be defined here */

/* Convert and pack the two channels in the format expected by DAC */
                        outData = (8192 - chan1) << 16;
                        outData |= (8192 - chan2) & 0x0000ffff;
/* Write to DAC */
                        write(fd, &outData, sizeof(unsigned int));
                }
        }
        return 0;
}
```

In this program three possible synchronization mechanisms for ADC readout can be defined:

- **Asynchronous**. In this mode, when the driver is requested to read ADC samples stored in the FIFO, and the FIFO is found empty, driver read returns soon reporting that no samples have been read. In this case, the read operation is continuously repeated until at least one sample has been read. This is in practice a polling mechanism and it has advantages and disadvantages, depending on the requirements. This method minimized latency because as soon as a sample has been acquired it is soon available to the program, without the overhead of the interrupt management. On the other side, such a program will take 100% of the processor time. You can easily convince yourself by issuing in a separate terminal the top command showing the CPU utilization when the program is running.

- **Synchronous, first sample interrupt**. In this case, when a read operation is performed and the ADC FIFO is still empty, the calling process is put on wait until a new sample is available. When such a sample arrives in the FIFO, it generates an interrupt and the associated driver interrupt routine will awake the calling process. Latency in this case is slightly worsened in respect of the asynchronous case, but no processor time is wasted in polling. As in asynchronous case, the read routine will return in most cases a single sample, unless the actual sampling rate becomes so high that more than one sample is acquired in the time the program takes to make a new read. This configuration is optimal for a feedback application where single samples must be processed as soon as possible in order to compute the control response and CPU power is not wasted but may prove sub-optimal for data acquisition tasks aiming at maximizing data throughput. In this case, batched acquisition is preferred since it reduces the average number of CPU cycles per acquired sample.

- **Synchronous, half FIFO interrupt**. This case differs from the previous one for the fact that a task that is waiting after issuing a read request is awakened only then 16k Samples are available in the ADC FIFO. Clearly such a configuration is not optimal for a feedback control aiming at minimizing latency, but it favors batch data transfer, possibly improving throughput.

A consequence of the synchronization mechanism is the choice of the data buffer dimension, that will be a single sample for feedback applications and a larger buffer for batch data acquisition. After opening the device, its configuration is carried out by some ioctl operations, defining the decimation rate and the interrupt mode (if synchronous). The symbols passed as first argument to ioctl are defined in the include file rp_adc_dac.h that is the same used as include file for the driver code.

In the following loop ADC samples are accessed via driver `read` and DAC samples sent via driver `write`, a quite natural choice in the driver organization. You may remember that in camera device driver presented in Chapter 2, `ioctl` and `mmap` operation were performed to access the acquire frames, but in that case the DMA mechanism was used.

17.4 SUMMARY

In this chapter we have presented a complete real-time application based on the popular RedPitaya STEMlab 125-14 board (`https://redpitaya.readthedocs.io/en/latest/intro.html`). This board hosts a two channel high speed Analog to Digital converted (ADC) and a two channels Digital to Analog Converter (DAC) making it the ideal hardware solution for several high speed real-time feedback systems. Even if several ready-to-use solutions for RedPitaya applications are available, in this chapter we have developed a new application from scratch in order to give the reader a comprehensive view of all the components that are involved in the complete data acquisition and control chain. In the first part of the chapter, an overview of the FPGA configuration is given, without entering into details as FPGA programming is outside the scope of this book. A more detailed description of the Linux driver and of a sample application reading and writing data samples is given afterward in order to present the reader a complete implementation of a Linux driver that can be used as a template for the development of Linux drivers in other applications. Most involved C code is presented in the chapter itself, but the reader will find the complete code sources and the other required files in a GitHub repository at `https://github.com/minimap-xl/RTOS_Book`. It is possible to replicate the driver and the application on a RedPitaya board by installing the standard RedPitaya Linux OS (Ubuntu) in a SD card, taking it from `https://redpitaya.readthedocs.io/en/latest/quickStart/SDcard/SDcard.html`, and then (1) copying the two support files `devicetree.dtb` and `bitstream.bin` available in the GitHub repository, (2) building the driver taking from the GitHub repository the source codes `rp_adc_dac.c` and `rp_adc_dac.h` as well as the `Makefile` required for the drive compilation. For the braves, willing to develop their own FPGA components, the complete Vivado Project including all the HDL sources is also available in the GitHub repository along with detailed instruction on how building the binaries and the device tree.

17.5 EXERCISES

EXERCISE 1

The RedPitaya board can be very useful in a larger environment, such as distributed data acquisition in an industrial application. In this context the ReadPitaya board can be configured remotely and acquired data sent over the network. When developing an application to be controlled remotely it is necessary to establish a communication protocol so that the client can properly configure the board and receive acquired data samples. If the `rp_adc_dac` driver is used the following protocol may be defined over TCP/IP:

```
Message              Response      Meaning

SYNC=[YES, NO]       OK            Defines synch or asynch mode
SYNC?                [YES, NO]     Query mode
FDIV=<division>      OK            Set frequency division
FDIV?                <division>    Get frequency division
BUFDIM=<dimension>   OK            Set databuffer dimension
BUFDIM?              <dimension>   Query current buffer dimension
SAMPLES=<nsamples>   OK            Set the number of samples to be acquired
SAMPLES?             <nsamples>    Query the number of samples
OPEN                 OK            Open and configures rp_adc_dac device
START                <see below>   Acquire and send back the acquired samples
CLOSE                OK            Close the current session
```

When the command START is issued, and the specified number of samples acquired, the number of samples (4 bytes, little endian) is sent, followed by the Channel 1 samples and then by the Channel 2 samples (2 bytes per sample, little endian).
Write a C program running on the RedPitaya board and listening for incoming connections at a given network port. When a client connects, then a session is tarted where the client can configure the board and acquire data. You have also to write a client program issuing commands and receiving answers and data samples. A C implementation of the server on the RedPitaya side and a Python implementation of the client are available at the GitHub repository.

EXERCISE 2

A method for measuring the round-trip delay, that is the time required to acquire an ADC sample, communicate it via the driver to the program and send it back to the DAC, is to generate a sawtooth wave with amplitude 1V (set both the jumpers close to the ADC inputs channels to LV) and connect it to the first ADC channel as well as the first channel of an oscilloscope. Connect then the second channel of the oscilloscope to the first DAC output and run such a program. The second waveform will appear shifted (delayed) on the oscilloscope and if you zoom the falling edge of the waveforms, you can easily measure the delay and see the jitter between the original wave and the reproduced one, discovering that the overall average delay is in the order of $10\mu s$. But what if you don't have an oscilloscope? No problem, you can use the RedPitaya board itself to measure the round trip delay. For this purpose you have to connect the used DAC output to the second ADC input. Then modify the program so that both ADC inputs are stored in memory. After running data acquisition and DAC generation for a given amount of samples in order to collect a meaningful sample history in memory, the program will compute the mean and the standard deviation of the delays for every falling edge in the two acquired channels (falling edges are easily detected in memory by computing the difference between every sample and the previous one). As the delay is in the order of $10\mu s$, in order to collect a meaningful statistics (the delay can only be measured in terms of sample count differences between the falling edge for the two ADC channels), it is necessary to acquire at the highest achievable sampling speed (around 200 kSamples/s). If the frequency of the sawtooth waveform is 1 kHz, after collecting 10 MSamples around 50000 edges are collected, providing an acceptable statistics.

As usual, a C implementation of this program is available in the GitHub repository.

18 Control Theory and Digital Signal Processing Primer

Two case studies will be presented in this chapter in order to introduce basic concepts in Control Theory and Digital Signal Processing, respectively. The first one will consist in a simple control problem to regulate the flow of a liquid in a tank in order to stabilize its level. Here, some basic concepts of control theory will be introduced to let the reader become familiar with the concept of transfer function, system stability, and the techniques for its practical implementation.

The second case study is the implementation of a low-pass digital filter and is intended to introduce the reader to some basic concepts in digital signal processing, using the notation introduced by the first example.

18.1 CASE STUDY 1: CONTROLLING THE LIQUID LEVEL IN A TANK

Consider a tank containing a liquid and a pump that is able to move the liquid back and forth in the tank, as shown in Figure 18.1. The pump is able to generate a liquid flow and receives the flow reference from a control system with the aim of maintaining the level of the liquid in the tank to a given reference value, possibly varying over time. A practical application of such a system is the regulation of the fuel level inside the aircraft wings housing the fuel reservoir. The level of the fuel within each wing has to be maintained controlled in order to ensure the proper weight balancing in the aircraft during the flight and at the same time provide engine fueling.

The system has one detector for the measurement of the level of the liquid, producing a time-dependent signal representing, at every time t, the level $h(t)$, and one actuator, that is, the pump whose flow at every time t depends on the preset value $f_p(t)$. In order to simplify the following discussion, we will assume that the pump represents an ideal actuator, producing a flow $f(t)$ at any time equal to the current preset value, that is, $f(t) = f_p(t)$. The tank-pump system of Figure 18.1 will then have an input, that is, the preset flow $f(t)$ and one output—the liquid level in the tank $h(t)$. A positive value of $f(t)$ means that the liquid is flowing into the tank, and a negative value of $f(t)$ indicates that the liquid is being pumped away the tank. The tank–pump system input and output are correlated as follows:

$$h(t) = h_0 + \frac{1}{B} \int_0^t f(\tau)d\tau \tag{18.1}$$

where h_0 is the level in the tank at time 0, and B is the base surface of the tank. $f(\tau)d\tau$ represents, in fact, the liquid volume change during the infinitesimal time $d\tau$. In order to control the level of the liquid, we may think of sending to the actuator

DOI: 10.1201/9781003593416-18   **385**

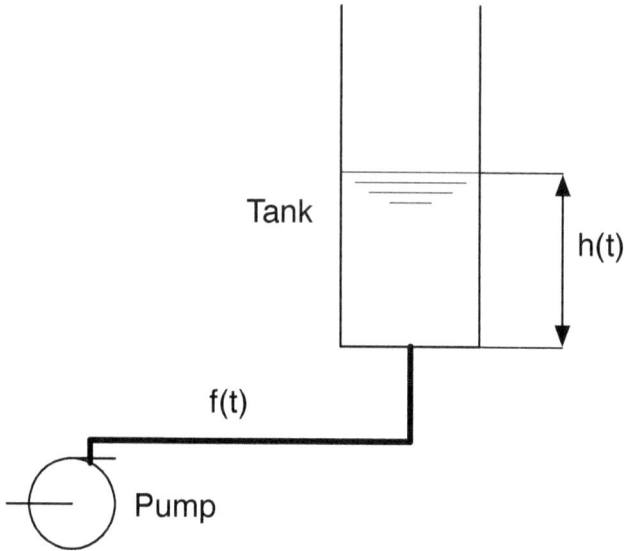

Figure 18.1   The tank–pump system.

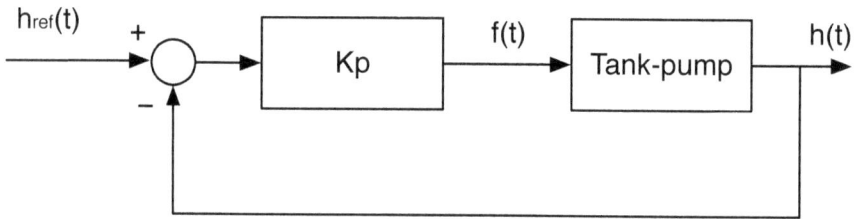

Figure 18.2   Tank–pump system controlled in feedback mode.

(the pump) a reference signal that is proportional to the difference of the measured level and the reference level value, that is,

$$f(t) = K_p[h_{ref}(t) - h(t)] \qquad (18.2)$$

corresponding to the schema shown in Figure 18.2. This is an example of feedback control where the reference signal depends also on its current output. Parameter K_p is called the *Proportional Gain*, and this kind of feedback control is called *proportional* because the system is fed by a signal which is proportional to the current error, that is, the difference between the desired output $h_{ref}(t)$ and the current one $h(t)$. This kind of control intuitively works. In fact, if the current level $h(t)$ is lower than the reference value $h_{ref}(t)$, the preset flow is positive, and therefore liquid enters the tank. Conversely, if $h(t) > h_{ref}(t)$, liquid is pumped away from the tank, and when the liquid level is ok, that is, $h(t) = h_{ref}(t)$, the requested flow is 0.

18.1.1   THE USE OF DIFFERENTIAL EQUATIONS TO DESCRIBE THE DYNAMICS OF THE SYSTEM

In order to compute the actual time evolution of the liquid level when, say, the reference value is set to h_{ref} at time $t = 0$ and is not changed afterward, and the initial level of the liquid is h_0, we must consider the input/output (I/O) relationship of the tank-pump system of Figure 18.1, where the actual input is given by (18.2). From (18.1) we obtain

$$h(t) = h_0 + \frac{1}{B}\int_0^t K_p[h_{ref} - h(\tau)]d\tau \tag{18.3}$$

If we consider the time derivative of both terms in (18.3), we get

$$\frac{dh(t)}{dt} = \frac{K_p}{B}[h_{ref} - h(t)] \tag{18.4}$$

that is, the differential equation

$$\frac{dh(t)}{dt} + \frac{K_p}{B}h(t) = \frac{K_p}{B}h_{ref} \tag{18.5}$$

whose solution is the actual evolution $h(t)$ of the liquid level in the tank.

The tank–pump system is an example of *linear system*. More in general, the I/O relationship for linear systems is expressed by the differential equation

$$a_n\frac{dy^n}{dt^n} + a_{n-1}\frac{dy^{n-1}}{dt^{n-1}} + \ldots + a_1\frac{dy}{dt} + a_0 = b_m\frac{du^m}{dt^m} + b_{m-1}\frac{du^{m-1}}{dt^{m-1}} + \ldots + b_1\frac{du}{dt} + b_0 \tag{18.6}$$

where $u(t)$ and $y(t)$ are the input and output of the system, respectively, and coefficients a_i and b_i are constant. In our tank–pump system, the input $u(t)$ is represented by the applied reference level $h_{ref}(t)$ and the output $y(t)$ is the actual liquid level $h(t)$. The general solution of the above equation is of the form

$$y(t) = y_l(t) + y_f(t) \tag{18.7}$$

where $y_l(t)$ is the *homogeneous* solution, that is, the solution of the same differential equation, where the term on the right is 0 and describes the *free* evolution of the system, and $y_f(t)$ is a particular solution of (18.6), also called *forced* evolution. In our case, the reference h_{ref} is constant, and therefore, a possible choice for the forced evolution is $y_f(t) = h_{ref}$. In this case, in fact, for $t > 0$, the derivative term of (18.5) is 0, and the equation is satisfied.

In order to find the homogeneous solution, we recall that the general solution of the generic differential equation

$$a_n\frac{dy^n}{dt^n} + a_{n-1}\frac{dy^{n-1}}{dt^{n-1}} + \ldots + a_1\frac{dy}{dt} + a_0 = 0 \tag{18.8}$$

is of the form

$$y(t) = \sum_{i=1}^{n'}\sum_{k=0}^{\mu_i} A_{ik}t^k e^{p_i t} \tag{18.9}$$

where A_{ik} are coefficients that depend on the initial system condition, and n' and μ_i are the number of different roots and their multiplicity of the polynomial

$$a_n p^n + a_{n-1} p^{n-1} + \dots + a_1 p + a_0 = 0 \qquad (18.10)$$

respectively. Polynomial (18.10) is called the *characteristic equation* of the differential equation (18.7). The terms $t^k e^{p_i t}$ are called the *modes* of the system. Often, the roots of (18.10) have single multiplicity, and the modes are then of the form $e^{p_i t}$. It is worth noting that the roots of polynomial (18.10) may be real values or complex ones, that is, of the form $p = a + jb$, where a and b are the *real* and *imaginary* parts of p, respectively. Complex roots (18.10) appear in conjugate pairs (the conjugate of a complex number $a + jb$ is $a - jb$). We recall also that the exponential of a complex number $p = a + jb$ is of the form $e^p = e^a[\cos b + j \sin b]$. The modes are very important in describing the dynamics of the system. In fact, if any root of the associated polynomial has a positive real part, the corresponding mode will have a term that diverges over time (an exponential function with an increasing positive argument), and the system becomes *unstable*. Conversely, if all the modes have negative real part, the system transients will become negligible after a given amount of time. Moreover, the characteristics of the modes of the system provide us with additional information. If the modes are real numbers, they have the shape of an exponential function; if instead they have a nonnull imaginary part, the modes will have also an oscillating term, whose frequency is related to the imaginary part and whose amplitude depends on the real part of the corresponding root.

The attentive reader may be concerned by the fact that, while the modes are represented by complex numbers, the free evolution of the system must be represented by real numbers (after all, we live in a real world). This apparent contradiction is explained by considering that the complex roots of (18.10) are always in conjugate pairs, and therefore, the imaginary terms elide in the final summation of (18.9). In fact, for every complex number $p = a + jb$ and its complex conjugate $\overline{p} = a - jb$, we have

$$e^{\overline{p}} = e^a[\cos(b) + j\sin(-b)] = e^a[\cos(b) - j\sin(b)] = \overline{(e^p)}; \qquad (18.11)$$

moreover, considering the common case in which solutions of the characteristic equation have single multiplicity, the (complex) coefficients A_i in (18.9) associated with $p_i = a + jb$ and $\overline{p_i} = a - jb$ are A_i and $\overline{A_i}$, respectively, and therefore

$$A_i e^{p_i t} + \overline{A_i} e^{\overline{p_i} t} = e^a[2\operatorname{Re}(A_i)\cos(bt) - 2\operatorname{Im}(A_i)\sin(bt)] \qquad (18.12)$$

where $\operatorname{Re}(A_i)$ and $\operatorname{Im}(A_i)$ are the real and imaginary parts of A_i, respectively. Equation (18.12) represents the contribution of the pair of conjugate roots, which is a real number.

Coming back to our tank–pump system, the polynomial associated with the differential equation describing the dynamics of the system is

$$p + \frac{K_p}{B} = 0 \qquad (18.13)$$

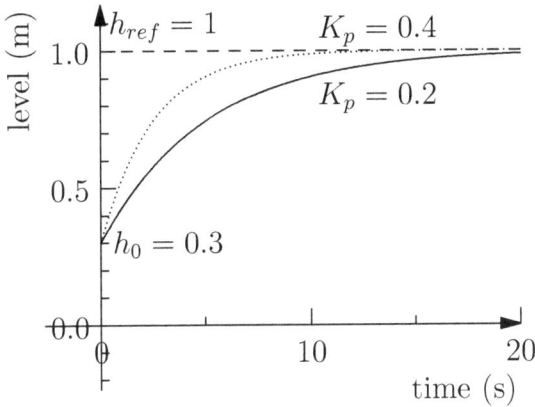

Figure 18.3   Tank–pump system response when controlled in feedback mode.

which yields the single real solution $p_0 = -\frac{K_p}{B}$. The homogeneous solution for (18.5) is then

$$h_l(t) = Ae^{-\frac{K_p t}{B}} \tag{18.14}$$

and the general solution for (18.5) is

$$h(t) = h_l(t) + h_f(t) = Ae^{-\frac{K_p t}{B}} + h_{ref} \tag{18.15}$$

where $h_l(t)$ and $h_f(t)$ are the free and forced solutions of (18.5).

Parameter A is finally computed considering the boundary condition of the system, that is, the values of $h(t)$ for $t = 0_-$. Just before the reference value h_{ref} has been applied to the system. For $t = 0_-$ (18.15) becomes

$$h_0 = Ae^0 + h_{ref} = A + h_{ref} \tag{18.16}$$

which yields the solution $A = h_0 - h_{ref}$, thus getting the final response of our tank–pump system

$$h(t) = (h_0 - h_{ref})e^{-\frac{K_p t}{B}} + h_{ref} \tag{18.17}$$

The system response of the tank–pump system controlled in feedback, that is, the time evolution of the level of the liquid in the tank in response to a step reference h_{ref} is shown in Figure 18.3 for two different values of the proportional gain K_p, with tank base surface $B = 1\,\text{m}^2$.

18.1.2   INTRODUCING AN INTEGRAL GAIN

We observe that the proportional gain K_p affects the readiness of the response, as shown by Figure 18.3, and larger proportional gains produce a faster response. We may therefore think that we can choose a proportional gain large enough to reach the desired speed in the response. However, in the choice of K_p we must consider

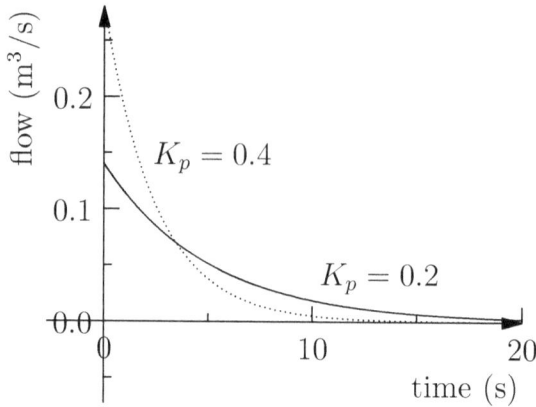

Figure 18.4 Flow request to the pump using feedback control with proportional gain.

the limit in the pump ability in generating the requested liquid flow. As an example, the flow request to the pump at the proportional gains considered in Figure 18.3, is shown in Figure 18.4.

From Figure 18.4 it can be seen that larger proportional gains imply larger flows requested to the pump. In practice, the maximum allowable value of the proportional gain is limited by the maximum flow the pump is able to generate. Moreover, the system will always approach the requested liquid level asymptotically. We may wonder if it is possible to find some other control schema that could provide a faster response. A possibility could be considering an additional term in the input for the pump that is proportional to the integral of the error, in the hope it can provide a faster response. In fact, in the evolution plotted in Figure 18.3, the integral of the error is always positive. We may expect that forcing the integral of the error to become null will provide a faster step response, possibly with an overshoot, in order to make the overall error integral equal to 0 (see Figure 18.9). To prove our intuition, we consider a reference signal for the pump of the form

$$f(t) = K_p[h_{ref}(t) - h(t)] + K_i \int_0^t [h_{ref}(\tau) - h(\tau)]d\tau \qquad (18.18)$$

where K_p and K_i are the *proportional* and *integral* gains in the feedback control, respectively. We obtain, therefore, the relation

$$h(t) = h_0 + \frac{1}{B}\int_0^t f(\tau)d\tau =$$

$$h_0 + \frac{1}{B}\int_0^t \{K_p[h_{ref}(\tau) - h(\tau)] + \frac{K_i}{B}\int_0^\tau [h_{ref}(\tau') - h(\tau')]d\tau'\}d\tau \qquad (18.19)$$

Derivating both terms, we obtain

$$\frac{dh(t)}{dt} = \frac{K_p}{B}[h_{ref}(t) - h(t)] + \frac{K_i}{B}\int_0^t [h_{ref}(\tau) - h(\tau)]d\tau \qquad (18.20)$$

and, by derivating again

$$\frac{d^2h(t)}{dt^2} = \frac{K_p}{B}[\frac{dh_{ref}(t)}{dt} - \frac{dh(t)}{dt}] + \frac{K_i}{B}[h_{ref}(t) - h(t)] \qquad (18.21)$$

that is, the differential equation

$$B\frac{d^2h(t)}{dt^2} + K_p\frac{dh(t)}{dt} + K_ih(t) = K_ih_{ref}(t) + K_p\frac{dh_{ref}(t)}{dt} \qquad (18.22)$$

whose solution is the time evolution $h(t)$ of the liquid level in the tank.

18.1.3 USING TRANSFER FUNCTIONS IN THE LAPLACE DOMAIN

At this point the reader may wonder if defining control strategies always means solving differential equations, which is possibly complicated. Fortunately, this is not the case, and control theory uses a formalism for describing linear systems that permits the definition of optimal control strategies without developing explicit solutions to the differential equations representing the I/O relationship.

Instead of dealing with time-dependent signals, we shall use *Laplace transforms*. Given a function on time $f(t)$, its Laplace transform is of the form

$$F(s) = L\{f(t)\} = \int_0^\infty f(t)e^{-st}dt \qquad (18.23)$$

where s and $F(s)$ are complex numbers. Even if at a first glance this approach may complicate things, rather than simplifying problems (we are now considering complex functions of complex variables), this new formalism can rely on a few interesting properties of the Laplace transforms, which turn out to be very useful for expressing I/O relationships in linear systems in a simple way. In particular we have

$$L\{Af(t) + Bg(t)\} = AL\{f(t)\} + BL\{g(t)\} \qquad (18.24)$$

$$L\{\frac{df(t)}{dt}\} = sL\{f(t)\} \qquad (18.25)$$

$$L\{\int f(t)dt\} = \frac{L\{f(t)\}}{s} \qquad (18.26)$$

Equation (18.24) states that the Laplace transform is a linear operator, and, due to (18.25) and (18.26), relations expressed by time integration and derivation become algebraic relations when considering Laplace transforms. In the differential equation in (18.22), if we consider the Laplace transforms $H(s)$ and $H_{ref}(s)$ in place of $h(t)$ and $h_{ref}(t)$, from (18.24), (18.25) and (18.26), we have for the tank–pump system

$$Bs^2H(s) + K_psH(s) + K_iH(s) = K_iH_{ref}(s) + K_psH_{ref}(s) \qquad (18.27)$$

that is,

$$H(s) = \frac{K_i + K_ps}{Bs^2 + K_ps + K_i}H_{ref}(s) \qquad (18.28)$$

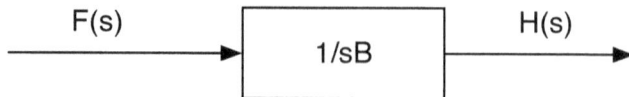

Figure 18.5 Graphical representation of the transfer function for the tank–pump system.

Observe that the I/O relationship of our tank–pump system, which is expressed in the time domain by a differential equation, becomes an algebraic relation in the Laplace domain. The term

$$W(s) = \frac{K_i + K_p s}{Bs^2 + K_p s + K_i} \tag{18.29}$$

is called the *Transfer Function* and fully characterizes the system behavior. Using Laplace transforms, it is not necessary to explicitly express the differential equation describing the system behavior, and the transfer function can be directly derived from the block description of the system. In fact, recalling the I/O relationship of the tank and relating the actual liquid level $h(t)$ and the pump flow $f(t)$ in (18.1), using property (18.26), we can express the same relationship in the Laplace domain as

$$H(s) = \frac{1}{sB} F(s) \tag{18.30}$$

where $H(s)$ and $F(s)$ are the Laplace transforms of $h(t)$ and $f(t)$, respectively. This relation can be expressed graphically as in Figure 18.5. Considering the control law involving the proportional and integral gain

$$f(t) = K_p[h_{ref}(t) - h(t)] + K_i \int_0^t [h_{ref}(\tau) - h(\tau)]d\tau \tag{18.31}$$

the same law expressed in the Laplace domain becomes:

$$F(s) = (K_p + \frac{K_i}{s})[H_{ref}(s) - H(s)] \tag{18.32}$$

and therefore, we can express the whole tank–pump system as in Figure 18.6. From that figure we can easily derive

$$H(s) = \frac{1}{sB}[K_p + \frac{K_i}{s}][H_{ref}(s) - H(s)] \tag{18.33}$$

that is

$$H(s) = H_{ref}(s)\frac{sK_p + K_i}{s^2 B + sK_p + K_i} \tag{18.34}$$

obtaining, therefore, the same transfer function of (18.29) directly from the graphical representation of the system without explicitly stating the differential equation describing the system dynamics.

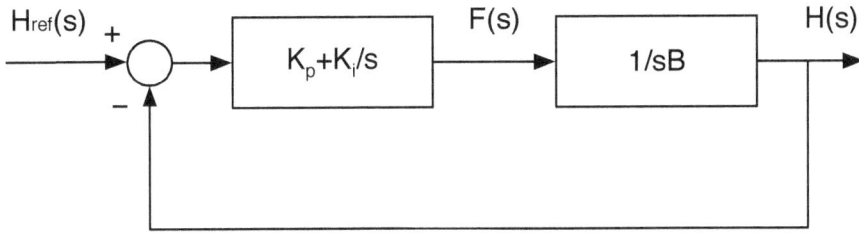

Figure 18.6   Graphical representation of tank–pump system controlled in feedback.

18.1.4   DERIVING SYSTEM PROPERTIES FROM ITS TRANSFER FUNCTION

The transfer function $W(s)$ of a linear system fully describes its behavior and, in principle, given any reference $h_{ref}(t)$, we could compute its Laplace transform, multiply it for $W(s)$, and then compute its antitransform to retrieve the system response $h(t)$. We do not report here the formula for the Laplace antitransform because, in practice, an analytical solution of the above procedure would be very difficult, if not impossible. Rather, computational tools for the numerical simulation of the system behavior are used by control engineers in the development of optimal control strategies. Several important system properties can, however, be inferred from the transfer function $W(s)$ without explicitly computing the system evolution over time. In particular, stability can be inferred from $W(s)$ when expressed in the form

$$W(s) = \frac{N(s)}{D(s)} \qquad (18.35)$$

The roots of $N(s)$ are called the *zeroes* of the transfer function, while the roots of $D(s)$ are called the *poles* of the transfer function. $W(s)$ becomes null in its zeroes, diverging when s approaches its poles. Recall that $W(s)$ is a complex value of complex variable. For a graphical representation of $W(s)$, usually its module is considered. Recall that the module of a complex number $a + jb$ is $\sqrt{a^2 + b^2}$, that is, the length of the segment joining the origin and the point of coordinates (a, b) in a two-dimensional cartesian system, where the x and y coordinates represent the real and imaginary parts, respectively. The module of $W(s)$ is therefore represented by a real function (the module) of two real values (the real and imaginary parts of the complex variable s), and can be represented in three-dimensional Cartesian system. As an example, Figure 18.7 shows the representation of the transfer function for the tank–pump system where the base area of the tank is $B = 1$, and the feedback gains are $K_p = 1$ and $K_i = 1$.

The function shown in Figure 18.7 diverges when s approaches the roots of $D(s) = s^2 + s + 1$, i.e., for $s_1 = -\frac{1}{2} + j\frac{\sqrt{3}}{2}$ and $s_2 = -\frac{1}{2} - j\frac{\sqrt{3}}{2}$, and becomes null when $N(s) = 0$, that is, for $s = -1 + j0$. There is normally no need to represent graphically in this way the module of the transfer function $W(s)$, and it is more useful to express graphically its poles and zeroes in the complex plane, as shown in Figure 18.8. By

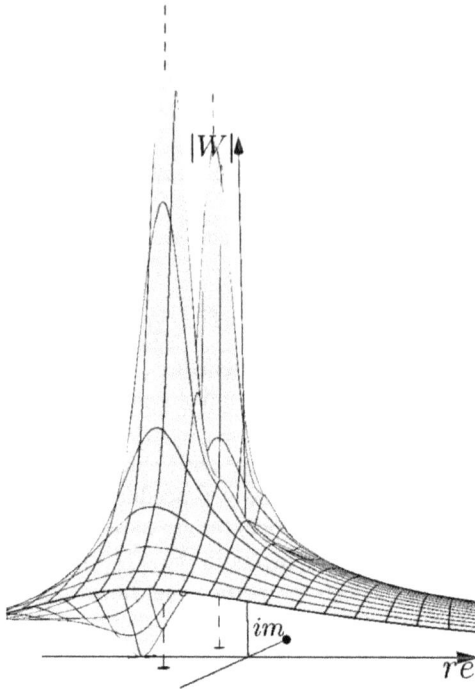

Figure 18.7   The module of the transfer function for the tank–pump system.

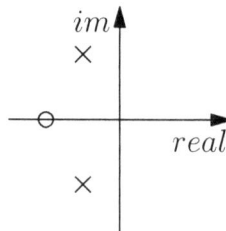

Figure 18.8   Zeroes and poles of the transfer function for the tank–pump system.

convention, zeroes are represented by a circle and poles by a cross. System stability can be inferred by $W(s)$ when expressed in the form $W(s) = \frac{N(s)}{D(s)}$, where the numerator $N(s)$ and the denominator $D(s)$ are polynomials in s. Informally stated, a system is stable when its natural evolution with a null input is toward quiet, regardless of its initial condition. The output of an unstable system may instead diverge even with

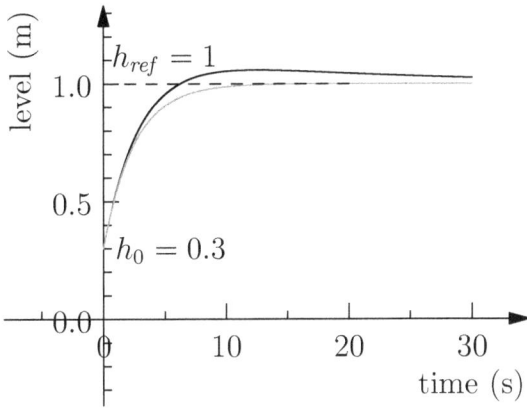

Figure 18.9 The response of the controlled tank–pump system with proportional gain set to 0.4 and integral gains set to 0.02 (black) and 0 (grey), respectively.

null inputs. If we recall how the expression of $W(s)$ has been derived from the differential equation expressing the I/O relationship of the system, we recognize that the denominator $D(s)$ corresponds to the characteristic equation of the linear system whose roots define the modes of the system. So, recalling the definition (18.9) of the modes of the system contributing to its free evolution, we can state that if the poles of the transfer function $W(s)$ have a positive real part, the system will be unstable. Moreover, if the poles of $W(s)$ have a non-null imaginary part, we can state that oscillations will be present in the free evolution of the system, and these oscillation will have a decreasing amplitude if the real part of the poles is negative, and increasing otherwise. The limit case is for poles of $W(s)$, which are pure imaginary numbers (i.e., with null real part); in this case, the free evolution of the system oscillates with constant amplitude over time.

Let us now return to the tank–pump system controlled in feedback using a proportional gain, K_p, and an integral gain, K_i. Recalling its transfer function in (18.34), we observe that its poles are the solution to equation

$$s^2 B + s K_p + K_i = 0 \qquad (18.36)$$

that is,

$$s_{1,2} = \frac{-K_p \pm \sqrt{K_p^2 - 4 K_i B}}{2B} \qquad (18.37)$$

(recall that B is the base surface of the tank). Figure 18.9 shows the controlled tank–pump response for $K_p = 0.4$ and $K_i = 0.02$, compared with the same response for a proportional gain $K_p = 0.4$ only. It can be seen that the response is faster, but an overshoot is present, as the consequence of the nonnull integral gain (recall that control tries to reduce the integral of the error in addition to the error itself). For $K_p > 2\sqrt{K_i B}$ Equation (18.36) yields two real solutions, and therefore, the modes of the system in its free evolution have an exponential shape and do not oscillate.

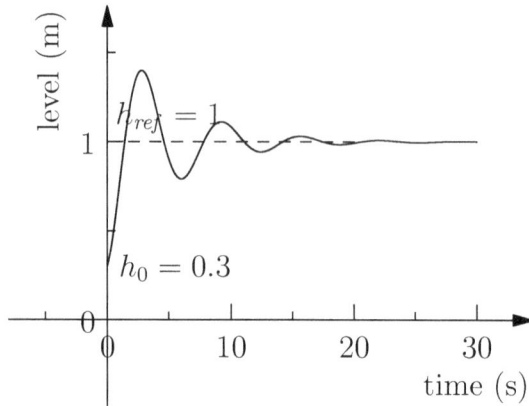

Figure 18.10 The response of the controlled tank–pump system with proportional gain set to 0.4 and integral gain set to 1.

Conversely, for K_i to be large enough, that is, $K_i > \frac{K_p^2}{4B}$ the two poles of $W(s)$ become complex and therefore the system response contains an oscillating term, as shown in Figure 18.10. In any case, the oscillations are smoothed since $K_p \geq 0$, and therefore, the real part of the poles is not positive. It is interesting to observe that if $K_p = 0$, that is, when considering only the integral of the error $h_{ref}(t) - h(t)$ in the feedback control, the system response contains an oscillating term with constant amplitude.

Before proceeding, it is worthwhile now to summarize the advantages provided by the representation of the transfer functions expressed in the Laplace domain for the analysis of linear systems. We have seen how it is possible to derive several system properties simply based on its block diagram representation and *without deriving the differential equation* describing the system dynamics. More importantly, this can be done *without deriving any analytical solution* of such differential equation. This is the reason why this formalism is ubiquitously used in control engineering.

18.1.5   IMPLEMENTING A TRANSFER FUNCTION

So far, we have learned some important concepts of control theory and obtained some insight in the control engineer's duty, that is, identifying the system, modeling it, and finding a control strategy. The outcome of this task is the specification of the control function, often expressed in the Laplace domain. It is worth noting that in the definition of the control strategy for a given system, we may deal with different transfer functions. For example, in the tank–pump system used throughout this chapter, we had the following transfer functions:

$$W_1(s) = \frac{1}{Bs} \tag{18.38}$$

which is the description of the tank I/O relationship

$$W_2(s) = K_p + \frac{K_i}{s} \qquad (18.39)$$

which represents the control law, and

$$W_3(s) = \frac{K_i + K_p s}{Bs^2 + K_p s + K_i} \qquad (18.40)$$

which describes the overall system response.

If we turn our attention to the implementation of the control, once the parameters K_p and K_i have been chosen, we observe that the embedded system must implement $W_2(s)$, that is, the controller. It is necessary to provide to the controller the input error $h_{ref}(t) - h(t)$, that is the difference between the current level of the liquid in the tank and the reference one. The output of the controller, $f(t)$ will drive the pump in order to provide the requested flow. The input to the controller is therefore provided by a *sensor*, while its output will be sent to an *actuator*. The signals $h(t)$ and $f(t)$ may be analog signals, such as a voltage level. This was the common case in the older times, prior to the advent of digital controllers. In this case the controller itself was implemented by an analog electronic circuit whose I/O law corresponded to the desired transfer function for control. Nowadays, analog controllers are rarely used and digital controllers are used instead. Digital controllers operate on the sampled values of the input signals and produce sampled outputs. The input may derive from analog-to-digital conversion (ADC) performed on the signal coming from the sensors, or taking directly the numerical values from the sensor, connected via a local bus, a Local Area Network or, more recently, a wireless connection. The digital controller's outputs can then be converted to analog voltages by means of a Digital to Analog converter (DAC) and then given to the actuators, or directly sent to digital actuators with some sort of bus interface.

When dealing with digital values, that is, the sampled values of the I/O signals, an important factor is the sampling period T. For the tank–pump system, the analog input $h(t)$ is then transformed into a sequence of sampled values $h(nT)$ as shown in Figure 18.11. Since sampling introduces unavoidable loss of information, we would like that the sampled signal could represent an approximation that is accurate enough for our purposes. Of course, the shorter the sampling period T, the more accurate the representation of the original analog signal. On the other side, higher sampling speed comes at a cost since it requires a faster controller and, above all, faster communication, which may become expensive. A trade-off is therefore desirable, that is, choosing a value of T that is short enough to get an acceptable approximation, but avoiding implementing an "overkill." Such a value of T cannot be defined a priori, but depends on the dynamics of the system; the faster the system response, the shorted must be the sampling period T. A crude method for guessing an appropriate value of T is to consider the step response of the system, as shown in (18.9), and to choose T as the rise time divided for 10, so that a sufficient number of samples can describe the variation of the system output. More accurate methods consider the

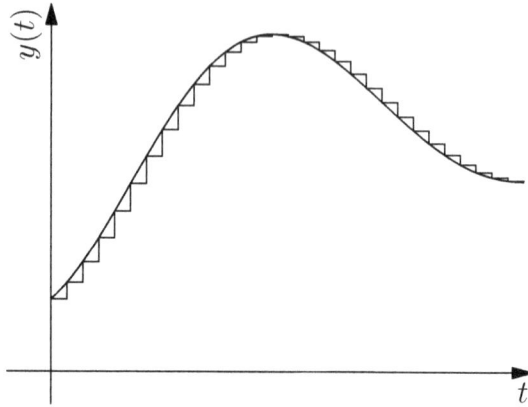

Figure 18.11 Sampling a continuous function.

poles of $W(s)$, which determine the modes of the free system evolution, that is, its dynamics. In principle, the poles with the largest absolute value of their real part, that is, describing the fastest modes of the system, should be considered, but they are not always significant for control purposes (for example, when their contribution to the overall system response is negligible). Moreover, what gets digitized in the system is only the controller, not the system itself. For this reason, normally the working range of frequencies for the controller is considered, and the sampling period chosen accordingly. The second test case presented in this chapter will describe more in detail how the sampling period is chosen based on frequency information.

Before proceeding further, it is necessary to introduce another mathematical formalism that turns out very useful in digitalizing transfer functions, and, in the end, in implementing digital controllers: the Z transform. Given a sequence of real values $y(n)$, its Z transform is a complex function of complex variable z defined as

$$Z\{y(n)\} = \sum_{n=-\infty}^{\infty} y(n)z^{-n} \qquad (18.41)$$

Even if this may seem another way to complicate the engineer's life (as for the Laplace transforms, we are moving from the old, familiar, real space into a complex one), we shall see shortly how Z transforms are useful for specifying the I/O relationship in digital controllers. The following important two facts hold:

$$Z\{Ay_1(n) + By_2(n)\} = AZ\{y_1(n)\} + BZ\{y_2(n)\} \qquad (18.42)$$

that is, the Z transform is linear, and

$$Z\{y(n-1)\} = \sum_{n=-\infty}^{\infty} y(n-1)z^{-n} = \sum_{n'=-\infty}^{\infty} y(n')z^{-n'}z^{-1} = z^{-1}Z\{y(n)\} \qquad (18.43)$$

The above relation has been obtained by replacing the term n in the summation with $n' = n - 1$. Stated in words, (18.43) means that there is a simple relation (the multiplication for z^{-1}) between the Z transform of a sequence $y(n)$ and that of the same sequence delayed of one sample. Let us recall the general I/O relationship in a linear system:

$$a_n \frac{dy^n}{dt^n} + a_{n-1} \frac{dy^{n-1}}{dt^{n-1}} + \ldots + a_1 \frac{dy}{dt} + a_0 =$$
$$b_m \frac{du^m}{dt^m} + b_{m-1} \frac{du^{m-1}}{dt^{m-1}} + \ldots + b_1 \frac{du}{dt} + b_0 \quad (18.44)$$

where $u(t)$ and $y(t)$ are the continuous input and output of the system, respectively. If we move from the continuous values of $y(t)$ and $u(t)$ to the corresponding sequence of sampled values $y(kT)$ and $u(kT)$, after having chosen a period T small enough to provide a satisfactory approximation of the system evolution, we need to compute an approximation of the sampled time derivatives of $y(t)$ and $u(t)$. This is obtained by approximating the first order derivative with a finite difference:

$$\frac{dy}{dt}(kT) \simeq \frac{y(kT) - y((k-1)T)}{T} \quad (18.45)$$

Recalling that the Z transform is linear, we obtain the following Z representation of the first-order time-derivative approximation:

$$Z\{\frac{y(kT) - y((k-1)T)}{T}\} = \frac{1}{T} Z\{y(kT)\} - \frac{z^{-1}}{T} Z\{y(kT)\} = \frac{1 - z^{-1}}{T} Z\{y(kT)\} \quad (18.46)$$

which is a relation similar to that of (18.25) relating the Laplace transforms of $y(t)$ and its time derivative expressing the I/O relationship for a linear system. Using the same reasoning, the second-order time derivative is approximated as

$$(\frac{1 - z^{-1}}{T})^2 Z\{y(kT)\} \quad (18.47)$$

The I/O relationship expressed using Z transforms then becomes

$$Y(z)[a_n \frac{(1 - z^{-1})^n}{T^n} + \ldots + a_1 \frac{(1 - z^{-1})}{T} + a_0] =$$
$$U(z)[b_m \frac{(1 - z^{-1})^m}{T^m} + \ldots + b_1 \frac{(1 - z^{-1})}{T} + b_0] \quad (18.48)$$

that is,

$$Y(z) = V(z)U(z) \quad (18.49)$$

where $Y(z)$ and $U(z)$ are the Z transform of $y(t)$ and $u(t)$, respectively, and $V(z)$ is the transfer function of the linear system in the Z domain. Again, we obtain an algebraic relationship between the transforms of the input and output, considering the sampled values of a linear system. Observe that the transfer function $V(z)$ can be derived directly from the transfer function $W(s)$ in the Laplace domain, by the replacement

$$s = \frac{1 - z^{-1}}{T} \tag{18.50}$$

From a specification of the transfer function $W(s)$ expressed in the form $W(s) = \frac{N(s)}{D(s)}$ with the replacement of (18.50), we derive a specification of $V(z)$ expressed in the form

$$V(z) = \frac{\sum_{i=0}^{m} b_i (z^{-1})^i}{\sum_{i=0}^{n} a_i (z^{-1})^i} \tag{18.51}$$

that is,

$$Y(z) \sum_{i=0}^{n} a_i (z^{-1})^i = U(z) \sum_{i=0}^{m} b_i (z^{-1})^i \tag{18.52}$$

Recalling that $Y(z)z^{-1}$ is the Z transform of the sequence $y((k-1)T)$ and, more in general, that $Y(z)z^{-i}$ is the Z transform of the sequence $y((k-i)T)$, we can finally get the I/O relationship of the discretized linear system in the form

$$\sum_{i=0}^{n} a_i y((k-i)T) = \sum_{i=0}^{m} b_i u((k-i)T) \tag{18.53}$$

that is,

$$y(kT) = \frac{1}{a_0} [\sum_{i=0}^{m} b_i u((k-i)T) - (\sum_{i=1}^{n} a_i y((k-i)T)] \tag{18.54}$$

Put in words, the system output is the linear combination of the $n-1$ previous outputs and the current input plus the $m-1$ previous inputs. This representation of the controller behavior can then be easily implemented by a program making only multiplication and summations, operations that can be executed efficiently by CPUs.

In summary, the steps required to transform the controller specification given as a transfer function $W(s)$ into a sequence of summations and multiplications are the following:

1. Find out a reasonable value of the sampling period T. Some insight into the system dynamics is required for a good choice.
2. Transform $W(s)$ into $V(z)$, that is, the transfer function of the same system in the Z domain by replacing variable s with $\frac{1-z^{-1}}{T}$.
3. From the expression of $V(z)$ as the ratio of two polynomials in z^{-1}, derive the relation between the current system output $y(kT)$ and the previous input and output history.

Returning to the case study used through this section, we can derive the implementation of the digital controller for the tank–pump system from the definition of its transfer function in the Laplace domain, that is,

$$W_2(s) = \frac{sK_p + K_i}{s} \tag{18.55}$$

representing the proportional and integral gain in the feedback control. Replacing s with $\frac{1-z^{-1}}{T}$ we obtain

$$V(z) = \frac{K_p + K_i T - z^{-1} K_p}{1 - z^{-1}} \tag{18.56}$$

from which we derive the definition of the algorithm

$$y(kT) = y((k-1)T) + (K_p + K_i T)u(kT) - K_p u((k-1)T) \tag{18.57}$$

where the input $u(kT)$ is represented by the sampled values of the difference between the sampled reference $h_{ref}(kT)$ and the actual liquid level $h(kT)$, and the output $y(kT)$ corresponds to the flow reference sent to the pump.

Observe that the same technique can be used in a simulation tool to compute the overall system response, given the transfer function $W(s)$. For example, the plots in Figures 18.9 and 18.10 have been obtained by discretizing the overall tank–pump transfer function (18.40).

An alternative method for the digital implementation of the control transfer function is the usage of the *Bilinear Transform*, that is, using the replacement

$$s = \frac{2}{T} \frac{1 - z^{-1}}{1 + z^{-1}} \tag{18.58}$$

which can be derived from the general differential equation describing the linear system using a reasoning similar to that used to derive (18.50).

18.1.6 WHAT WE HAVE LEARNED

Before proceeding to the next section introducing other important concepts for embedded system development, it is worthwhile to briefly summarize the concepts that have been presented here. First of all, an example of linear system has been presented. Linear systems represent a mathematical representation of many practical control applications. We have then seen how differential equations describe the dynamics of linear systems. Even using a very simple example, we have experienced the practical difficulty in finding solutions to differential equations. The definition and some important concepts of the Laplace transform have been then presented, and we have learned how to build a transfer function $W(s)$ for a given system starting from its individual components. Moreover, we have learned how it is possible to derive some important system characteristics directly from the transfer function $W(s)$, such as system stability. Finally, we have learned how to implement in practice the I/O relationship expressed by a transfer function.

It is useful to highlight what we have not learned here. In fact, control theory is a vast discipline, and the presented concepts represent only just a very limited introduction. For example, no technique has been presented for finding the optimal proportional and integral gains, nor we have explored different control strategies. Moreover, we have restricted our attention to systems with a single-input and a single-output (SISO systems). Real-world systems, however, may have multiple inputs and multiple outputs (MIMO systems), and require more sophisticated mathematical techniques involving linear algebra concepts.

It is also worth stressing the fact that even the most elegant and efficient implementation can fail if the underlying algorithm is not correct, leading, for example, to an unstable system. So, it is often very important that an accurate analysis of the system is carried out in order to find a good control algorithm. The transfer function of the digital controller should also be implemented in such a way that its parameters can be easily changed. Very often, in fact, some fine tuning is required during the commissioning of the system.

As a final observation, all the theory presented here and used in practice rely on the assumption that the system being controlled can be modeled as a linear system. Unfortunately, many real-world systems are not linear, even simple ones. In this case, it is necessary to adopt techniques for approximating the nonlinear systems with a linear one in a restricted range of parameters of interest.

18.2   CASE STUDY 2: IMPLEMENTING A DIGITAL LOW-PASS FILTER

In the previous section we have introduced the Laplace transform, that is, a transformation of a real function defined over time $y(t)$, representing in our examples the time evolution of the signals handled by the control system, into a complex function $Y(s)$ of a complex variable. In this section we present another powerful and widely used transformation that converts the time evolution of a signal into its representation in the *frequency domain*.

18.2.1   HARMONICS AND THE FOURIER TRANSFORM

In order to better understand what we intend for frequency domain, let us introduce the *harmonics* concept. A harmonic function is of the form

$$y(t) = A \cos(2\pi f t + \vartheta) \qquad (18.59)$$

that is, a sinusoidal signal of amplitude A, frequency f, and phase ϑ. Its period T is the inverse of the frequency, that is, $T = \frac{1}{f}$. Under rather general conditions, every periodic function $y(t)$ can be expressed as a (possibly infinite) sum of harmonics to form *Fourier series*. As an example, consider the square function $f_{square}(t)$ with period T shown in Figure 18.12. The same function can be expressed by the following Fourier series:

$$f_{square}(t) = \frac{4}{\pi} \sum_{k=1}^{\infty} \frac{\cos(2\pi(2k-1)t - \frac{\pi}{2})}{2k-1} \qquad (18.60)$$

Each harmonic is represented by a sinusoidal signal of frequency $f = 2k - 1$, phase $\vartheta = -\frac{\pi}{2}$ and amplitude $4/(\pi(2k-1))$. Figure 18.13 shows the approximation provided by (18.60) when considering 1 and 10 harmonics in the summation, respectively. The first harmonic in this case is a sine function with the same period and amplitude of the square wave, and considering more and more harmonics makes

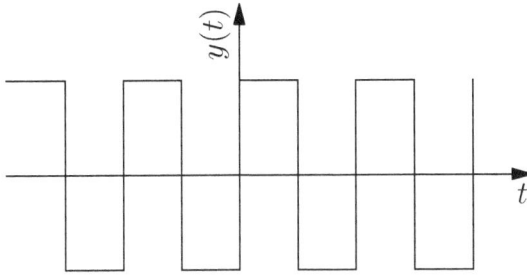

Figure 18.12   A square function.

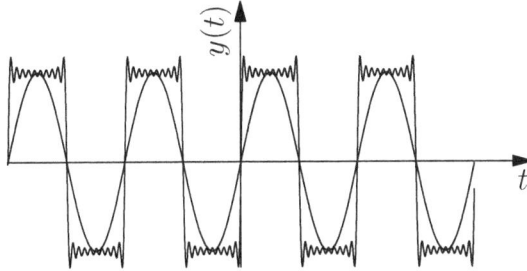

Figure 18.13   The approximation of a square function considering 1 and 10 harmonics.

the approximation closer and closer to the square function. The possibility of representing a periodic function $y(t)$ as a summation of a (possibly infinite) series of harmonics suggests a possible graphical representation of $y(t)$ different from its time evolution, that is, plotting the amplitude of its harmonics components against their frequency value. If we do this for the square function of Figure 18.12 we get the plot shown in Figure 18.14. The example we have considered so far is, however, somewhat simplified: in fact, all the harmonics have the same phase, so only the amplitude depends on their frequency. Moreover, the set of frequencies in the harmonic expansion is discrete, so that the periodic function can be represented by a summation of harmonics. When the function is not periodic, this does not hold any more, and a continuous range of frequencies must be considered in the representation of the function in the frequency domain.

In the general case, the representation of a function $y(t)$ in the frequency domain is mathematically formalized by the *Fourier transform*, which converts a signal $y(t)$ expressed in the time domain into its representation in the frequency domain, that is, a complex function of real variable of the form

$$F\{y(t)\} = Y(f) = \int_{-\infty}^{\infty} y(t)e^{-j2\pi ft}\,dt \qquad (18.61)$$

Recall that, due to Euler's formula,

$$e^{-j2\pi ft} = \cos(2\pi ft) - j\sin(2\pi ft) \qquad (18.62)$$

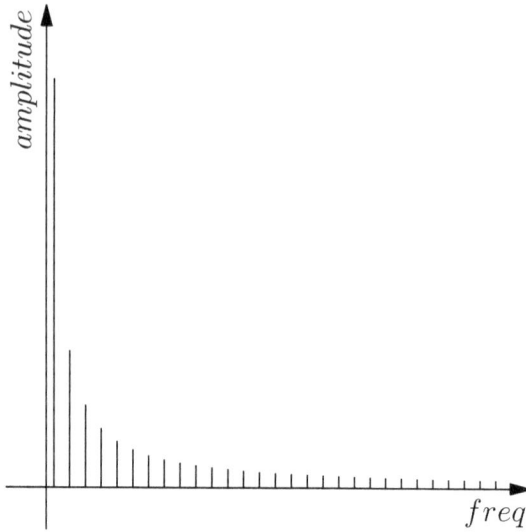

Figure 18.14 The components (amplitude vs. frequency) of the harmonics of the square function.

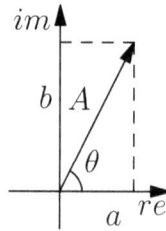

Figure 18.15 Representation of a complex number in the re–im plane.

The reader may wonder why the Fourier transform yields a complex value. Intuitively, this is because for every frequency f, the corresponding harmonics is characterized by two parameters: its amplitude and its phase. Given a frequency value f_1, we shall see that the corresponding value of the Fourier transform $Y(f_1)$ is a complex number whose module and phase represent the amplitude and the phase of the corresponding harmonic. Recall that for a complex number $a + jb$, its module is $A = \sqrt{a^2 + b^2}$, and its phase is $\theta = \arctan(b/a)$. Module and phase are also graphically represented in the cartesian plane (Re, Im) as shown in Figure 18.15. The module is the length of the segment joining the origin with the point representing the number; the phase is the angle between the real axis and such segment.

Informally stated, the Fourier transform represents, for every given frequency f, the (infinitesimal) contribution of the corresponding harmonic to function $y(t)$. We can better understand this concept considering the inverse Fourier transform, that is, the transformation from $Y(f) = F\{y(t)\}$ into $y(t)$:

$$y(t) = \int_{-\infty}^{\infty} F(f)e^{j2\pi ft}df \tag{18.63}$$

In words, every infinitesimal contribution $F(f)e^{j2\pi ft}df$ represents the contribution of the harmonic at frequency f whose amplitude and phase correspond to the module and phase of $F(f)$, respectively. This may seem not so intuitive (we are considering the product between two complex numbers), but it can be easily proven as follows.

Consider a given frequency value f_1. For this value, the Fourier transform yields $Y(f_1)$, which is a complex number of the form $a + jb$. From Figure 18.15 we can express the same complex number as

$$Y(f_1) = a + jb = A[\cos\theta + j\sin\theta] = Ae^{j\theta} \tag{18.64}$$

Where $A = \sqrt{a^2 + b^2}$ and $\theta = \arctan(b/a)$ are the module and phase of $F(f_1)$, respectively. Consider now the expression of the inverse Fourier transform (18.63). In particular, the contribution due to frequencies f_1 and $-f_1$ (the integral spans from $-\infty$ to ∞) is the following:

$$Y(f_1)e^{j2\pi f_1 t} + Y(-f_1)e^{-j2\pi f_1 t} \tag{18.65}$$

A property of the Fourier transform $Y(f)$ of a real function $y(t)$ is that $Y(-f)$ is the complex conjugate of $Y(f)$, and therefore, for the Euler's formula, if $Y(f) = Ae^{j\theta}$, then $Y(-f) = Ae^{-j\theta}$. We can then rewrite (18.65) as

$$Ae^{j\theta}e^{j2\pi f_1 t} + Ae^{-j\theta}e^{-j2\pi f_1 t} = A[e^{j2\pi f_1 t + \theta} + e^{-j2\pi f_1 t + \theta}] \tag{18.66}$$

The imaginary terms in (18.66) elide, and therefore,

$$Y(f_1)e^{j2\pi f_1 t} + Y(-f_1)e^{-j2\pi f_1 t} = 2A\cos(2\pi f_1 t + \theta) \tag{18.67}$$

That is, the contribution of the Fourier transform $Y(f)$ at the given frequency f_1 is the harmonic at frequency f_1 whose amplitude and phase are given by the module and phase of the complex number $Y(f_1)$.

Usually, the module of the Fourier transform $Y(f)$ is plotted against frequency f to show the frequency distribution of a given function $f(t)$. The plot is symmetrical in respect of the Y axis. In fact, we have already seen that $Y(-f) = \overline{Y(f)}$, and therefore, the modules of $Y(f)$ and of $Y(-f)$ are the same.

The concepts we have learned so far are can be applied to a familiar concept, that is, sound. Intuitively, we expect that grave sounds will have a harmonic content mostly containing low frequency components, while acute sounds will contain harmonic components at higher frequencies. In any case, the sound we perceive will have no harmonics over a given frequency value because our ear is not able to perceive sounds over a given frequency limit.

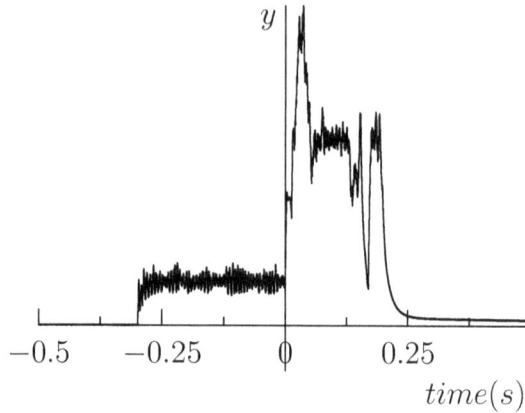

Figure 18.16   A signal with noise.

18.2.2   LOW-PASS FILTERS

A low-pass filter operates a transformation over the incoming signal so that frequencies above a given threshold are removed. Low-pass filters are useful, for example, to remove noise from signals. In fact, the noise has a frequency distribution where most components are above the frequencies of interest for the signal. A filter able to remove high frequency components will then remove most of the noise from the signal. As an example, consider Figure 18.16 showing a noisy signal. Its frequency distribution (spectrum) is shown in Figure 18.17, where it can be shown that harmonics are present at frequencies higher than 5Hz. Filtering the signal with a low-pass filter that removes frequencies above 5Hz, we get the signal shown in Figure 18.18,

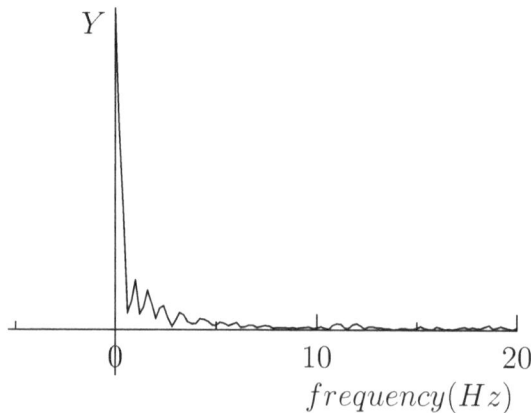

Figure 18.17   The spectrum of the signal shown in Figure 18.16.

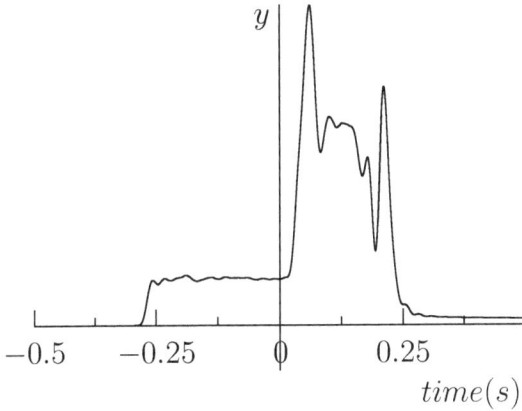

Figure 18.18   The signal of Figure 18.16 after low-pass filtering.

whose spectrum is shown in Figure 18.19. It can be seen that the frequencies above 5Hz have been removed thus removing the noise superimposed to the original signal.

The ideal low-pass filter will completely remove all the harmonics above a given cut-off frequency, leaving harmonics with lower frequency as they are. So the frequency response of the ideal low-pass frequency filter with cut-off frequency f_c would have the form shown in Figure 18.20, where, in the Y axis, the ratio between the amplitudes of the original and filtered harmonics is shown. In practice, however, it is not possible to implement low-pass filters with such a frequency response. For example, the frequency response of the low-pass filter used to filter the signal shown in Figure 18.16 is shown in Figure 18.21. The gain of the filter is normally expressed in *decibel* (dB), corresponding to $20\log_{10}(A_1/A_0)$, where A_0 and A_1 are

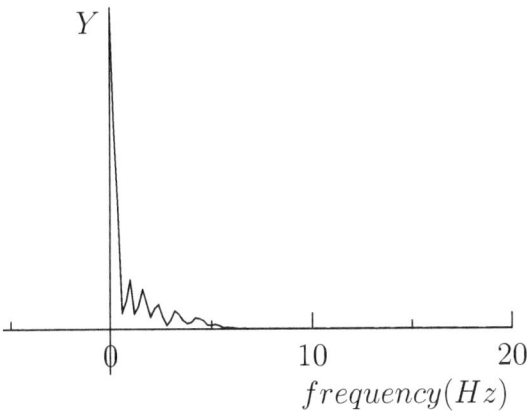

Figure 18.19   The spectrum of the signal shown in Figure 18.18.

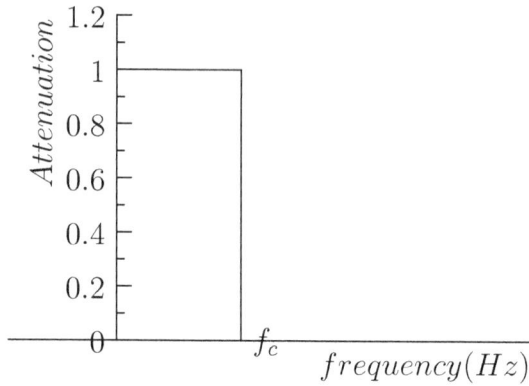

Figure 18.20 Frequency response of an ideal filter with cut-off frequency f_c.

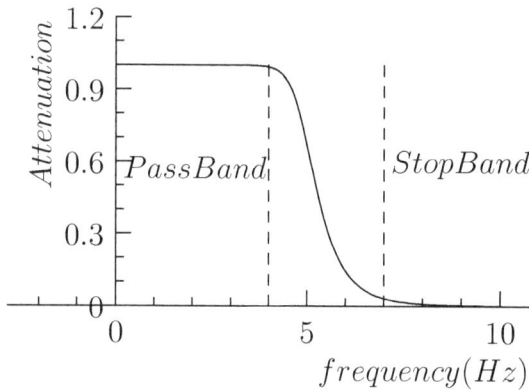

Figure 18.21 Frequency response of the filter used to filter the signal shown in Figure 18.16.

the amplitude of the original and filtered harmonic, respectively. Since the gain is normally less than or equal to one for filters, its expression in decibel is normally negative. The frequency response shown in Figure 18.21 is shown expressed in decibel in Figure 18.22. Referring to Figure 18.21, for frequencies included in the *Pass Band* the gain of the filter is above a given threshold, normally -3dB (in the ideal filter the gain in the pass band is exactly 0 dB), while in the *Stop Band* the gain is below another threshold, which, depending on the application, may range from -20 dB and -120 dB (in the ideal filter the gain in decibel for these frequencies tends to $-\infty$). The range of frequencies between the pass band and the stop band is often called transition band: for an ideal low-pass filter there is no Transition Band, but in practice the transition band depends on the kind of selected filters, and its width is never null.

A low-pass filter is a linear system whose relationship between the input (unfiltered) signal and the output (filtered) one is expressed by a differential function in the

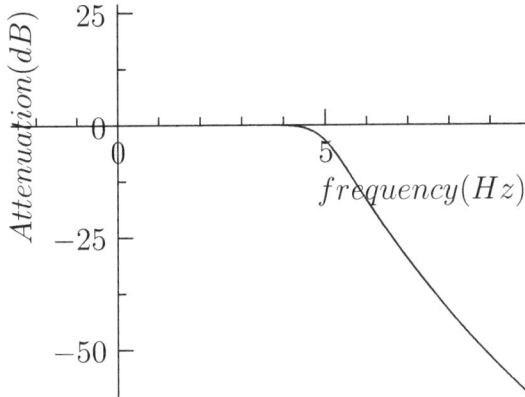

Figure 18.22 Frequency response shown in Figure 18.21 expressed in decibel.

form of (18.6). We have already in hand some techniques for handling linear systems and, in particular, we know how to express the I/O relationship using a transfer function $W(s)$ expressed in the Laplace domain. At this point, we are able to recognize a very interesting aspect of the Laplace transform. Recalling the expression of the Laplace transform of function $w(t)$

$$W(s) = \int_0^\infty w(t)e^{-st}dt \qquad (18.68)$$

and of the Fourier transform

$$W(f) = \int_{-\infty}^\infty w(t)e^{-2\pi ft}dt \qquad (18.69)$$

and supposing that $f(t) = 0$ before time 0, the Fourier transform corresponds to the Laplace one, for $s = j2\pi f$, that is, when considering the values of the complex variable s corresponding to the imaginary axis. So, information carried by the Laplace transform covers also the frequency response of the system. Recalling the relationship between the input signal $u(t)$ and the output signal $y(t)$ for a linear system expressed in the Laplace domain

$$Y(s) = W(s)U(s) \qquad (18.70)$$

we get an analog relationship between the Fourier transform $U(f)$ and $Y(f)$ of the input and output, respectively:

$$Y(f) = W(f)U(f) \qquad (18.71)$$

In particular, if we apply a sinusoidal input function $u(t) = \cos(2\pi ft)$, the output will be of the form $y(t) = A\cos(2\pi ft + \theta)$, where A and θ are the module and phase of $W(s), s = e^{j2\pi f}$.

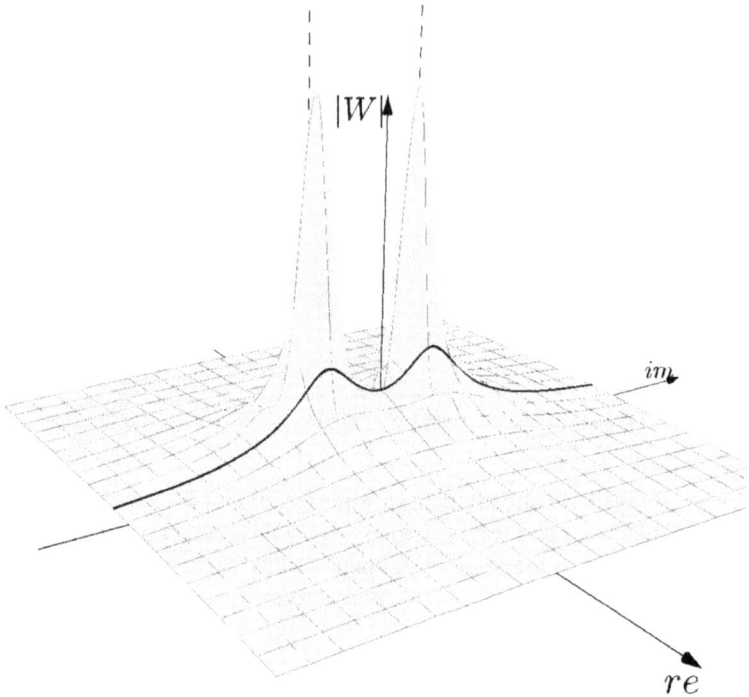

Figure 18.23   The module of the transfer function for the tank–pump system controlled in feedback highlighting its values along the imaginary axis.

Let us consider again the transfer function of the tank–pump with feedback control we defined in the previous section. We recall here its expression

$$W(s) = \frac{sK_p + K_i}{s^2 B + sK_p + K_i} \tag{18.72}$$

Figure 18.23 shows the module of the transfer function for $B = 1$, $K_p = 1$, and $K_i = 1$, and, in particular, its values along the imaginary axis. The corresponding Fourier transform is shown in Figure 18.24. We observe that the tank–pump system controlled in feedback mode behaves somewhat like a low-pass filter. This should not be surprising: We can well expect that, if we provide a reference input that varies too fast in time, the system will not be able to let the level of the liquid in the tank follow such a reference. Such a filter is far from being an optimal low-pass filter: The decrease in frequency response is not sharp, and there is an amplification, rather than an attenuation, at lower frequencies. In any case, this suggests us a way for implementing digital low-pass filters, that is finding an analog filter, that is, a system with the desired response in frequency, and then digitalizing it, using the technique

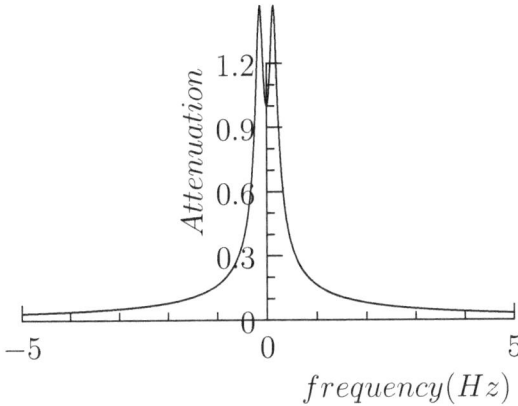

Figure 18.24   The Fourier representation of the tank–pump transfer function.

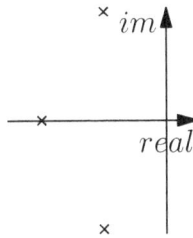

Figure 18.25   The poles of a Butterworth filter of the third-order.

we have learned in the previous section.

A widely used analog low-pass filter is the *Butterworth filter*, whose transfer function is of the form

$$W(s) = \frac{1}{\prod_{k=1}^{n}(s - s_k)/(2\pi f_c)} \tag{18.73}$$

where f_c is the cut-off frequency, and n is called the *order* of the filter and s_k, $k = 1, \ldots, n$ are the poles, which are of the form

$$s_k = 2\pi f_c e^{(2k+n-1)\pi/(2n)} \tag{18.74}$$

The poles of the transfer function lie, therefore, on the left side of a circle of radius $2\pi f$. Figure 18.25 displays the poles for a Butterworth filter of the third order. Figure 18.26 shows the module of the transfer function in the complex plane, highlighting its values along the imaginary axis corresponding to the frequency response shown in (18.27). The larger the number of poles in the Butterworth filter, the sharper the frequency response of the filter, that is, the narrower the Transition Band. As an example, compare the frequency response of Figure 18.27 corresponding to a Butterworth filter with 3 poles, and that of Figure 18.21 corresponding to a Butterworth filter with 10 poles.

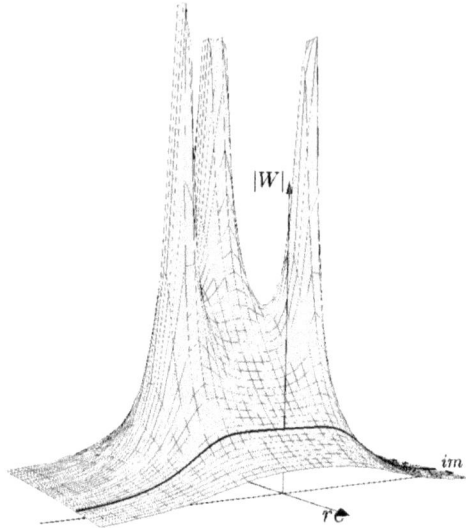

Figure 18.26   The module of a Butterworth filter of the third-order, and the corresponding values along the imaginary axis.

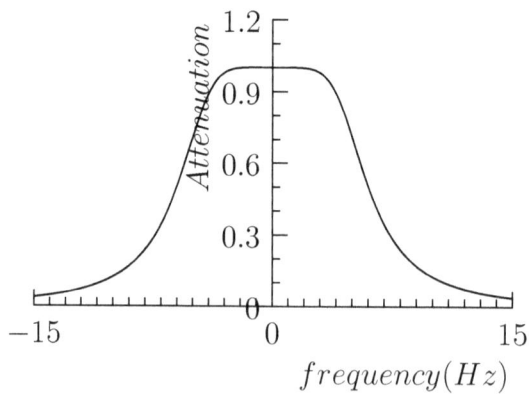

Figure 18.27   The module of the Fourier transform of a third-order Butterworth filter with 5 Hz cutoff frequency.

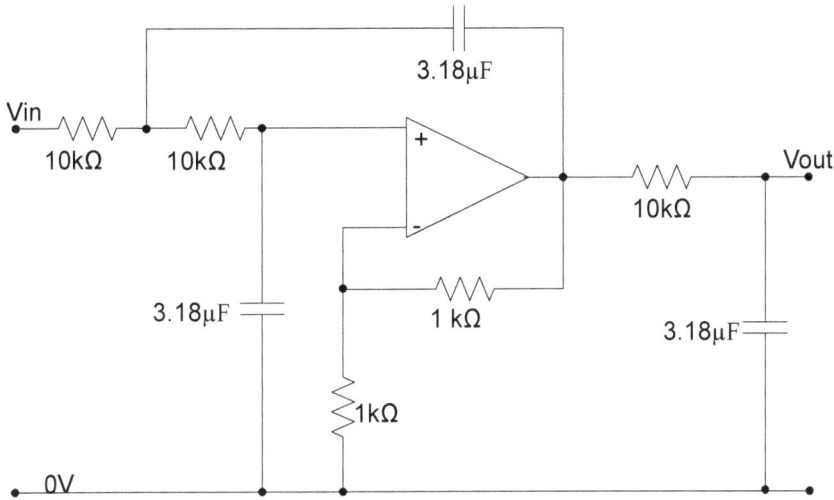

Figure 18.28 An electronic implementation of a Butterworth filter of the third order with 5 Hz cutoff frequency.

An analog Butterworth filter can be implemented by an electronic circuit, as shown in in Figure 18.28. We are, however, interested here in its digital implementation, which can be carried out using the technique introduces in the previous section, that is

1. Find an appropriate value of the sampling period T.
2. Transform the transfer function $W(s)$ expressed in the Laplace domain into the corresponding transfer function $W(z)$ expressed in the Z domain by replacing s with $(1 - z^{-1})/T$, or using the bilinear transform $s = 2(1 - z^{-1})/T(1 + z^{-1})$.
3. Implement the transfer function as a linear combination of previous samples of the input and of the output and the current input.

Up to now we have used informal arguments for the selection of the most appropriate value of the sampling period T. We have now the necessary background for a more rigorous approach in choosing the sampling period T.

18.2.3 THE CHOICE OF THE SAMPLING PERIOD

In the last section, discussing the choice of the sampling period T for taking signal samples and using them in the digital controller, we have expressed the informal argument that the value of T should be short enough to avoid losing significant information for that signal. We are now able to provide a more precise statement of this: We shall say that the choice of the sampling period T must be such that the continuous signal $y(t)$ *can be fully reconstructed from its samples* $y(nT)$. Stated in other words, if we are able to find a mathematical transformation that, starting from the sampled values $y(nT)$ can rebuild $y(t)$, for every time t, then we can ensure that no information has been lost when sampling the signal. To this purpose, let us recall the expression of the Fourier transform for the signal $y(t)$

$$Y(f) = \int_{-\infty}^{\infty} y(t)e^{-2\pi ft}dt \qquad (18.75)$$

which transforms the real function of real variable $y(t)$ into a complex function of real variable $Y(f)$. $Y(f)$ maintains all the information of $y(t)$, and, in fact, the latter can be obtained from $Y(f)$ via the Fourier antitransform:

$$y(t) = \int_{-\infty}^{\infty} Y(f)e^{j2\pi ft}df \qquad (18.76)$$

Now, suppose we have in hand only the sampled values of $y(t)$, that is, $y(nT)$ for a given value of the sampling period T. An approximation of the Fourier transform can be obtained by replacing the integral in (18.75) with the summation

$$Y_T(f) = T \sum_{n=-\infty}^{\infty} y(nT)e^{-j2\pi fnT} \qquad (18.77)$$

(18.77) is a representation of the *discrete-time Fourier transform*. From $Y_T(f)$ it is possible to rebuild the original values of $y(nT)$ using the inverse transform

$$y(nT) = \int_{-T/2}^{T/2} Y_T(f)e^{j2\pi fnT}df \qquad (18.78)$$

Even if from the discrete-time Fourier transform we can rebuild the sampled values $y(nT)$, we cannot yet state anything about the values of $y(t)$ at the remaining times. The following relation between the continuous Fourier transform $Y(f)$ of (18.75) and the discrete time version $Y_T(f)$ of (18.77) will allow us to derive information on $y(t)$ also for the times between consecutive samples:

$$y_T(f) = \frac{1}{T} \sum_{r=-\infty}^{\infty} Y(f - \frac{r}{T}) \qquad (18.79)$$

Put in words, (18.79) states that the discrete time Fourier representation $Y_T(f)$ can be obtained by considering infinite terms, being the rth term composed of the continuous Fourier transform shifted on the right of $rf_c = r/T$. The higher the sampling frequency f_c, the more separate will be the terms of the summation. In particular, suppose that $Y(f) = 0$ for $|f| < f_1 < f_c/2$. Its module will be represented by a curve similar to that shown in Figure 18.29. Therefore the module of the discrete time Fourier will be of the form shown in Figure 18.30, and therefore, for $-f_c/2 < f < f_c/2$, the discrete time transform $Y_T(f)$ will be *exactly the same* as the continuous one $Y(f)$.

This means that, if the sampling frequency is *at least twice* the highest frequency of the harmonics composing the original signal, *no information is lost in sampling*. In fact, in principle, the original signal $y(t)$ could be derived by antitrasforming using (18.76), the discrete-time Fourier transform, built considering only the sampled

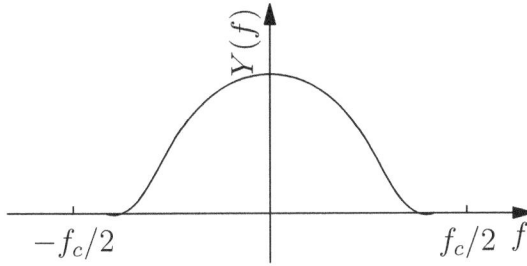

Figure 18.29 A frequency spectrum limited to $f_c/2$.

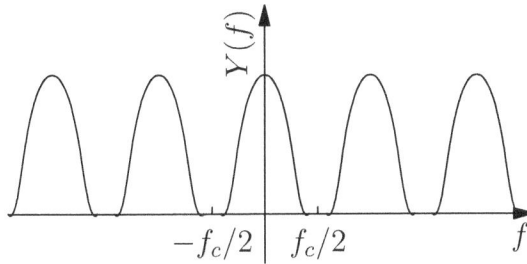

Figure 18.30 The discrete-time Fourier transform corresponding to the continuous one of Figure 18.29.

values $y(nT)$. Of course, we are not interested in the actual computation of (18.76), but this theoretical result gives us a clear indication in the choice of the sampling period T.

This is a fundamental result in the field of signal processing, and is called the Nyquist–Shannon sampling theorem, after Harry Nyquist and Claude Shannon, even if other authors independently discovered and proved part of it. The proof by Shannon was published in 1949 [84] and is based on an earlier work by Nyquist [69].

Unfortunately, things are not so bright in real life, and normally, it is not possible for a given function $y(t)$ to find a frequency f_0 for which $Y(f) = 0, |f| > f_0$. In this case we will have an *aliasing* phenomenon, as illustrated in Figure 18.31, which shows how the spectrum is distorted as a consequence of sampling. The effect of the aliasing when considering the sampled values $y(nT)$ is the "creation" of new harmonics that do not exists in the original continuous signal $y(t)$. The aliasing effect is negligible for sampling frequencies large enough, and so the amplitude of the tail in the spectrum above $f_c/2$ becomes small, but significant distortion in the sampled signal may occur for a poor choice of f_c.

The theory presented so far provides us the "golden rule" of data acquisition, when signals sampled by ADC converters are then acquired in an embedded system, that is, *choosing a sampling frequency which is at least twice the maximum frequency of any significant harmonic of the acquired signal*. However, ADC converters cannot provide an arbitrarily high sampling frequency, and in any case, this may be limited

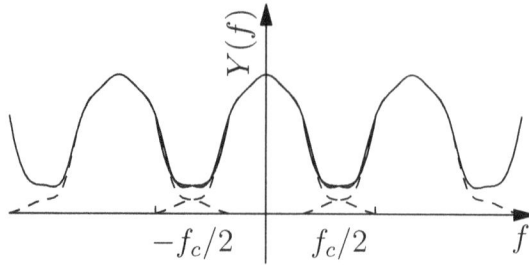

Figure 18.31   The Aliasing effect.

by the overall system architecture. As an example, consider an embedded system that acquires 100 signals coming from sensors in a controlled industrial plant, and suppose that a serial link connects the ADC converters and the computer. Even if the single converter may be able to acquire the signal at, say, 1 kHz (commercial ADC converters can have a sampling frequency up to some MHz), sending 100 signals over a serial link means that a data throughput of 100 KSamples/s has to be sustained by the communication link, as well as properly handled by the computer. Moving to a sampling frequency of 10 kHz may be not feasible for such a system because either the data link is not able to sustain a higher data throughput, or the processing power of the computer becomes insufficient.

Once a sampling frequency f_c has been chosen, it is mandatory to make sure that the conversion does not introduce aliasing, and therefore it is necessary to filter the input signals *with an analog low-pass filter* whose cut-off frequency is at least $f_c/2$, before ADC conversion. Butterworth filters, whose electrical schema is shown in Figure 18.28, are often used in practice, and are normally implemented inside the ADC boards themselves.

18.2.4   BUILDING THE DIGITAL LOW-PASS FILTER

We are now ready to implement the digital low-pass filter with a cut-off frequency of 5 Hz similar to that which has been used to filter the signal shown in Figure 18.16, but using, for simplicity, 3 poles instead of 10 (used to obtain the filtered signal of Figure 18.18). We suppose a sampling frequency of 1 kHz, that is, $T = 10^{-3}s$, assuming therefore that the spectrum of the incoming analog signal is negligible above 500 Hz.

Using (18.74), we obtain the following values for the three poles:

$$p_1 = (-15.7080 + j27.2070) \tag{18.80}$$

$$p_2 = (-31.4159 + j0) \tag{18.81}$$

$$p_3 = (-15.7080 - j27.2067) \tag{18.82}$$

and, from (18.73), we derive the transfer function in the Laplace domain

$$W(s) =$$
$$\frac{31006.28}{(s-(-15.708+j27.207))(s-(-31.416+j0))(s-(-15.708-j27.207))}$$
(18.83)

From the above transfer function, with the replacement $s = (1 - z^{-1})/T, T = 10^{-3}$, we derive the transfer function in the Z domain

$$W(z) = \frac{3.1416 \times 10^{-5}}{-z^{-3} + 3.0628z^{-2} - 3.1276z^{-1} + 1.06483} \qquad (18.84)$$

and finally the actual computation for the desired low-pass digital filter:

$$y(nT) =$$
$$\frac{y((n-3)T) - 3.063y((n-2)T) + 3.128y((n-1)T) + 3.142 \times 10^{-5}x(nT)}{1.065}$$
(18.85)

We recognize that, in the general case, the implementation of a low-pass filter (and more in general of the discrete time implementation of a linear system) consists of a linear combination of the current input, its past samples and the past samples of the output. It is therefore necessary to keep in memory the past samples of the input and output. A common technique that is used in order to avoid unnecessary copies in memory and therefore minimize the execution time of digital filters is the usage of *circular buffers*. A circular buffer is a data structure that maintains the recent history of the input or output. When new samples arrive, instead of moving all the previous samples in the buffer array, the pointer to the head of the buffer is advanced instead. Then, when the head pointer reaches one end of the history array, it is moved to the other end: If the array is large enough to contain the required history the samples on the other end of the array are no more useful and can be discarded. Below is the C code of a general filter implementation. Routine initFilter() will build the required data structure and will return a pointer to be used in the subsequent call to routine doFilter(). No memory allocation is performed by routine doFilter(), which basically performs a fixed number of sums, multiplications and memory access. In fact, this routine is to be called run time, and for this reason it is important that the amount of time required for the filter computation is bounded. Conversely, routine initFilter() has to be called during system initialization since its execution time may be not predictable due to the calls to system routines for the dynamic memory allocation. An alternative implementation would have used statically allocated buffers and coefficient array, but would have been less flexible. In fact, the presented implementation allows the run-time implementation of a number of independent filters.

```
/* Filter Descriptor Structure: fully describes the filter and
   its current state.
   This structure contains the two circular buffers and the
   current index within them.
   It contains also the coefficients for the previous input
   and output samples for the filter computation
   y(nT) = aN*y((n-N)T)+...+a1*y((n-1)T)
         + bMu((n-M)T)+...+b1*u((n-1)T + b0*u(nT) */

typedef struct {
  int currIndex;        //Current index in circular buffers
  int bufSize;   //Number of elements in the buffers
  float *yBuf;   //Output history buffer
  float *uBuf;   //Input history buffer
  float *a;             //Previous output coefficients
  int aSize;            //Number of a coefficient
  float *b;             //Previous input coefficients
  int bSize;            //Number of b coefficients
}FilterDescriptor;

/* Filter structure initialization.
   To be called before entering the real-time phase
   Its arguments are the a and b coefficients of the filter */
FilterDescriptor *initFilter(float *aCoeff, int numACoeff,
  float *bCoeff, int numBCoeff)
{
  int i;
  FilterDescriptor *newFilter;
  newFilter = (FilterDescriptor *)malloc(sizeof(FilterDescriptor));
/* Allocate and copy filter coefficients */
  newFilter->a = (float *)malloc(numACoeff * sizeof(float));
  for(i = 0; i < numACoeff; i++)
    newFilter->a[i] = aCoeff[i];
  newFilter->aSize = numACoeff;
  newFilter->b = (float *)malloc(numBCoeff * sizeof(float));
  for(i = 0; i < numBCoeff; i++)
    newFilter->b[i] = bCoeff[i];
  newFilter->bSize = numBCoeff;
/* Circular Buffer dimension is the greatest between the number
   of a and b coefficients */
  if(numACoeff > numBCoeff)
    newFilter->bufSize = numACoeff;
  else
    newFilter->bufSize = numBCoeff;
/* Allocate circularBuffers, initialized to 0 */
  newFilter->yBuf = (float *)calloc(newFilter->bufSize, sizeof(float));
  newFilter->uBuf = (float *)calloc(newFilter->bufSize, sizeof(float));
/* Buffer index starts at 0 */
  newFilter->currIndex = 0;
  return newFilter;
}


/* Run time filter computation.
   The first argument is a pointer to the filter descriptor
   The second argument is the current input
   It returns the current output */
float doFilter(FilterDescriptor *filter, float currIn)
{
  float currOut = 0;
  int i;
  int currIdx;
/* Computation of the current output based on previous input
   and output history */
  currIdx = filter->currIndex;
  for(i = 0; i < filter->aSize; i++)
  {
```

```
   currOut += filter->a[i]*filter->yBuf[currIdx];
/* Go to previous sample in the circular buffer */
   currIdx--;
   if(currIdx < 0)
      currIdx += filter->bufSize;
   }
   currIdx = filter->currIndex;
   for(i = 0; i < filter->bSize-1; i++)
   {
      currOut += filter->b[i+1]*filter->uBuf[currIdx];
/* Go to previous sample in the circular buffer */
      currIdx--;
      if(currIdx < 0)
         currIdx += filter->bufSize;
   }
/* b[0] contains the coefficient for the current input */
   currOut += filter->b[0]*currIn;
/* Upate input and output filters */
   currIdx = filter->currIndex;
   currIdx++;
   if(currIdx == filter->bufSize)
      currIdx = 0;
   filter->yBuf[currIdx] = currOut;
   filter->uBuf[currIdx] = currIn;
   filter->currIndex = currIdx;

   return currOut;
}


/* Filter deallocation routine.
   To be called when the filter is no longer used outside
   the real-time phase */
void releaseFilter(FilterDescriptor *filter)
{
   free((char *)filter->a);
   free((char *)filter->b);
   free((char *)filter->yBuf);
   free((char *)filter->uBuf);
   free((char *)filter);
}
```

18.2.5   SIGNAL TO NOISE RATIO (SNR)

Up to now we have seen the main components of a digital low-pass filter, that is, antialiasing analog filtering, data sampling and filter computation. The resulting data stream can be used for further computation or converted to an analog signal using a DAC converter if filtering is the only function of the embedded system. We have seen how information may be lost when sampling a signal, and the techniques to reduce this effect. There is, however, another possible reason for information loss in the above chain. The ADC converter, in fact, converts the analog value of the input into its digital representation using a *finite* number of bits. This unavoidably introduces an error called the *quantization error*. Its effect can be considered as a superimposed noise in the input signal. If N bits are used for the conversion, and A is the input range for the ADC converter, the quantization interval is $\Delta = A/2^N$, and therefore, the quantization error $e(nT)$ is included in the interval $[-\Delta/2, \Delta/2]$. The sequence $e(nT)$ is therefore a sequence of random values, which is often assumed to be a *strictly stationary process*, where the probability distribution of the random samples is not related to the original signal $y(nT)$. It does not change over time, and the distribution is uniform in the interval $[-\Delta/2, \Delta/2]$. Under these assumptions, the probability distribution of the error is

$$f_e(a) = \frac{1}{\Delta}, -\frac{\Delta}{2} \le a \le \frac{\Delta}{2} \qquad (18.86)$$

For such a random variable e, its power, that is, the expected value $E[e^2]$ is

$$P_e = E[e^2] = \int_{-\infty}^{\infty} a^2 f_e(a) da = \int_{-\Delta/2}^{\Delta/2} \frac{a^2}{\Delta} da = \frac{\Delta^2}{12} \qquad (18.87)$$

An important parameter in the choice of the ADC device is the *signal-to-noise ratio* (SNR), which expresses the power ratio between the input signal and its noise after sampling, that is,

$$SNR = 10 \log_{10} \frac{P_{signal}}{P_{noise}} \qquad (18.88)$$

using the expression of the power for the quantization error (18.87), we have

$$SNR = 10 \log_{10} \frac{P_{signal}}{P_{noise}} = 10 \log_{10} P_{signal} - 10 \log_{10} \frac{\Delta^2}{12} \qquad (18.89)$$

If we use B bits for the conversion, and input range of A, the quantization interval Δ is equal to $A/(2^B)$ and therefore

$$SNR = 10 \log_{10} P_{signal} + 10 \log_{10} 12 - 20 \log_{10} A + 20B \log_{10} 2 =$$
$$CONST + 6.02B \qquad (18.90)$$

that is, for every additional bit in the conversion, the SNR is incremented of around 6dB for every additional bit in the ADC conversion (the other terms in (18.90) are constant). This gives us an estimation of the effect of the introduced quantization error and also an indication on the number of bits to be considered in the ADC conversion. Nowadays, commercial ADC converters use 16 bits or more in conversion, and the number of bits may reduce for very high-speed converters.

18.3   SUMMARY

In this section we have learned the basic concepts of control theory and the techniques that are necessary to design and implement a digital low-pass filter. The presented concepts represent facts that every developer of embedded systems should be aware of. In particular, the effects of sampling and the consequent harmonic distortion due to the aliasing effect must always be taken into account when developing embedded systems for control and data acquisition. Another important aspect that should always be taken in consideration when developing systems handling acquired data, is the choice of the appropriate number of bits in analog-to-digital conversion. Finally, once the parameters of the linear system have been defined, an accurate implementation of the algorithm is necessary in order to ensure that the system will have a deterministic execution time, less than a given maximum time. Most of this book will be devoted to techniques that can ensure real-time system responsiveness. A precondition to every technique is that the number of machine instructions required for the execution of the algorithms is bounded. For this reason, in the presented example, the implementation of the filter has been split into two sets of routine: offline routines for the creation and the deallocation of the required data structures, and an online routine for the actual run-time filter computation. Only the latter one will be executed under real-time constraints, and will consist of a fixed number of machine instructions.

18.4   EXERCISES

EXERCISE 1

Implement in C a Proportional/Integrative/Derivative (PID) controller using bilinear transform. You may then use it in a closed loop control using the RedPitaya board presented in the previous chapter.

Hint: starting from the Laplace transform of the transfer function $W(s) = K_p + K_i/s + K_d s$, obtain the transfer function in Z replacing variable s with $s = \frac{2}{T}\frac{1-z^{-1}}{1+z^{-1}}$. The conversion into C code is then straightforward, recalling that z^{-n} means the nth element in the history. You can maintain the history for both input and output in a structure that is passed to the `pid()` routine along with the current input and the PID parameters.

EXERCISE 2

Implement in C a low pass filter that can then be used for real-time noise reduction in the signal acquired in the RedPitaya board presented in the previous chapter.

Hint: Some considerations about the frequency side effects of defining the sampling frequency as described in the previous chapter are worth first. The RedPitaya STEM board performs ADC sampling at a frequency of 125MHz. For this reason, an anti-aliasing analog low-pass filter is provided in the board before the ADC stage, cutting frequencies above 62MHz. However, in the application presented in the case study the signal sampled at 125MHz is then directly subsampled in the FPGA in order to read data at a lower frequency. This introduces the risk of aliasing in case frequencies above half the selected sampling speed were present in the signal. In order to keep the presented case study simple enough, no digital filtering is performed in the FPGA, but a real-world application should consider this and implement in the FPGA the required low-pass filter with a cutoff frequency that depends on the selected sampling speed.

The software implementation of a further low pass filter may use the approach presented in this chapter, i.e. select the Laplace transfer function of a low pass filter (e.g. Butterworth) and then transform it in a discrete implementation using the bilinear transform. Other methods exist, such as the direct implementation of digital Finite Response (FIR) filters, not described here. In any case, for the Nyquist–Shannon theorem, the selected cutoff frequency cannot be larger than half the sampling speed selected in the software driver.

Answers to Selected Exercises

The source code of some other solutions is available in the online repository at `https://github.com/minimap-xl/RTOS_Book` under the GPL-2.0 license.

CHAPTER 3

EXERCISE 1

Sequences 2 and 4 produce correct results. Sequence 1 does not because D is executed before B. Sequence 3 is also incorrect because B is executed before A. Note that in sequence 4 the interposition of C between A and B has no adverse effect.

EXERCISE 2

Both *Blocked* and *Ready* are process state diagram states that have only involuntary outgoing transitions. As a consequence, a task may not get out of these states unless an event *external to the task* occurs. For *Blocked*, the event is generated by another process, while for *Ready* it depends on the scheduler. In other words, since a task that is *Blocked* or *Ready* does not execute, it is completely unable to perform any actions to get out of these states.

CHAPTER 4

EXERCISE 1

There is no deadlock in the system because there are no cycles in the resource allocation graph. The path $P_1 \to R_3 \leftarrow P_3 \to R_2 \to P_1 \to R_3$ is not a cycle because the arc $R_3 \leftarrow P_3$ is oriented in the wrong way.

EXERCISE 2

There is a deadlock in the system because there is at least one cycle in the resource allocation. In fact, there are two:

1. $P_1 \to R_1 \to P_2 \to R_2 \to P_1$, and
2. $P_1 \to R_1 \to P_2 \to R_4 \to P_3 \to R_3 \to P_1$.

The two arcs $P_1 \to R_1$ and $R_1 \to P_2$ are common to all cycles, so removing either of them breaks all cycles, thus resolving the deadlock. Working on $R_1 \to P_2$ is relatively complex because it is an ownership arc. Hence, removing it would imply notifying P_2 that a resource previously allocated to it is no longer available for use. In turn, this would waste any partial work P_2 may have already performed on R_1 and would require R_1 to be rolled back to its original state. $P_1 \to R_1$ is a request arc, instead. It can be removed by simply denying the resource allocation request being performed by P_1. From a practical standpoint, this means the system must force the function that P_1 invoked to request R_1 to return an error indication instead of waiting for R_1 to become available.

EXERCISE 3

- The system deadlocks when the following sequence of events occurs: 1) P acquires resource A; 2) Q acquires resource B; 3) P waits for B (currently assigned to Q); 4) R acquires resource C; 5) Q waits for C (currently assigned to R); 6) R waits for A (currently assigned to P). Other sequences exist that also lead to a deadlock.
- A possible total resource ordering is $f(B) > f(A) > f(C)$. No changes are needed to processes P and R because they already acquire the resources they need in the right order. Process Q must be modified to request C first, and then B.

EXERCISE 4

From the information given in the exercise, we can calculate N and $\mathbf{r}$:

$$N = \begin{pmatrix} 1 & 1 & 0 \\ 2 & 0 & 1 \\ 1 & 1 & 0 \end{pmatrix}, \quad \mathbf{r} = \begin{pmatrix} 0 \\ 2 \\ 0 \end{pmatrix}$$

Moreover, we initially set $\mathbf{w} = \mathbf{r}$ to track resource availability as the safety check proceeds. The first process suitable for termination is P_3 because:

$$\mathbf{n_3} = \begin{pmatrix} 0 \\ 1 \\ 0 \end{pmatrix} \leq \mathbf{w}.$$

Then, we can update $\mathbf{w}$ accordingly:

$$\mathbf{w} = \begin{pmatrix} 0 \\ 2 \\ 0 \end{pmatrix} + \mathbf{c_3} = \begin{pmatrix} 4 \\ 2 \\ 3 \end{pmatrix}$$

After this update to $\mathbf{w}$, P_1 becomes suitable for termination because:

$$\mathbf{n_1} = \begin{pmatrix} 1 \\ 2 \\ 1 \end{pmatrix} \leq \mathbf{w}.$$

The termination of P_1 leads to a further update to $\mathbf{w}$:

$$\mathbf{w} = \begin{pmatrix} 4 \\ 2 \\ 3 \end{pmatrix} + \mathbf{c_1} = \begin{pmatrix} 7 \\ 3 \\ 5 \end{pmatrix}$$

Finally, P_2 is able to terminate as well, because it is:

$$\mathbf{n_2} = \begin{pmatrix} 1 \\ 0 \\ 1 \end{pmatrix} \leq \mathbf{w}.$$

Since all three processes in the system are able to terminate after reaching their worst-case resource allocation requirements, we can conclude that the system state is safe. In the solution, we could have chosen to terminate P_2 before P_1, but the end result would have been the same.

EXERCISE 5

1. The request coming from P_2 is not legitimate because it is not $\mathbf{q_2} \leq \mathbf{n_2}$. The banker must immediately return an error indication to P_2.
2. The request coming from P_1 is legitimate because $\mathbf{q_1} \leq \mathbf{n_1}$. However, it is not $\mathbf{q_1} \leq \mathbf{r}$, meaning that the request cannot be satisfied immediately because the available resources are insufficient. The banker must block P_2 until sufficient resources become available and then evaluate its request again.
3. The request coming from P_3 is legitimate (because $\mathbf{q_3} \leq \mathbf{n_3}$) and can be satisfied immediately (because $\mathbf{q_3} \leq \mathbf{r}$). Before granting it, we must ensure that the update system state is still safe. To this purpose, we calculate the new system state using (4.16) obtaining:

$$C' = \begin{pmatrix} 3 & 3 & 4 \\ 1 & 0 & 1 \\ 2 & 2 & 3 \end{pmatrix}, \quad N' = \begin{pmatrix} 1 & 1 & 0 \\ 2 & 0 & 0 \\ 1 & 1 & 0 \end{pmatrix}, \quad \mathbf{r'} = \begin{pmatrix} 0 \\ 1 \\ 0 \end{pmatrix}$$

Proceeding like in the Exercise 4, we conclude that the new state is safe because at least one safe termination sequence, for instance, P_3, P_1, P_2, exists. The banker must grant the request and update the actual system state accordingly.
4. The request coming from P_1 is legitimate and can be satisfied. If we granted the request the new system state would be:

$$C' = \begin{pmatrix} 3 & 3 & 4 \\ 3 & 0 & 0 \\ 2 & 2 & 3 \end{pmatrix}, \quad N' = \begin{pmatrix} 1 & 1 & 0 \\ 0 & 0 & 1 \\ 1 & 1 & 0 \end{pmatrix}, \quad \mathbf{r'} = \begin{pmatrix} 0 \\ 0 \\ 0 \end{pmatrix}$$

However, this new state would be unsafe because $\mathbf{r'}$ is zero whereas all $\mathbf{n_j}$ are non-zero. The banker must therefore force P_1 to wait to avoid bringing the system into an unsafe state from the deadlock point of view.

CHAPTER 5

EXERCISE 1

- The P() and V() primitives on mutual exclusion semaphore `mutex` surround the increments of a and b performed by P_1. P_2 also performs P() and

V() around its access to these shared variables. Hence, P_2 always gets valid values for them.

- Process P_1 releases mutex after incrementing a and acquires it again before incrementing b. Therefore, P_2 may enter its own mutual exclusion region in between the two increments. When this happens, the values of a and b seen by P_2 will differ by one and c will be odd.

All in all, this exercise highlights that, even though the mutual exclusion regions present in the code guarantee that variable values are valid when *taken individually*, there may still be inconsistencies when another process considers *multiple variables* together. The most immediate way to solve the issue is to use only one critical region in P_1, which comprises both increments.

EXERCISE 2

In order to determine which kind of monitor the code implements, we must first understand how it works.

- Semaphore m guarantees the mutual exclusion among processes that enter the monitor. Variable nw counts how many processes are currently waiting on the condition variable.
- When a process waits on the condition variable, it releases m to allow other processes to enter the monitor and waits on semaphore c. Before releasing m, the process also increments nw. The fact that the process still holds m when it performs the increment ensures the consistency of nw.
- The decrement of nw performed after P(c) may look suspicious at first sight, but we will soon see that, even at this point, mutual exclusion is indeed guaranteed.
- When a process signals the condition variable, it first checks whether or not there are other processes waiting on it, by looking at nw. This raises no consistency issues because the signaling process holds m, and hence, is the only one within the monitor.
- If at least one process is waiting on the condition variable, the process unblocks exactly one of them by means of V(c) *without releasing* the mutual exclusion semaphore m. This ensures that no other processes will be allowed into the monitor until the process just unblocked resumes execution and either exits from the monitor or performs another wait().
- If no processes are waiting on the condition variable, the signaling process simply releases m.

Besides convincing us that the code is indeed correct, this short analysis also helps us realize that it implements a Brinch Hansen's monitor. This is because:

- The signaling process performs V(m) when it finds no processes to unblock. This prevents it from continuing within the monitor without violating the mutual exclusion constraint.

- When the signaling process unblocks a process that was waiting on the condition variable, it transfers ownership of the mutual exclusion semaphore to the unblocked process, which then becomes responsible of releasing it at a later time. As before, the signaling process would violate the mutual exclusion constraint if it continued within the monitor.

CHAPTER 6

EXERCISE 1

- We can use a mailbox that allows $Q_1, \ldots, Q_n$ to send P a message containing an integer, assuming that it is large enough to hold the values $1, \ldots, n$.
- To generate an event, process $Q_i, 1 \leq i \leq n$ sends a message containing i to the mailbox.
- To wait for an event, process P waits on the mailbox until a message arrives. When this happens, it can inspect its content and get to know the sender identity.
- For what concerns the synchronization model, two alternative approaches are possible, each with its own advantages and disadvantages:
    1. A synchronous transfer blocks Q_i when it generates an event until P waits for the event. This ensures that no events are ever lost but couples the timings of Q_i with those of P and, for instance, may lead to an indefinite wait if P malfunctions.
    2. An asynchronous transfer lets Q_i generate an event and continue even though P is not currently waiting for it. The mailbox size must be determined based on a trade-off between the probability of being unable to generate an event because the mailbox is full— the higher the size, the lower the probability—and the timeliness of event generation—the higher the size, the older an event may be when P eventually becomes aware of it.

EXERCISE 2

- A message is inserted into message queue M upon initialization. The actual message content is unimportant.
- The number of messages in M represents the value of the mutual exclusion semaphore. The presence of a message corresponds to the value 1 and no messages means 0.
- To perform a P(), a process extracts a message from M, waiting if there is none.
- To perform a V(), a process inserts a message into M. Also in this case, message content is unimportant.

Note that, if the semaphore is used properly, M will never overflow and any process executing V() will never be blocked because M was full.

EXERCISE 3

We may proceed by induction:

- If there is $k = 1$ message in the queue, that message must have been generated at a time $t_S - t_G \leq t < t_S$. The constraint $t < t_S$ is trivial because it simply states that a message cannot be received before it has been sent. Since G inserts a message into the queue every t_G, the presence of a message generated at a certain time $t < t_S - t_G$ would imply that at least a second message must also be the queue, contrary to out hypothesis. It must therefore be $t \geq t_S - t_G$.
- If there are $k > 1$ messages in the queue, the oldest message must have been generated at a time $t_S - k t_G \leq t < t_S - (k-1)t_G$ because G generates one message every t_G. Assuming that the queuing discipline is first-in, first-out (FIFO), as is usually the case, the message retrieved by S at time t_S will be the oldest one.

CHAPTER 10

EXERCISE 1

- The minor cycle length must be:

$$T_m = \gcd(T_1, \ldots, T_3) = \gcd(60, 20, 30) = 10\,\text{ms}.$$

- The major cycle length must be:

$$T_M = \text{lcm}(T_1, \ldots, T_3) = \text{lcm}(60, 20, 30) = 60\,\text{ms}.$$

- Each major cycle consists of $T_M/T_m = 6$ minor cycles. It must contain $T_M/T_1 = 1$ instance of τ_1, $T_M/T_2 = 3$ instances of τ_2, and $T_M/T_3 = 2$ instances of τ_3.
- In the absence of any other constraints or requirements, we can assign task instances to minor cycles in the most intuitive way and devise the following cyclic executive. Other solutions are also possible.

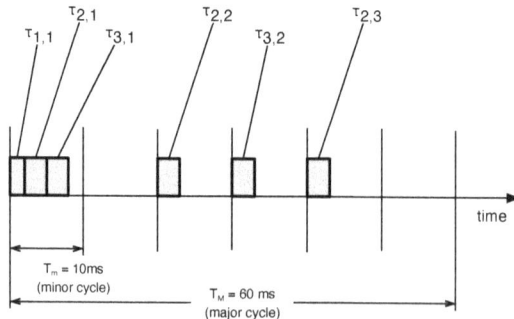

EXERCISE 2

No adjustments to the major and minor cycle are needed. We must simply add to the cyclic executive $T_M/T_4 = 3$ instances of τ_4. Since each instance has an execution time $C_4 = 4$ ms we cannot put them into the first, third, and fifth minor cycle because this would overflow the first minor cycle.

However, since there are no constraints on the phase relationships among tasks, we can use the second, fourth, and sixth minor cycle, as shown in the following scheduling diagram, without violating the requirement $T_4 = 20$ ms.

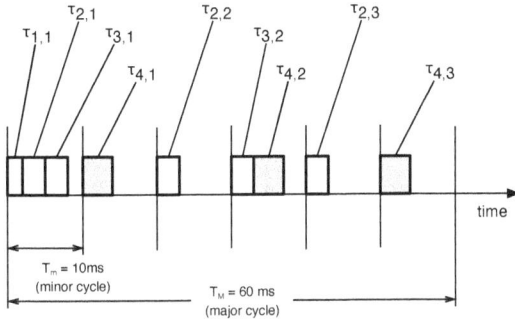

EXERCISE 3

Accommodating τ_5 without using a secondary schedule would not change T_m, but would increase T_M. Namely, it would be:

$$T'_M = \text{lcm}(T_1, \ldots, T_5) = \text{lcm}(60, 20, 30, 20, 600) = 600\,\text{ms}.$$

However, we observe that T_5 is an integral multiple of the previous major cycle length T_M, because $T_5/T_M = 600/60 = 10$. We can therefore use a secondary schedule with $k = T_5/T_M = 600/60 = 10$ and place it, for instance, in the second minor cycle, as shown in the scheduling diagram that follows. Any other minor cycle, except the first, would work equally well.

It is worth noting that we must reserve an executing time of $C_5 = 2$ ms in the minor cycle, although τ_5 will actually use it only once every k major cycles.

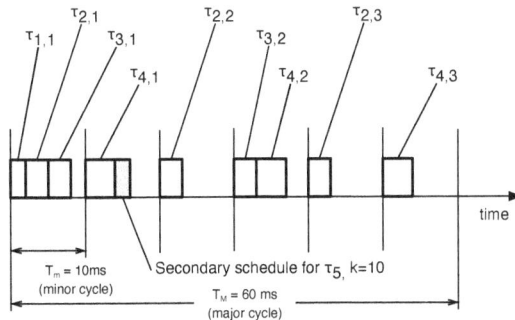

EXERCISE 4

As in the previous exercise, the period T_6 of task τ_6 suggests we should use a secondary schedule with $k = 10$ for it, unless we want to significantly increase the major cycle length T_M.

Moreover, the execution time C_6 of task τ_6 is as big as T_m, the minor cycle length. Since no minor cycles are completely empty, because other tasks have already been allocated to them, there is no obvious way to place τ_6 in any of them.

To accommodate τ_6 we must therefore resort to a combination of a secondary schedule (to keep T_M unchanged) and task splitting (to properly spread C_6 across multiple minor cycles).

The most appropriate way of splitting a task heavily depends on its internal structure. For the sake of this exercise, let us assume that τ_6 is a background communication task that performs 5 network transactions, each requiring up to 2 ms of execution time.

Under this assumption, we can split the task into 5 parts, one for each transaction, and place them in all minor cycles except the first, as shown in the following diagram. Each part is executed by a secondary schedule with $k = 10$.

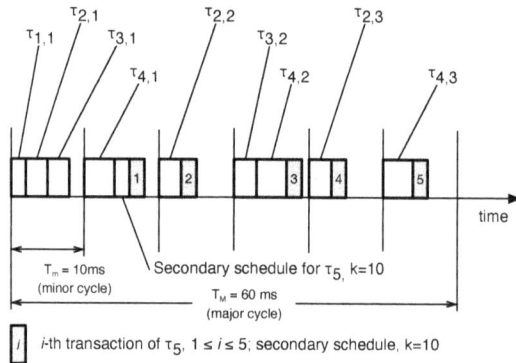

CHAPTER 11

EXERCISE 1

The RM algorithm assigns to the tasks a priority inversely proportional to their period. Since $T_2 = T_4$, τ_2 would have the same priority as τ_4 but, due to our assumption, we assign to τ_2 a priority higher than the priority of τ_4. As a result, the final task priorities are the ones listed in the table that follows. Higher values correspond to higher priorities.

Task τ_i	Period T_i (ms)	Execution time C_i (ms)	Priority
τ_1	60	2	1
τ_2	20	3	4
τ_3	30	3	2
τ_4	20	4	3

The corresponding scheduling diagram is:

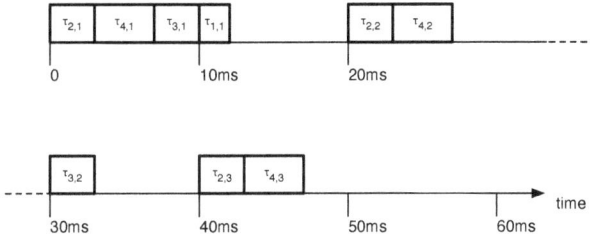

The diagram has been built based on the following reasoning:

- At $t = 0$ all tasks are ready for execution. They are executed in priority order, that is, τ_2, τ_4, τ_3, and τ_1. No tasks are preempted because there are no further task releases before the first instance of τ_1 completes, at $t = 12$ ms.
- The system is idle until $t = 20$ ms when the second instance of τ_2 and τ_4 are released. They are executed in priority order with no preemption.
- At $t = 30$ ms the second instance of τ_3 is released and immediately executed to completion.
- At $t = 40$ ms the scheduler behaves as it did at $t = 20$ ms. The system then remains idle until $t = 60$ ms, the hyperperiod of its tasks.

There is only one critical instant in the diagram, at $t = 0$.

When using EDF, priorities are dynamically assigned to tasks based on their absolute deadline. Namely, the task with the earliest deadline receives the highest priority. The following figure shows the EDF scheduling diagram and, above it, the current distance from the deadline for all tasks whenever a scheduling decision is made. The earliest deadline, which gives to its task the highest priority, is highlighted in bold. As before, when two tasks are at the same distance from their deadline, higher priority is given to the task with the lower index.

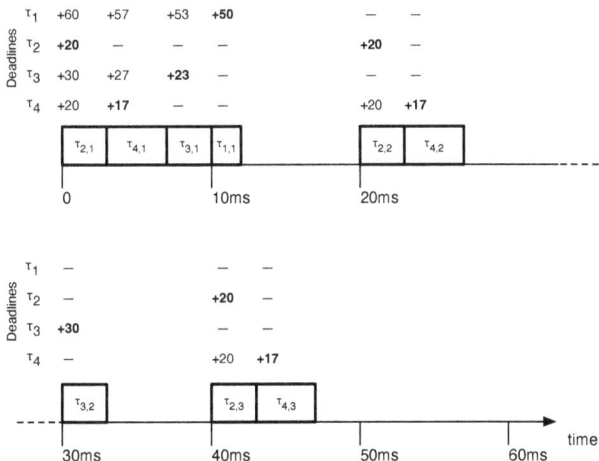

In this particular case, both RM and EDF took exactly the same scheduling decision, although the underlying algorithms are very different, leading to two identical scheduling diagrams.

EXERCISE 2

The priorities that RM assigns to the tasks are listed in the rightmost column of the following table:

Task τ_i	Period T_i (ms)	Execution time C_i (ms)	Priority
τ_1	40	8	1
τ_2	20	3	4
τ_3	30	12	2
τ_4	20	5	3

The RM scheduling diagram is:

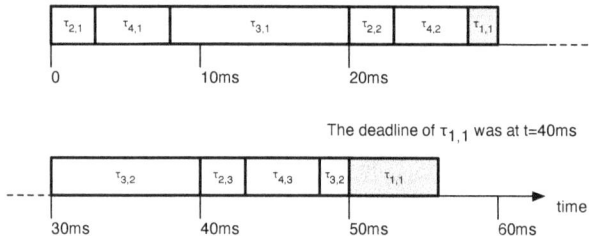

The following aspects of the RM algorithm are important to understand the difference with respect to EDF:

- At $t = 20$ ms tasks τ_2, τ_4, and τ_1 are all ready for execution. Since RM task priorities are static, the algorithm executes τ_2, τ_4 and, eventually τ_1.
- At $t = 30$ ms task τ_3 becomes ready for execution. Since its priority is higher than the priority of τ_1, the scheduler preempts τ_1 to execute τ_3.
- At $t = 40$ ms task τ_3 concludes its execution, but tasks τ_2 and τ_4 become ready for execution again. They keep the processor busy until $t = 50$ ms. At that time, the execution of τ_1 becomes possible again, but its deadline has already expired.

As a consequence, RM is unable to schedule the task set in the given time interval. The EDF algorithm makes different scheduling decisions in this case, as depicted in the following diagram:

Deadlines

τ1	+40	+37	+32		+20		–
τ2	+20	–	–		+20		+12
τ3	+30	+27	+22		–		–
τ4	+20	+17	–		+20		+12

| τ2,1 | τ4,1 | τ3,1 | τ1,1 | τ2,2 |

0   10ms   20ms

Deadlines

τ1	–	–	–	+40	+37	+29
τ2	+10	–	–	+20	–	–
τ3	+30 +29	+24	+20	+17	–
τ4	+10 +9	–	+20	+17	+9

τ2,2 | τ4,2 | τ3,2 | τ2,3 | τ3,2 | τ4,3 | ...

30ms   40ms   50ms   60ms   time

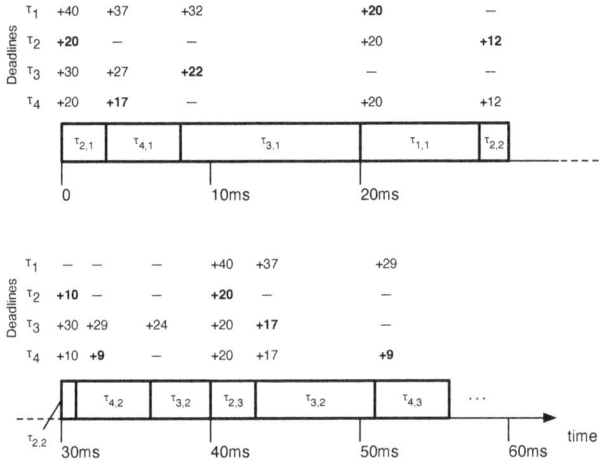

The most notable difference takes place at $t = 20\,\text{ms}$, when EDF executes τ_1 first, followed by τ_2 and τ_4. This makes the EDF scheduling of the task set possible in the given time interval.

CHAPTER 12

EXERCISE 1

- For the task set of Exercise 1, Chapter 11:

$$U = 2/60 + 3/20 + 3/30 + 4/20 \simeq 0.48.$$

- For the task set of Exercise 2, Chapter 11:

$$U = 8/40 + 3/20 + 12/30 + 5/20 = 1.$$

EXERCISE 2

It is:
$$U = 20/60 + 2/10 + 3/30 \simeq 0.63$$

- Since $U \leq \ln 2$ the RM algorithm is certainly able to schedule the task set.
- We can come to the conclusion that EDF is able to schedule the task set in two different ways:
  1. EDF can schedule any task set schedulable with RM.
  2. It is $U \leq 1$.

EXERCISE 3

- It is:
$$U = 20/60 + 5/10 + 3/30 = 0.93.$$

- Since $U > \ln 2$ the sufficient schedulability test for RM does not provide any useful information unless we consider the number of tasks in the system. Even if we take into account that there are only $N = 3$ tasks in this case, it is still $U > N(2^{1/N} - 1) \simeq 0.779$. To obtain conclusive results, we have to draw the scheduling diagram at a critical instant. Assuming that all tasks are ready for execution at $t = 0$ we obtain the following diagram:

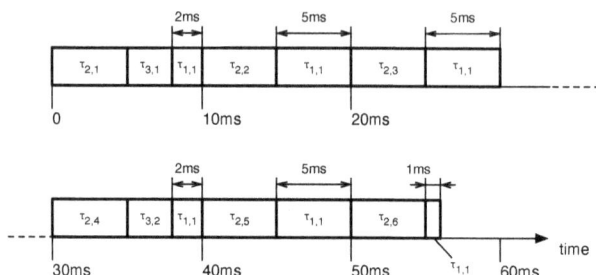

Looking at the diagram, it is evident that RM is able to schedule the task set, although τ_1 undergoes a number of preemptions before concluding its execution.

- Since $U < 1$ we can be certain that EDF is able to schedule the task set, without drawing any scheduling diagram.

CHAPTER 13

EXERCISE 1

The worst-case response times of tasks $\tau_1, \ldots, \tau_4$ are:

$$
\begin{aligned}
R_2 &= 3\,\text{ms}, &&\text{task } \tau_2 \\
R_4 &= 7\,\text{ms}, &&\text{task } \tau_4 \\
R_3 &= 10\,\text{ms}, &&\text{task } \tau_3 \\
R_1 &= 12\,\text{ms}, &&\text{task } \tau_1
\end{aligned}
$$

Tasks are listed in decreasing priority order. These response times coincide with what the first part of the scheduling diagram, starting at the critical instant $t = 0$, shows.

EXERCISE 2

The worst-case response times of tasks $\tau_1, \ldots, \tau_4$ are:

$$
\begin{aligned}
R_2 &= 3\,\text{ms}, &&\text{task } \tau_2 \\
R_4 &= 8\,\text{ms}, &&\text{task } \tau_4 \\
R_3 &= 20\,\text{ms}, &&\text{task } \tau_3 \\
R_1 &= 56\,\text{ms}, &&\text{task } \tau_1 \text{ (beyond the deadline)}
\end{aligned}
$$

Results agree with the corresponding scheduling diagram for all tasks. It is worth noting that RTA was able to accurately calculate the worst-case response time of task τ_1 even though it exceeded its deadline.

CHAPTER 14

EXERCISE 1

On a single-processor system the two tasks may indirectly affect the response time of τ_M in two different ways:

- Task τ_H may preempt τ_M, thus increasing the amount of *interference* that τ_M endures.
- Task τ_L may not preempt τ_M as long as it stays at its baseline priority.
- However, τ_L may inherit the priority of τ_H and induce some *push-through blocking* to τ_M.
- Task τ_M does not suffer from *direct blocking* due to τ_L or τ_H because it does not share any resources with them.

EXERCISE 2

1. Since τ_{L1} and τ_{L2} do not use any mutual exclusion semaphore, the priority inheritance algorithm will never raise their priority above the baseline. As a consequence, they may not affect the response time of τ_M, whose baseline priority is higher.
2. The priority inheritance algorithm may temporarily raise the priority of τ_{L1} and τ_{L2} to the maximum between their two baseline priorities. However, both of them are lower than the baseline priority of τ_M. Hence, as in the previous scenario, τ_{L1} and τ_{L2} may not affect the response time of τ_M.
3. Due to semaphore S, τ_{L1} may temporarily inherit the priority of τ_H, which is higher than the priority of τ_M. Depending on whether and how the critical regions involving S and S' are nested within τ_{L1}, τ_{L2} may also inherit the priority of τ_H, due to semaphore S' and transitive priority inheritance. As a consequence, not only τ_{L1} but also τ_{L2} may induce push-through blocking on τ_M.

In neither scenario τ_{L1} and τ_{L2} induce any interference or direct blocking on τ_M because they do no share any resources with it.

CHAPTER 15

EXERCISE 1

Cases 2 and 4 count as self-suspension. In case 2, the self-suspension duration depends on the amount of time needed to retrieve the data the process intends to read. In case 4, assuming no other tasks are also waiting on the same semaphore, the self-suspension ends when another task performs a V() on the semaphore.

In case 1, the calculation may take a long time to complete, but this is irrelevant from the point of view of self-suspension because the task does make use of the processor during the calculation. In case 3, the task may or may not have to wait while acquiring the mutual exclusion semaphore, depending on the semaphore state, but any amount of waiting is better treated as blocking instead of self-suspension.

EXERCISE 2

- Task T must wait for at most $t = \min(t_1, t_2) = 40\,\text{ms}$ for the arrival of a message. We must take the minimum between the two periods because T unblocks—and, in other words, ends its self-suspension—when a message arrives from *any* of the two sources. Therefore, t is the worst-case self-suspension time of T.
- Tasks V_1 and V_2 may have to wait while sending a message to the message queue if they use a blocking primitive to do so, the message queue has a limited capacity, and is currently full. This wait counts as self-suspension and, besides on t_1 and t_2, it also depends on the worst-case time task T needs to process a message, as well as the maximum number of messages the queue can hold.

References

1. ACTORS—adaptivity and control of resources in embedded systems. Project web site at http://www.actors-project.eu/.
2. J. H. Anderson and M. Moir. Universal constructions for large objects. *IEEE Transactions on Parallel and Distributed Systems*, 10(12):1317–1332, 1999.
3. J. H. Anderson and S. Ramamurthy. A framework for implementing objects and scheduling tasks in lock-free real-time systems. In *Proc. 17th IEEE Real-Time Systems Symposium*, pages 94–105, December 1996.
4. J. H. Anderson, S. Ramamurthy, and K. Jeffay. Real-time computing with lock-free shared objects. In *Proc. 16th IEEE Real-Time Systems Symposium*, pages 28–37, December 1995.
5. ARM Ltd. *ARM Architecture Reference Manual*, July 2005. DDI 0100I.
6. ARM Ltd. *ARMv7-M Architecture Reference Manual*, February 2010. DDI 0403D.
7. ARM Ltd. *Cortex™-M3 Technical Reference Manual, rev. r2p0*, February 2010. DDI 0337H.
8. N. C. Audsley, A. Burns, M. Richardson, and A. J. Wellings. Hard real-time scheduling: the deadline monotonic approach. In *Proc. 8th IEEE Workshop on Real-Time Operating Systems and Software*, pages 127–132, 1991.
9. N. C. Audsley, A. Burns, and A. J. Wellings. Deadline monotonic scheduling theory and application. *Control Engineering Practice*, 1(1):71–78, 1993.
10. M. J. Bach. *The Design of the UNIX Operating System*. Prentice-Hall, Upper Saddle River, NJ, 1986.
11. T. P. Baker and A. Shaw. The cyclic executive model and Ada. In *Proc. IEEE Real-Time Systems Symposium*, pages 120–129, December 1988.
12. J. H. Baldwin. Locking in the multithreaded FreeBSD kernel. In *BSDC'02: Proceedings of the BSD Conference 2002*, Berkeley, CA, 2002. USENIX Association.
13. R. Barry. The FreeRTOS.org project. Available online, at http://www.freertos.org/.
14. R. Barry. *The FreeRTOS™ Reference Manual*. Real Time Engineers Ltd., 2011.
15. P. Brinch Hansen. Structured multiprogramming. *Communications of the ACM*, 15(7):574–578, 1972.
16. P. Brinch Hansen. *Operating System Principles*. Prentice-Hall, Englewood Cliffs, NJ, 1973.
17. P. Brinch Hansen, editor. *The Origin of Concurrent Programming: from Semaphores to Remote Procedure Calls*. Springer-Verlag, New York, 2002.
18. A. Burns and A. Wellings. *Real-Time Systems and Programming Languages*. Pearson Education, Harlow, England, 3rd edition, 2001.
19. G. C. Buttazzo. *Hard Real-Time Computing Systems. Predictable Scheduling Algorithms and Applications*. Springer-Verlag, Santa Clara, CA, 2nd edition, 2005.
20. Cadence Design Systems, Inc. Xtensa LX processor platform. Available online, at https://www.cadence.com/en_US/home/tools/silicon-solutions/compute-ip/tensilica-xtensa-controllers-and-extensible-processors/xtensa-lx-processor-platform.html, 2024.

21. E. G. Coffman, M. Elphick, and A. Shoshani. System deadlocks. *ACM Computing Surveys*, 3(2):67–78, 1971.

22. F. J. Corbató, M. Merwin Daggett, and R. C. Daley. An experimental time-sharing system. In *Proc. of the AFIPS Spring Joint Computer Conference (SJCC)*, volume 21, pages 335–344, May 1962.

23. Robert Davis, Alan Burns, Reinder Bril, and Johan Lukkien. Controller Area Network (CAN) schedulability analysis: Refuted, revisited and revised. *Real-Time Systems*, 35(3):239–272, 2007.

24. R. Devillers and J. Goossens. Liu and Layland's schedulability test revisited. *Information Processing Letters*, 73(5-6):157–161, 2000.

25. E. W. Dijkstra. Cooperating sequential processes. Technical Report EWD-123, Eindhoven University of Technology, 1965. Published as [27].

26. E. W. Dijkstra. The multiprogramming system for the EL X8 THE. Technical Report EWD-126, Eindhoven University of Technology, June 1965.

27. E. W. Dijkstra. Cooperating sequential processes. In F. Genuys, editor, *Programming Languages: NATO Advanced Study Institute*, pages 43–112. Academic Press, Villard de Lans, France, 1968.

28. E. W. Dijkstra. The structure of the "THE"-multiprogramming system. *Communications of the ACM*, 11(5):341–346, 1968.

29. A. Dunkels. lwIP—a lightweight TCP/IP stack. Available online, at `http://savannah.nongnu.org/projects/lwip/`.

30. Espressif Systems. *ESP32 ESP-IDF Programming Guide, release v5.1*, June 2023. Available online, at `https://docs.espressif.com/projects/esp-idf/en/stable/esp32/`.

31. Espressif Systems. *ESP32-WROOM-32E, ESP32-WROOM-32UE Datasheet, version 1.6*, 2023. Available online, at `https://espressif.com/documentation/esp32-wroom-32e_esp32-wroom-32ue_datasheet_en.pdf`.

32. K. Etschberger. *Controller Area Network Basics, Protocols, Chips and Applications*. IXXAT Press, Weingarten, Germany, 2001.

33. D. Faggioli, F. Checconi, M. Trimarchi, and C. Scordino. An EDF scheduling class for the Linux kernel. In *Proc. 11th Real-Time Linux Workshop*, pages 197–204, Dresden, Germany, 2009.

34. M. D. Godfrey and D. F. Hendry. The computer as von Neumann planned it. *IEEE Annals of the History of Computing*, 15(1):11–21, 1993.

35. Google LLC. Protocol buffers documentation. Available online, at `https://protobuf.dev`.

36. A. N. Habermann. Prevention of system deadlocks. *Communications of the ACM*, 12(7):373–377, 1969.

37. J. W. Havender. Avoiding deadlock in multitasking systems. *IBM Systems Journal*, 7(2):74–84, 1968.

38. M. Herlihy. A methodology for implementing highly concurrent data objects. *ACM Trans. on Programming Languages and Systems*, 15(5):745–770, November 1993.

39. M. P. Herlihy and J. M. Wing. Axioms for concurrent objects. In *Proc. 14th ACM SIGACT-SIGPLAN symposium on Principles of programming languages*, pages 13–26, New York, 1987.

40. C. A. R. Hoare. Towards a theory of parallel programming. In *Proc. International Seminar on Operating System Techniques*, pages 61–71, 1971. Reprinted in [17].

41. C. A. R. Hoare. Monitors: An operating system structuring concept. *Communications of the ACM*, 17(10):549–557, 1974.

42. Richard C. Holt. Comments on prevention of system deadlocks. *Communications of the ACM*, 14(1):36–38, 1971.

43. Richard C. Holt. Some deadlock properties of computer systems. *ACM Computing Surveys*, 4(3):179–196, 1972.

44. IEC. *Industrial Communication Networks—Fieldbus specifications—Part 3-3: Data-Link Layer Service Definition—Part 4-3: Data-link layer protocol specification—Type 3 elements*, December 2007. Ed 1.0, IEC 61158-3/4-3.

45. *IEEE Std 1003.13™-2003, IEEE Standard for Information Technology—Standardized Application Environment Profile (AEP)—POSIX® Realtime and Embedded Application Support*. IEEE, 2003.

46. *IEEE Std 1003.1™-2008, Standard for Information Technology—Portable Operating System Interface (POSIX®) Base Specifications, Issue 7*. IEEE and The Open Group, 2008.

47. *IEEE Std 802.3™-2008, Standard for Information technology—Telecommunications and information exchange between systems—Local and metropolitan area networks—Specific requirements Part 3: Carrier sense multiple access with Collision Detection (CSMA/CD) Access Method and Physical Layer Specifications*. IEEE, 2008.

48. *Industrial Communication Networks—Fieldbus specifications—Part 3-12: Data-Link Layer Service Definition—Part 4-12: Data-link layer protocol specification—Type 12 elements*. IEC, December 2007. Ed 1.0, IEC 61158-3/4-12.

49. Intel Corp. *Intel® 64 and IA-32 Architectures Software Developer's Manual*, 2007.

50. *International Standard ISO/IEC 8802-2:1998, ANSI/IEEE Std 802.2, Information technology—Telecommunications and information exchange between systems—Local and metropolitan area networks—Specific requirements Part 2: Logical Link Control*. IEEE, New York, 1988.

51. *International Standard ISO/IEC 7498-1, Information Technology—Open Systems Interconnection—Basic Reference Model: The Basic Model*. ISO/IEC, 1994.

52. *International Standard ISO/IEC/IEEE 9945, Information Technology—Portable Operating System Interface (POSIX®) Base Specifications, Issue 7*. IEEE and The Open Group, 2009.

53. ISO. *ISO 11898-1 – Road vehicles – Controller area network (CAN) – Part 1: Data link layer and physical signalling*. International Organization for Standardization, 3rd edition, May 2024.

54. *ISO 11898-2—Road vehicles—Controller area network (CAN)—Part 2: High-speed medium access unit*. International Organization for Standardization, 3rd edition, March 2024.

55. *ITU-T Recommendation H.323—Packet-based multimedia communications systems*. International Telecommunication Union, December 2009.

56. M. Khiszinsky. *CDS: Concurrent Data Structures library*. Available online, at `http://libcds.sourceforge.net/`.

57. S. Rao Kosaraju. Limitations of Dijkstra's semaphore primitives and Petri nets. *Operating Systems Review*, 7(4):122–126, January 1973.

58. K. Lakshmanan and R. Rajkumar. Scheduling self-suspending real-time tasks with rate-monotonic priorities. In *Proc. 16th IEEE Real-Time and Embedded Technology and Applications Symposium*, pages 3–12, April 2010.

59. L. Lamport. A new solution of Dijkstra's concurrent programming problem. *Communications of the ACM*, 17(8):453–455, August 1974.

60. L. Lamport. The mutual exclusion problem: part I—a theory of interprocess communication. *Journal of the ACM*, 33(2):313–326, 1986.

61. L. Lamport. The mutual exclusion problem: part II—statement and solutions. *Journal of the ACM*, 33(2):327–348, 1986.

62. J. Y.-T. Leung and J. Whitehead. On the complexity of fixed-priority scheduling of periodic, real-time tasks. *Performance Evaluation*, 2(4):237–250, 1982.

63. C. L. Liu and J. W. Layland. Scheduling algorithms for multiprogramming in a hard-real-time environment. *Journal of the ACM*, 20(1):46–61, 1973.

64. J. W. S. Liu. *Real-Time Systems*. Prentice Hall, Upper Saddle River, NJ, 2000.

65. C. Douglass Locke. Software architecture for hard real-time applications: cyclic executives vs. fixed priority executives. *Real-Time Systems*, 4(1):37–53, 1992.

66. M. K. McKusick, K. Bostic, M. J. Karels, and J. S. Quarterman. *The Design and Implementation of the 4.4BSD Operating System*. Addison-Wesley, Reading, MA, 1996.

67. P. Mockapetris. *Domain Names—-Implementation and Specification, RFC 1035*. ISI, November 1987.

68. P. Mockapetris. *Domain Names—Concepts and Facilities, RFC 1034*. ISI, November 1987.

69. H. Nyquist. Certain topics in telegraph transmission theory. *Transactions of the AIEE*, pages 617–644, February 1928. Reprinted in [70].

70. H. Nyquist. Certain topics in telegraph transmission theory. *Proceedings of the IEEE*, 90(2):280–305, February 2002.

71. J. Oberg. Why the Mars probe went off course. *IEEE Spectrum*, 36(12):34–39, December 1999.

72. Olimex Ltd. ESP32-EVB revision K1 board schematics. Available online, at `https://www.olimex.com/Products/IoT/ESP32/ESP32-EVB/open-source-hardware`, 2025.

73. Olimex Ltd. MOD-LCD2.8RTP revision B board schematics. Available online, at `https://www.olimex.com/Products/Modules/LCD/MOD-LCD2-8RTP/open-source-hardware`, 2025.

74. S. Owicki and L. Lamport. Proving liveness properties of concurrent programs. *ACM Trans. Program. Lang. Syst.*, 4(3):455–495, July 1982.

75. G. L. Peterson. Myths about the mutual exclusion problem. *Information Processing Letters*, 12(3):115–116, 1981.

76. D. C. Plummer. *An Ethernet Address Resolution Protocol—or—Converting Network Protocol Addresses to 48.bit Ethernet Address for Transmission on Ethernet Hardware, RFC 826*. MIT, November 1982.

77. J. Postel. *User Datagram Protocol, RFC 768*. ISI, August 1980.

78. J. Postel, editor. *Internet Protocol—DARPA Internet Program Protocol Specification, RFC 791*. USC/Information Sciences Institute, September 1981.

79. J. Postel, editor. *Transmission Control Protocol—DARPA Internet Program Protocol Specification, RFC 793*. USC/Information Sciences Institute, September 1981.

80. R. Rajkumar, L. Sha, and J. P. Lehoczky. Real-time synchronization protocols for multiprocessors. In *Proc. 9th IEEE Real-Time Systems Symposium*, pages 259–269, December 1988.

81. M. H. Schimek, B. Dirks, H. Verkuil, and M. Rubli. *Video for Linux Two API Specification Revision 0.24*, March 2008. Available online, at `http://v4l2spec.bytesex.org/`.

82. L. Sha, T. Abdelzaher, K.-E. Årzén, A. Cervin, T. Baker, A. Burns, G. Buttazzo, M. Caccamo, J. Lehoczky, and A. K. Mok. Real time scheduling theory: A historical perspective. *Real-Time Systems*, 28(2):101–155, 2004.

83. L. Sha, R. Rajkumar, and J. P. Lehoczky. Priority inheritance protocols: an approach to real-time synchronization. *IEEE Transactions on Computers*, 39(9):1175–1185, September 1990.

84. C. E. Shannon. Communication in the presence of noise. *Proceedings of the IRE*, 37(1):10–21, January 1949. Reprinted in [85].

85. C. E. Shannon. Communication in the presence of noise. *Proceedings of the IEEE*, 86(2):447–457, February 1998.

86. A. Shoshani and E. G. Coffman. Prevention, detection, and recovery from system deadlocks. In *Proc. 4th Princeton Conference on Information Sciences and Systems*, March 1970.

87. A. Silberschatz, P. B. Galvin, and G. Gagne. *Operating System Concepts*. John Wiley & Sons, New York, 7th edition, 2005.

88. A. S. Tanenbaum and A. S. Woodhull. *Operating Systems Design and Implementation*. Pearson Education, Upper Saddle River, NJ, 3rd edition, 2006.

89. J. von Neumann. First draft of a report on the EDVAC. *IEEE Annals of the History of Computing*, 15(4):27–75, 1993. Reprint of the original typescript circulated in 1945.

Index